TH4811 .D5 1974

Dietz, Albert George Henry, 1908-

Dwelling house construction

Dwelling House Construction

The MIT Press
Cambridge, Massachusetts, and London, England

Dwelling House Construction

Fourth Edition, Revised

Albert G. H. Dietz

Original edition published by
D. Van Nostrand Company, Inc.,
Princeton, New Jersey,
and copyright © 1946, 1954 by
Walter C. Voss and Albert G. H. Dietz

New material copyright © 1971, 1974 by
The Massachusetts Institute of Technology

All rights reserved. No part of this book may be reproduced in any form or by any means, electronic or mechanical, including photocopying, recording, or by any information storage and retrieval system, without permission in writing from the publisher.

Fourth edition, revised (second MIT Press edition) 1974
First MIT Press paperback edition, 1977
Eighth printing, 1988

Library of Congress Cataloging in Publication Data

Dietz, Albert George Henry, 1908–
   Dwelling house construction.

   1. House construction. I. Title.
TH4811.D5 1974   690.8   73-22321
ISBN 0-262-04044-1 (hardcover)
ISBN 0-262-54033-9 (paperback)

This book is dedicated to
Ross Francis Tucker,
whose interest in young men and construction
left an indelible mark
on the personnel of the industry.

Contents

Preface ix

Acknowledgments xi

1 Inspection of the Site 1

2 Building Layout 10

3 Excavation. Sanitary Systems 16

4 Foundations 35

5 Framing 62

6 Chimneys and Fireplaces 147

7 Windows 164

8 Roofing and Flashing 194

9 Cornices, Gutters, and Leaders 218

10 Exterior Finishes. Water Tables 231

11 Insulation 246

12 Wallboard. Lath and Plaster 263

13 Interior Finish 313

14 Hardware 356

15 Coatings 381

16 Plastics 390

17 Manufactured Housing 401

18 Mobile Homes 415

Index 429

Preface

Enough changes have occurred in dwelling house construction since this book was last revised (1954) to make another revision desirable. A careful review of the text, however, reveals that changes are mainly in details; the basic principles have altered little since the book first appeared. The plan of the book, therefore, is essentially unchanged.

Plastics and mobile homes have developed so rapidly since the last revision that entirely new chapters have been added. Although the total volume of plastics in dwelling house construction is relatively small, the number of uses is large and constantly expanding. It is, consequently, desirable to have some understanding of what these materials are and how they are used. Mobile homes represented a minor part of the housing market as late as a few years ago; now they constitute a large part of the total and are, therefore, treated in a separate new chapter.

Considerable changes have occurred in manufactured homes, i.e., homes in which large components are factory-produced for assembly at the site. The material on panel construction has, therefore, been largely rewritten to reflect changing practice.

The use of dry-wall construction for wall surfacing has increased to a great degree, but plaster and stucco remain important, and veneer plaster promises to become significant. Chapter 12 has, consequently, been rewritten to reflect the present situation. Similarly, the advent of synthetic polymeric materials has brought about far-reaching changes in the technology of coatings, and Chapter 15 has accordingly been rewritten.

As I have indicated, basic principles change slowly. Much of the text, therefore, has been reused with only such alterations and additions as are necessary to reflect the present situation. Another reason for retaining as much of the old text and illustrations as possible is that these reflect the construction of millions of existing houses, and the old text is useful when such structures are altered or rehabilitated. For example, although much new material on windows has been added, a great deal of the old text is still valid, and many of the old illustrations of wood windows have been retained to present principles as well as for their relevance to the large stock of existing windows. Masonry firestopping is found in old houses and hence in some of the illustrations in this book. Similarly, the principles of chimney and fireplace construction are essentially unaltered, and I have mingled traditional and contemporary details. Millwork and cabinet work have changed in details but not in essentials; I have used illustrations that picture both the old and the new.

Some of the old text has been eliminated or shortened to make room for the new. For example, half-timber construction has virtually disappeared since its heyday, and is gone from this text. The description of rubble masonry has been curtailed. Some treatment of traditional roofing has been shortened to make room for the description of newer methods. Similar changes have taken place throughout the text.

The present edition, then, is an amalgam of new material and those parts of the old that remain pertinent. Its basic premise is unchanged: to present those principles of dwelling house construction that underlie good construction and do not change with whim and fashion. Although I originally intended *Dwelling House Construction* to be a textbook, it has found a use among builders, homeowners, architects, engineers, and others involved in the building or rehabilitation of dwelling houses, and, in this fourth edition, should continue to do so.

A. G. H. D.
Cambridge, Massachusetts

Acknowledgments

Many individuals and organizations have been called upon for assistance in this revision. It is manifestly impossible to take note of them all; I must content myself by gratefully acknowledging the contributions of the following.

Dushan Stankovich prepared the new drawings and revised many of the old ones, skillfully blending the new and the old. Marvin Goody generously made architectural drawings and details available from his files of house plans. Ralph Johnson of the National Association of Home Builders' Building Research Institute carefully reviewed the old text, made many suggestions, and supplied much information that has been incorporated in the new text. Malcolm Hope of the Public Health Service, Department of Health, Education, and Welfare, was instrumental in obtaining material about sanitary systems. Eliot Snider of Massachusetts Lumber Company supplied a great deal of information. The National Forest Products Association granted permission to draw extensively upon its publications. David Countryman of the American Plywood Association was similarly helpful. Professor E. George Stern of Virginia Polytechnic Institute made available an extensive literature about nails. Clifford O'Brien of Plasticrete Co. supplied information on concrete block and similar units.

W. R. Haverkampf of Sargent & Co. undertook the arduous task of rewriting the material on finish hardware, originally prepared by R. G. Salaman of the same company. Dean Paul Witherell of Wentworth Institute made many suggestions and was particularly helpful in the revision of the chapter on windows, supplying much of the text. Professor William Litle and Anthony Herrey of M.I.T. made available the originals of the sketches showing different types of industrialized housing. D. E. Brackett of the Gypsum Association sent much information and made many suggestions regarding wallboard and plaster, particularly veneer plaster, for which he supplied part of the text. George Dandrow supplied much of the information on plaster, with the assistance of the Northeast Plasterers Bureau. The drawing of the hypothetical house showing where plastics could be used was developed in cooperation with Marvin Goody. The drawing was originally commissioned by the Society of the Plastics Industry. Francis Fleetwood, a graduate student at M.I.T., drew upon his Master's thesis for much of the text and illustrations in the chapter on manufactured houses. Professor Arthur Bernhardt of M.I.T. and his

Acknowledgments

associates wrote the chapter on mobile homes. The drawing was adapted from material originating with Broadline.

Manufacturers' literature appearing in *Sweet's Catalog* and elsewhere was drawn upon for illustrations and text as appropriate. Among them should be mentioned Alcoa Building Products, Inc., Anaconda Aluminum Co., Andersen Corp., Aflas Aluminum Co., Bird & Son, Inc., Bliss Steel Products Co., Dayton Sure-Grip and Shore Co., F. H. Maloney Co., PPG Industries, The Redwood Bureau, Republic Steel Corp., Rohm and Haas Co., and Rolscreen Co.

Special mention must be made of Professor Edward Allen, my colleague in the teaching of dwelling house construction. Special thanks also go to Christine Pufhal, who typed and revised often undecipherable manuscript, caught errors, and was, in general, of incalculable help in preparing this edition.

Finally, the whole project would have been impossible without the warm encouragement and monumental patience of my wife, who immeasurably aided the original writing of this book and each succeeding revision.

Dwelling House Construction

1 Inspection of the Site

1.1 General

a Before construction proceeds, it is customary, in the case of contract or custom building, for the owner and architect to invite a selected number of builders to prepare estimates of cost and to submit bids, i.e., their proposals for building the structure for a specified sum of money in accordance with plans and specifications. The builder must be acquainted with the site of the proposed structure, and must investigate the various factors which will influence costs of construction before he can make an intelligent estimate. The builder who builds for sale must be equally familiar with the site and other factors affecting costs.

b In order to place responsibility upon the builder for careful inspection of the site, architects customarily insert into specifications clauses similar to the following example.

1.2 Specification Clauses

EXAMINATION OF SITE

a The Contractor shall visit the site of the building, examine for himself the condition of the lot, and satisfy himself as to the nature and bearing power of the soil. He shall furnish all the labor and materials necessary to prepare the site for the execution of this contract.

RELATION TO GRADES

b The drawings show the approximate present natural as well as the finished grades, but the Contractor shall include in his proposal all work necessary to carry the foundations of all parts to sound footings and below the influence of frost.

c Features the builder must closely investigate are:
Building Site—location; physical features (e.g., soil and water table); and availability of utilities.
Legal Restrictions—zoning ordinances; building code; and labor and materials.

d There are many more considerations which affect the owner's selection of a site; these do not concern the custom builder directly, but they do affect the builder who builds to sell. They include community facilities such as schools and shopping facilities, transportation, the immediate neighborhood, orientation of the site, and many others.

BUILDING SITE

1.3 Location

a. General The following general questions immediately arise: What is the location of the site in the community? On what street or streets does it face? Which direction will the house face with respect to the points of the compass and with respect to the site itself? Are the boundaries well marked by surveyors' stakes, bounds, or some other means? If not, has provision been made to have the boundaries clearly delineated?

1.4 Roads

a What is the elevation of the site with reference to the road, and is the road private or public? Private roads are often narrow, little improved, and likely to follow existing surface contours. If taken over by public authority, such roads are likely to be straightened, widened, and levelled, thereby probably altering the position of a house with respect to the road. It is wise to ascertain whether plans have been made to alter the roadway which a house is to face, and to locate the house in such a manner as to avoid being adversely affected. This may or may not be up to the custom builder, depending upon whether or not a plot plan is provided. It is of considerable importance to the developer.

b Of immediate concern to the builder is the condition of the roads. If roads are good, he can count on easy access to the site at all times; if poor, he may have to bring in and store considerable amounts of material when the roads are passable.

1.5 Abutting Properties

a What is the general elevation or "lie of the land" of abutting properties? This is important because of drainage. If abutting properties are higher, they may drain onto the plot in question, and the builder may be called upon to correct this condition by filling, grading, and sloping the finished ground level so as to divert drainage elsewhere. A matter of law may be involved that varies in different localities. Although a property owner may not be liable for the drainage from his land onto his neighbor's if it is natural drainage, he is not permitted deliberately to alter existing topography so as to cause a bad flow from his property to his neighbor if that flow did not exist before.

Inspection of the Site

PROTECTION OF ADJACENT PROPERTIES

b. Specification Clause The builder . . . shall adequately protect adjacent property as provided by law and the contract documents.

c If the custom builder follows instructions regarding grading which may result in lawsuits from adjoining owners because of altered drainage, he can be held liable unless he protects himself by securing a written acceptance of responsibility by the owner. Therefore, to avoid later complications, the prudent builder determines from his inspection of the site and the finished grades indicated on the drawings whether such a contingency can arise. The developer of land and the builder for sale must keep such contingencies in mind when developing and building upon a site.

d This is only one example of the care which must be taken to protect adjacent properties. In general, it is required by law that no permanent injury to abutting properties may result from building operations, and that any temporary injury must be made good.

PHYSICAL FEATURES

1.6 General Topography

a Is the site on a hill, in a valley, or on flat land? If it slopes, how much is the slope and in what direction? Are the new grades to be the same as the old, or will there be a considerable amount of cutting, filling, and regrading? If there is to be much change, is an accurate plot plan available showing both existing and final grades? Will the final grades be such as to pose the problem respecting drainage toward and away from the house? (An excellent time to inspect a site is immediately after a heavy rain. Low spots and drainage features are then plainly visible, whereas they might escape notice when the site is dry.) Will walks and curbs have to be protected, or replaced because of breakage when heavy loads pass over them?

1.7 Soil

a. Topsoil What is the depth and composition of the topsoil, or surface layer of earth? The owner is interested because of its gardening possibilities. The builder is primarily interested because he should strip the topsoil and stack it to one side for future finish grading. Usually the stripped area is about 20' greater each way than the size of the house.

Dwelling House Construction　　4

REMOVAL OF EARTH

b. Specification Clause The builder shall remove and stack at a place directed by the architect, all loam within an area 20' each way in excess of the area to be covered by the building.

c. Subsoil The best way to examine the character of the soil is to dig a pit on the site of the house. Sometimes cuts or excavations in the neighborhood give sufficient information, and frequently builders or excavators know subsoil conditions by experience. However, it is necessary to be careful, because the character of the subsoil may change over a short distance.

d With the exception of rock, clay is the most troublesome material with which to deal. Since clay is both relatively compressible and impermeable, it may allow settlement and make quick drainage difficult. As a consequence, water may be held against foundation walls and eventually find its way through, thereby causing leaks that call for waterproofing. Heaving brought about by freezing in winter may cause considerable damage unless all foundations are carried below frost line. In some parts of the country, expansive soils cause uneven swelling and settlement. Special measures may be needed to deal with them.

1.8 Rock and Ledge

a Are there any outcroppings of rock? The answer is often obvious, but sometimes the rock barely penetrates the surface or is slightly below it. If there is reason to suspect the presence of rock it is wise to drive a sharpened iron bar into the ground at a sufficient number of points to determine its depth and location. These investigations should go at least as deep as the proposed excavation.

b As a rule, a rocky site is undesirable, not only because it is expensive to excavate but because ledge often carries subsurface water which may cause wet basements unless it is brought under control.

c Architects who have reason to suspect that hidden rock may exist frequently insert a clause in their specifications calling for rock above a certain size, such as one-half cubic yard or more, to be removed at a specified cost per cubic yard.

ROCK AND BLASTING

d. Specification Clause Although no ledge is expected, should any be encountered in areas bounded by portions of the work enclosed by trench walls, the rock need not be removed except where it may interfere with

Inspection of the Site

the running of drains or service lines. All rock or ledge of over one-half cubic yard shall be removed at the rate of $ _____ per cubic yard. At the option of the owner or architect, interfering ledge may be left intact, and the basement space reduced.

e This approach allows the excavation cost estimates to be based upon ordinary soil; otherwise they may be generously increased to take care of an unknown quantity of rock excavation at an increase in price which may not be justified by subsequent actual excavation.

1.9 Water

a Are there any low or damp spots which may indicate the presence of springs? It is wise to dig into such spots to see if springs actually exist. Sometimes springs do not appear until excavation has commenced, and then they present difficulties of a serious character, such as the necessity for continuous pumping or for complete waterproofing of the foundation at additional cost. The contract builder may or may not be able to secure extra remuneration for this.

b Water cannot be ignored or neglected; it always asserts itself and often causes trouble. It frequently occurs in rock or clay formations and usually presents a problem that is expensive to solve. The presence of water is one of the most important items to look for during the inspection of the site. If a water condition exists, the most prudent course is not to build on the site. If construction must proceed, adequate steps to remove the water by drainage must be undertaken. In extreme cases, this may require a permanent pumping installation.

1.10 Trees

a Trees are a valuable addition to any home building site. As few as possible should be felled and those remaining should be carefully protected. Every effort should be made to avoid needless destruction by judiciously relating the house to the trees. In any event, the position, size, species, and condition of trees on the site should be noted on the plot plan. Trees may be killed by altering the grade of the soil surrounding them, especially by heaping soil around the trunks or by cutting important roots.

1.11 Utilities

a Are services such as sewer, water, gas, telephone, and electricity available?

b. Sewer Is there a sewer in the street? If so, what is its size, depth below the surface, and the direction of its flow? It is important to know the depth, if only to determine whether plumbing fixtures in the lowest part of the house can drain by gravity flow into the sewer. If not, pumps are required. It is also important to know the direction of the flow of the sewer, so that the connection of house sewer to street sewer may be made in the direction of the flow and not against it. In communities which have sewage disposal systems, rain water cannot ordinarily be discharged into the sewer, as this may interfere with the sewage disposal process or increase its cost. Other means, usually dry wells, must be found for the disposal of rain water.
c If there is no sewer, and a household disposal system such as a septic tank is required, the proper soil conditions are essential. Soil must be permeable to allow suitable leaching of the effluent from the system. The size and arrangement of the system are strongly influenced by the soil (Chapter 3).
d. Water Where is the water main, what is its size, pressure, and the location of the water connection for the site? If no water main exists, wells have to be dug. A careful survey should be made by a well drilling expert to determine at what depth an adequate supply of water may be found, and what the cost will be. The quality of the water must be determined.
e. Gas If gas is to be used, ascertain the location, depth, size, and pressure of the main and the location of the house connection. If there is no main, the cost and means of supplying gas must be determined.
f. Electricity Ascertain whether electric power and telephone lines are in or whether they will have to be brought in and at what cost. Are lines overhead or underground? Will house lines be overhead or underground?

1.12 Plot Plan

a Figure 1.1 illustrates a plot plan; a house to be built on rather heavily wooded, sloping terrain. A topographical survey can be a decided help in locating and planning the house, as well as to the contract builder in making his estimate. As the contours show, the long dimensions of the house follow the topography rather than the lot lines. Cutting and filling are both necessary to meet the requirements for a level drive and a terrace on the downhill side of the house. House corners are situated to conform to zoning requirements. Trees which may have to be removed or protected during construction are marked, as are outstanding individual trees. Lot lines have been surveyed and located by stone bounds or pipe stakes. Notes

Inspection of the Site

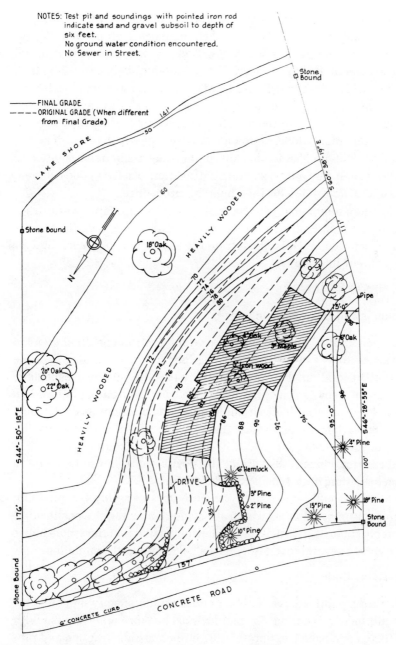

Figure 1.1 Plot Plan of a House on a Sloping Site. Pertinent Information for Construction of the Building Is Given.

respecting the subsoil have been made. Any changes in the topography are not of such a nature as to affect the neighboring properties. Water, gas, and electrical lines are available. There is no sewer; either it will have to be extended to the property, if it is available at all, or a septic tank installation will be required. Subsoil and natural drainage are ideal for the latter, but the installation will have to be located carefully to avoid spoiling the wooded lower slope, and must be far enough from the lake shore to conform to sanitary regulations. The owner must decide which to use, if he has a choice. With this plan and a checklist of the other major items mentioned in this discussion, the builder should be in position to make a good estimate of costs and to determine his final building procedure.

b If more than one house is to be built, as in the operations of a tract builder, a plot plan is drawn for the entire tract and shows the streets and utilities, individual lots, the location of each house on its lot, and all other pertinent information.

LEGAL RESTRICTIONS

1.13 Zoning and Deed Restrictions

a Most communities have zoning laws which restrict specified areas to certain uses, such as industrial, business, general residential, and single residential. In addition, they usually require that the buildings, especially residences, be kept back a certain distance from the front lot line and almost always require set-backs from the side and back lines as well. Sometimes the percentage of total area of the lot which may be covered by buildings is restricted.

b Deed restrictions may also affect the position of a house on a site in addition to limiting cost, type, and other features.

c The builder should check into these matters carefully because any infringements of ordinances which are his fault may require costly alterations at his expense. It is most disconcerting to find that the whole house must be moved several feet, after it is well along toward completion.

1.14 Building Code

a Most municipalities have building codes which specify minimum requirements for construction. The builder must be thoroughly familiar with the building code because it profoundly influences his practice and procedure, as well as the costs of construction. A builder going into a new locality should first of all familiarize himself with the building code, and to do so is logically a part of his general inspection of the site.

b It by no means follows that because two communities are adjacent, they have identical building codes. The opposite is often true, and what is considered excellent practice in one municipality may be expressly prohibited in its neighbor. Furthermore, state codes of practice must be considered, and if the provisions of the state code are more stringent than those of the municipality, the state code must be observed.

c Fees for the issuance of building permits and for inspection by the building commissioner's office are usually required. The builder should ascertain these costs as a part of his general inspection.

2 Building Layout

2.1 General

a The first step in actual construction is to locate the house or houses upon the land. This is known as the "stake-out." Generally this takes place once, but occasionally the building is laid out twice, once for the excavation and again for the foundations.

b All buildings of whatever kind must be built with "lines." The outside faces of the foundation walls are the building lines, and the chief outside walls of the structure are known as the main building lines. These lines are used as reference or base lines from which any subsidiary portions of the buildings are laid out. The dimensions between building lines are shown on the plans, especially the foundation plans, and these figures must be transferred and located on the ground so that the construction of the building may follow them with precision.

c It cannot be emphasized too soon, too strongly, or too often that all good building is "precision" building, which means that floors must be level, walls must be plumb, lines must be straight, corners must be square, and dimensions must be correct, precisely as called for by the plans.

d In order to lay out the building precisely, a system of stakes, lines, and batter boards or offset stakes is customarily used. These are placed after topsoil has been stripped. Surveyors' instruments, especially the transit, are convenient for making the layout, but they are not essential, and accurate work can be done without them. Here will be considered only methods which do not require instruments.

2.2 Establishing Building Lines (Figure 2.1)

a The front building line is in many ways the most important because it is the starting point for all others. Furthermore, in most municipalities zoning ordinances establish minimum distances from the front lines of buildings to front lot lines, and if through error the building is not back sufficiently far, it may entail costly tearing down and rebuilding to correct the error. It is permissible to go back farther than the minimum, but often the owner wants to stay directly on the line. Almost always the "front line" refers to the outside face of the foundation wall under the front wall of the house proper, exclusive of porches, stoops, areas, and so forth, but local zoning peculiarities must be checked to make sure.

b The builder must have clearly defined lot boundaries from which to work. Generally these are defined by surveyors' stakes or bounds of some

Building Layout

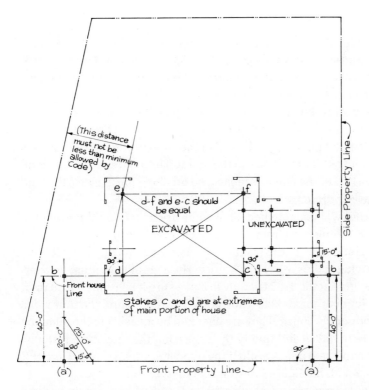

Figure 2.1 The Procedure for Layout of a House on its Site. The Building Lines Are Established First. Batter Boards for Excavated and Unexcavated Portions Are Located to Determine the Building Lines during Excavation.

kind situated at the corners of the lot. If such stakes or other clear markers are not present, the builder should require that they be given him; otherwise he cannot be responsible for errors in establishing house lines.

c Suppose a plot plan shows a house to be 40′ from the front line and 15′ from one side line, and that this conforms with zoning regulations (Figure 2.1). To locate and lay out this house the builder would proceed about as follows. A pair of stakes, *(a), (a)*, is driven on the front property line and a mason's line is stretched between them. These stakes are farther apart than the width of the house. At each stake another line is erected perpendicular to the base line, measured back 40′ with the tape, and two more stakes, *(b), (b)*, are driven. The right angle between the perpendicular and the front property line can be laid off by using the familiar 3-4-5 right triangle. This is accomplished by measuring off from each front stake, along the front line, some convenient multiple of 3′, say 15′, next measuring off on the perpendicular the same multiple times 4'—20' in this instance — and finally making the hypotenuse of the right triangle 25'. If the front lot line is on a curved street it is necessary to work from a tangent to the curve, or to work from a set-back such as a chord of the curve. Here regular surveyors' instruments such as a transit are most useful.

d A line is now stretched between the two house line stakes, and the extreme edge of the house is located by measuring in 15' from the side lot line, as shown on the plot plan. The side wall of the main portion of the house is next located, and a stake *(c)* driven at that point. The length of the front wall of the main portion of the house is measured off along the mason's line and another stake *(d)* driven.

e With the front wall established, side walls are next laid off perpendicular to it by stretching lines from these stakes (that is, from nails driven in the tops), at right angles to the front line, measuring back the correct distance, and driving stakes *(e)*, *(f)*, as before. Here great care must be exercised when laying out the 3-4-5 triangle to make sure it contains a true right angle. These four corner stakes, if correctly located, define the four sides of the main body of the house. They are checked by measuring the distance between the rear stakes and finally by measuring the diagonals **(df)** and **(ec),** the most important check of all. If the opposite sides are equal and the diagonals are equal, the figure is a true rectangle.

f Any ells, wings, areas, etc., which may be attached to the main body of the house are laid off in the same way, using one of the sides of the main body as a base line. In each instance the diagonals of the rectangle are measured as a final check.

g The stakes establish the corners of the house itself, but the excavation must be somewhat larger in order to allow room for the masons to lay up the foundation walls, or for the carpenters to build their forms. Also, when the footings are built, they extend beyond the foundation wall. Finally, only in very rare instances will an earth bank stand vertically; it must be sloped back varying amounts, depending upon the soil. The excavation lines, therefore, must be set out from the house lines 1' or more, depending upon the depth of the excavation.

2.3 Batter Boards and Offset Stakes (Figure 2.2)

a Batter boards or offset stakes are employed to fix the building lines during excavation, since the corner stakes themselves are lost. Batter

Building Layout

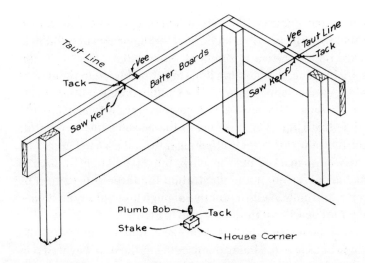

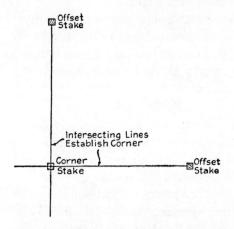

Figure 2.2 Batter Boards. Lines Drawn Taut at Vee, Saw Kerf, and Nail Establish Several Building Lines. Top of the Batter Board Establishes Grades.
Figure 2.3 Offset Stakes Used to Establish Building Lines.

boards, as illustrated in Figure 2.2, consist either of a pair of boards nailed to three uprights so as to form a right angle at exterior corners of buildings, or a single board nailed to two uprights at intermediate points. When batter boards are in place, the building lines are transferred to them from the corner stakes, as shown in Figure 2.2, by holding plumb bobs over the tacks or nails in the tops of the stakes and moving a taut mason's line into contact with the plumb line. The crossing of the mason's line and batter board is marked by a saw kerf or some other means. As shown in Figure 2.2, several lines may be so marked but the marks must be distinct enough to avoid any possible confusion. In the illustration the three sets consist of "vees," saw kerfs, and nails, each distinctive enough to avoid any mistakes.
b Positions of batter boards for the house plan in Figure 2.1 are shown. For the main portion of the house, which is to be excavated, batter boards are placed far enough back to avoid being disturbed. For the unexcavated portion they are placed much closer to the building lines.
c Offset stakes are illustrated in Figure 2.3. These are simply stakes driven far enough away from the corner stakes to avoid being disturbed by the excavation but so situated that when mason's lines are drawn between pairs of stakes, the crossings of the lines locate the house corners.

2.4 Establishing Grades

a The excavator now is enabled to place the cellar hole properly, but he must also know how deep to go. This means establishing the grade of the house and from it the finish grade of the lot, particularly if (as is often true) the finish grade is not to be the same as the natural grade.
b Any point in the house could be used as a reference point for the establishment of all other grades, but for many reasons the top of the foundation wall is the most convenient, although the finished first floor line is also much used. In any event, either the owner or architect must tell the builder what the height of the finished first floor or some other point in the house is to be, or information must be given which will enable the builder to determine it. For instance, in level country, the owner may wish to have his first floor line at the same elevation as those of his neighbors. In that event, the builder can level across from the neighboring house to the site of the new house and establish the first floor line on a stake driven there. This point can also be located on the batter boards by placing the tops of the boards at the desired level. When a line is stretched across the boards, it establishes both building line and reference elevation.

Building Layout

c The plan may show a uniform pitch of grade from front lot line to house. If this were ½″ per ft, for example, and the house were 40′ back, the grade would be 20″ higher at the house than at the front lot line. If then the drawings indicated that the top of the foundation wall was, for example, 8″ above finished grade, the foundation line would be established at 28″ above the front lot line. By setting a stake at the front lot line with its top 28″ above finished grade, the top of the foundation wall could be established by levelling back from front lot line to batter boards or grade stake.

d Once the house has been located on the lot and the grade established, the excavator is ready to proceed, because he can easily check the depth of his excavation from the grade marker and from the depths indicated on the drawings. Similarly, if the house is to be built on a slab at grade level, such excavation, grading, levelling, and other preparation as may be necessary can proceed.

2.5 Tract Layout

a If more than one house is to be laid out, as in a housing tract, standard surveying instruments, especially the transit, level, and tape, are much faster and more efficient in laying out the individual houses than the method described above. An experienced surveying team should be employed for such a project.

3 Excavation. Sanitary Systems

3.1 Specification Clauses

WORK TO BE DONE

a The contractor shall do all excavating, cutting, filling, blasting, and work of similar character necessary for the preparation of the site and the construction of the building. He shall remove all loam 20' each way in excess of the area of the building, and stack where directed. He shall excavate for the entire area of the building as shown, and for all piers, walls, areas, footings, dry wells, tanks, sewer, water supply, gas, electric power, telephone conduit, conductors and other drains and supplies, and whatever excavation is required by the drawings.

FROST, SOLID BED, BACKFILL

b All excavation shall extend at least 2' outside of walls, 6" below finished basement floor, and below frost. If any excavation is dug too low, no filling will be allowed, but masonry and floor foundations shall be started on solid bed. After masonry work is inspected and accepted, and only after such acceptance, trenches shall be filled to a depth of 3' with small stones and then filled with gravel, carefully wetted and thoroughly tamped.

REMOVAL OF SUPERFLUOUS SOIL, GRADING

c Before completion, all superfluous earth shall be removed from the site or carefully spread for grading as may be directed. All clay shall be removed from the site and shall not be used for backfill in any event. No filling shall be done which will cover any portion of the work until it has been inspected and accepted by the architect.

SPRINGS AND RUNNING WATER

d If a spring or running water is encountered which must be permanently provided for, the architect shall be notified, and the owner will pay the contractor at current prices for such work as is performed on the order of the architect. The contractor shall be responsible for any damage.

SPECIAL FILL

e Preparatory to laying concrete floor, level up excavation bottom and bring to proper grade with broken stone or coarse gravel, thoroughly wetted and tamped. Fill areas to proper grade with 1' of small stone and 4" of gravel topping, sloped slightly away from the building. Fill any other places to receive concrete or tile as directed.

Source: *Manual of Septic-Tank Practice*, US Department of Health, Education, and Welfare, Public Health Service. Publication No. 526.

EXCAVATION

3.2 General

a Most excavations, even small ones, are now made with power tools, but if a power shovel or other excavator is not available or the conditions are not favorable for its use, excavation by hand tools and light power equipment becomes necessary. Besides, whether power tools are used or not, on every building operation there is always a certain amount of hand excavation, such as trimming banks and so on.

b Knowledge and understanding of the kinds of soil to be excavated and of the working conditions are essential in order to determine how many men to employ and what equipment to use.

c Excavation is the most variable of all operations associated with building. Once a building is "out of the ground" job conditions are fairly well under control and can be predicted with some degree of certainty. With excavation, however, the story is different. Soil conditions vary with almost every site. Working conditions vary with every operation. Even the best knowledge gained from test pits and other sources gives only a general idea of what is to be expected, and many factors may enter to affect the final result. Good management and good judgment will accomplish much and are indispensable, but the best of these may be upset by unforeseen circumstances such as bad weather and a variety of other unanticipated possibilities, any or all of which may call for unexpected measures to be taken.

3.3 Classification of Soils

a Much of the procedure on any excavation job depends on the nature of the soil. Although excavators frequently lump soils into convenient classifications such as light, medium, or heavy, these classifications are often oversimplified and may lead to faulty decisions. It is not possible to provide a detailed classification of soils in this book, but a classification developed by the Corps of Engineers and the Bureau of Reclamation is summarized in Table 3.1, which also shows the relative suitability of the various soils for foundations of dwelling houses and for domestic sewage systems. The relative numerical ratings are applicable within the individual columns only and not from column to column. The numeral 1 indicates the best condition, and larger numerals indicate progressively poorer conditions. The symbol NS indicates the soil is not suitable.

Table 3.1 Classification of Soils

	Relative Desirability for Various Purposes				
	Foundations			Sewage Disposal	
	Undisturbed		Disturbed (Fill)	Undisturbed	Disturbed (Fill)
	Dense	Loose			

1. Coarse-grained soils					
Gravels					
Clean (little or no fines)					
Well-graded gravel, gravel-sand mixture	1[a]	1	1	1	1
Poorly-graded gravels or gravel-sand mixtures	1	2	2	1	1
With fines in appreciable amounts					
Silty gravels, gravel-sand-silt mixtures	2	2	3	2	2
Clayey gravels, gravel-sand-clay mixtures	3	1	4	2	2
Sands					
Clean (little or no fines)					
Well-graded sands, gravelly sands	1	1	2	1	1
Poorly-graded sands or gravelly sands	1	2	4	1	1
With fines in appreciable amounts					
Silty sands, sand-silt mixtures	2	2	4	2	2
Clayey sands, sand-clay mixtures	3	2	5	2	NS[b]
2. Fine-grained soils					
Silts and clays, liquid limit less than 50[c]					
Inorganic silts and very fine sands, rock flour, silty or clayey fine sands or clayey fine sands or clayey silts with slight plasticity,	3	3	7	2	NS
Inorganic clays of low to medium plasticity, gravelly clays, sandy clays, silty clays, lean clays,	3 Expansion dangerous if dry	3–5	6	2	NS
Organic silts and organic silty clays of low plasticity	4 Expansion dangerous	4	8	2	NS
Silts and clays, liquid limit greater than 50[c]					
Inorganic silts, micaceous or diatomaceous fine sandy or silty soils, elastic silts	5	4	9	2	NS
Inorganic clays of high plasticity, fat clays	5 Expansion dangerous if dry	4 Expansion may be dangerous	8	NS	NS
Organic clays of medium to high plasticity, organic silts	6 Expansion dangerous	5	10	NS	NS
3. Highly organic soils					
Peat and other highly organic soils	7	NS	NS	NS	NS

[a] The numeral 1 indicates best conditions; progressively larger numerals indicate progressively poorer conditions in any one vertical column.
[b] NS = not suitable.
[c] An arbitrary limit between the liquid and plastic states of consistency of a soil. Measured in a standard liquid limit apparatus.
Source: *Engineering Soil Classification for Residential Developments*, Federal Housing Administration, Washington, D.C., 1959.

Excavation. Sanitary Systems 19

3.4 Excavation Equipment

a Equipment for hand excavation is quite simple and consists of shovel (short or long handle depending upon the locality), pick, and means of dirt removal, usually motor truck.

b The difference in type of shovel used is just one of many instances which show that building has not yet fully emerged from its former intensely local character. Although good work can be and is done with both types of shovels, certain regions prefer a short handle; others, the long handle.

c The pick, used to loosen soils too heavy to dig directly, is often aided by the mattock, or grub hoe, which is similar to the pick except that the points are replaced by shorter, flat blades, one at right angles to the other.

d Hand tools are used primarily for digging trenches and trimming excavations otherwise principally performed by power equipment.

e Power equipment used for house excavation usually consists of tractor-driven bulldozer, dipper-type power shovel, or clam shell mounted on a power crane. Dipper and clam shell are usually used if excavations are too deep for convenient maneuvering by bulldozer, otherwise the bulldozer is used for simultaneous loosening and removal of the soil to one side. Large or long trenches are commonly dug by trenching machines.

3.5 Procedure

a Once the building lines are located, the job is ready for excavation to begin. No two excavations are alike, but the general procedure is more or less the same for all. Any job is divided into two classes of work: first, the main excavation, including trimming of banks; and second, trenches of all kinds, including footings, piers, foundation trenches for basementless portions of the building, and trenches for services such as sewer, water, and gas. Portions of these two classes of work, such as the general excavation and trenches for services, can proceed simultaneously; others of necessity have to follow each other.

b. Removal of Topsoil As mentioned in Chapter 1, it is generally necessary to remove the topsoil and store it at a point where it will not be disturbed during construction, and subsequently to replace it as the last step in the finish grading. Topsoil ordinarily is 8" to 12" deep. It is moved to the designated point, the area cleared usually being considerably larger than the ground area of the house. Because the topsoil is removed over a fairly large area, batter boards or offset stakes should not be placed before this operation is completed. It is necessary, in this event, to make a preliminary rough layout so that the topsoil will be stripped over the correct area. For a basementless house, the area of topsoil removal is smaller, and this may complete the general excavation, leaving only trenches to be excavated.

c. General Excavation When excavation is performed by bulldozer, the machine simply moves back and forth over the site to be excavated, loosening and pushing the earth to the ends of the excavation, digging deeper on successive passes. The direction of motion is generally the long direction of the excavation. This procedure results in a more or less dish-shaped excavation whose edges may have to be trimmed to proper dimensions.

d. Trenches and Miscellaneous Holes (Figure 3.1) As has been noted, a fairly large amount of hand work is necessary on any excavation job, and in the dwelling house, this takes the form of trimming, as well as the digging of small trenches, pier holes and, frequently, fairly deep holes for dry wells and septic tanks. This work is seldom very complicated and only occasionally is much bracing required to keep banks from caving in.

e Most trenches are fairly shallow, and excavated earth from the bottom can easily be thrown directly up onto the bank. The chief exception to this rule is the sewer trench, which frequently is quite deep and may require excavated earth to be rehandled from an intermediate level before it is thrown onto the bank. This can be accomplished by digging the trench in steps; the material from the lower level is thrown onto the higher step and then thrown from there onto the bank. Staging may also be built into the trench as it goes down, so that earth from the lower level may be thrown onto the staging, to be rehandled. Detailed discussion of the various methods of bracing and sheet-piling deep trenches and other deep excavations will be omitted here, except to mention that trenches may be continuously sheeted if soil is very loose (Figure 3.1a), or braced with horizontal lagging at intervals if soil is fairly firm (Figure 3.1b).

f Usually separate trenches are required for the various services because they come in at different depths, and street connections for the house are at different points. Occasionally several services can be placed in the same trench. Pipes should not be placed directly over each other because subsequent repairs to the lower ones are made difficult if others are in the way. Water and sewer lines should be separated as far as possible

Excavation. Sanitary Systems

if in the same trench because simultaneous leaks in the two lines might pollute the water supply. Water lines should be placed above sewer lines, preferably not directly above but offset to one side.

g Certain precautions must be observed when digging trenches or deep holes for dry wells, septic tanks, and so forth. Earth is treacherous and a bank that looks perfectly safe may slide. Sometimes warning is given by cracks appearing in the surface of the earth along the top of the trench. Earth trickling down the side of the bank also indicates that it may give way soon. In any event, all excavated material should be thrown well back and not piled close to the trench. If bracing is not to be used, the banks should be sloped sufficiently to avoid a cave-in; the looser the soil, the gentler the slope. Finally, once the excavation is made, it should receive whatever it is to contain—pipes, tanks, crushed stone, etc.—as rapidly as possible. Steep banks stand for a short time but with time, rain, and vibration they may slide.

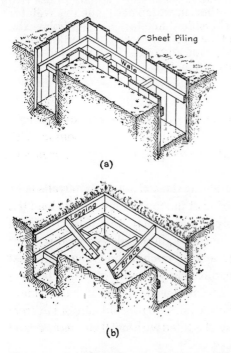

Figure 3.1 Bracing for Trenches. (a) Vertical Sheet Piling. (b) Horizontal Lagging.

DRAINS AND DRY WELLS

3.6 Specification Clauses

LAND OR WALL DRAIN

a Fill in around all outside walls with 1' of small stone above the pipe, and from the top of stone to grade with coarse gravel. Lay the wall drain without mortar upon treated planks and put asphalt paper over all joints.

CONDUCTOR DRAINS

b Furnish and lay 3" glazed earthenware pipe from each downspout to a dry well, each pipe to turn up with a quarter bend at the foundation wall and to be brought 6" above the finished grade with one length of 3" heavy cast iron soil pipe, with cast hub to receive the foot of the downspout. Make connections to dry wells.

DRY WELLS

c Excavate for and build dry wells, one for each downspout, unless two downspouts are within 10' of each other, in which case, one dry well for both will suffice. Wells are to be 4' inside diameter, 4' deep below the basement bottom, 3' below the inlet, and 10' away from the building. Refill with broken stone to within 3' of grade and cover with pressure-creosoted plank (or inverted sod in place of planks).

3.7 Wall Drains (Figure 3.2)

a When foundations are built in impermeable soils such as clay or mixtures of clay and sand, underground water is apt to be held in contact with the walls and to cause leakage, particularly after heavy rains, or in wet seasons.

b One of the simplest methods of draining the soil adjacent to walls is to lay a line of land tile all around the base of the foundation, and to lead the line into a large dry well, preferably at a low point, and at least several feet from the house. Land tile are porous cylindrical clay pipe without hubs. They are laid end to end with a small opening between individual pieces to allow water to seep in readily. They may be carefully bedded in a layer of small stones or laid on a treated wood plank. It is customary to lay strips of tarred felt loosely over the joints to prevent clay or silt from entering. The space adjacent to the wall is filled with small stones or gravel

Excavation. Sanitary Systems   23

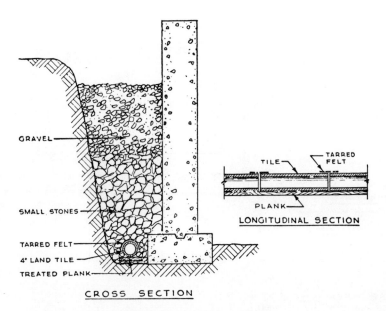

Figure 3.2 Footing Drains around Foundation Wall. Burned Clay Tile Covered with Porous Fill to Carry off Ground Water.

to within a foot or two of the finished grade. Finish grading with topsoil subsequently covers the gravel fill. Backfill should be compacted to avoid future settlement (Figure 3.2).

3.8 Roof Drains (Figures 3.3–5)

a Roof drains, called conductors, leaders, or downspouts, must be led into underground drainage basins, or dry wells, unless roof water can be led into the house sewer and into the municipal sewerage system, or unless the water can be discharged upon the ground. The underground drain, leading from downspout to dry well, must be watertight, unlike the land tile wall drains discussed in the previous section. Glazed earthenware hub and spigot drain tile or various composition pipes are customarily employed for the horizontal underground portion. This is connected by an elbow to a vertical soil pipe into which the downspout is led at or just above the grade line. Joints in earthenware pipe are tightly caulked. A typical section is shown in Figure 3.3.

b If rain water is permitted to run into the house sewer, an encircling drain, connecting all conductors, may be built as shown in Figure 3.4.

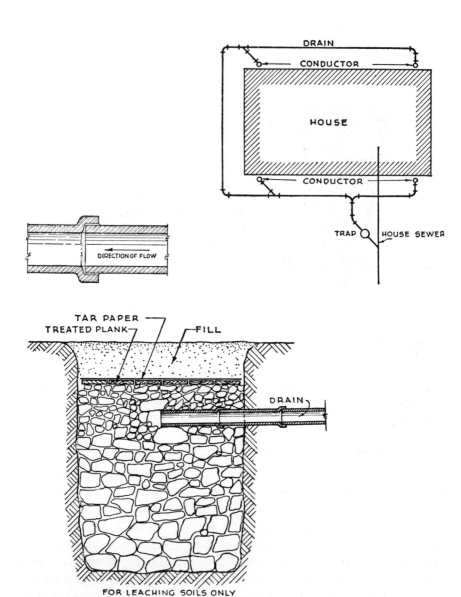

Figure 3.3 Section through Hub and Spigot Drain Pipe.
Figure 3.4 Plan of Encircling Drain for Foundations.
Figure 3.5 Section through Dry Well.

Excavation. Sanitary Systems

c A rule for establishing proper grading for a drain is usually stated: To 1' of fall in a drain, allow a length of 10' for each inch of diameter of the pipe.

d To make a formula of this rule, we would have, if f = fall in feet, l = length of drain in feet, and d = diameter of pipe in inches:

$$f = \frac{l}{10d}$$

Thus, for a 6" drain, 60' long, the fall required would be $60/(10 \times 6) = 1'$.

3.9 Dry Wells (Figure 3.5)

a Dry wells are devices for disposing of excess water conducted from roofs or carried away from foundation walls by drain tile. They are most effective in sandy or gravelly soils because their function is to distribute the waste water into the subsoil, and in impermeable soils such as clay this cannot be accomplished satisfactorily.

b Dry wells are simply fairly deep, fairly large excavations filled with crushed stone, gravel, broken brickbats, or other broken masonry, into which waste water is led for distribution to the surrounding soil. For roof water, dry wells must be away from the house a sufficient distance to prevent the water seeping into the soil from building up a hydrostatic head against the foundation walls. Usually there are several dry wells for roof water, one for each principal downspout. They may be anywhere from 4' to 6' or 7' in diameter and as deep; the size depends upon the amount of roof water to be disposed of and upon the permeability of surrounding soil.

c The dry well for the collection of water around the foundation walls, served by the wall drains, is usually considerably larger than the others, and its top is at the elevation of footing drains. Its size, which may be from 6' to 8' each way, also depends upon the size of the house and the permeability of the soil, as well as upon the condition in the site. A section through a typical dry well is shown in Figure 3.5.

SANITARY SYSTEMS

3.10 Sanitary Waste Disposal

a Household and human wastes must be disposed of in such a way as not to contaminate drinking water, not be accessible to insects, rodents, or

Dwelling House Construction 26

other disease carriers, not be accessible to children, not violate laws governing pollution, not pollute beaches and streams, and not create a nuisance due to odor or appearance. By far the best way is to discharge such wastes into an adequate community sewerage system. If this is impossible, a carefully designed and constructed and properly maintained septic tank system can be satisfactory.

b A cesspool, rarely used today except where such installations already exist, is simply a sewage receptacle built underground, of sufficient size to accommodate the house, and with walls constructed of porous material such as open-jointed masonry to allow seepage into surrounding soil — which must be permeable if the cesspool is to be effective. Cesspools are only partially satisfactory at best and may be positive menaces to health. If an individual sewage disposal system must be used, the septic tank is much to be preferred.

c A septic tank system consists essentially of two parts: a tank into which wastes are discharged, digested, solids separated from liquids, and the liquids prepared for disposal; and the disposal field in which the liquids are allowed to percolate into the soil. Anaerobic bacteria (bacteria not dependent on free oxygen) accomplish much of the digestive action in the tank; aerobic bacteria (bacteria requiring free oxygen) in the soil attack the liquid effluent, pathogens are removed by percolation, and disease bacteria eventually die out in the unfavorable environment of the soil. Although detergents in the quantity usually used do not affect the system as such, they are often highly resistant to digestion and may not be removed. If they penetrate into water supplies, they may cause foaming and other undesirable effects. Special care should therefore be taken to avoid penetration of potable water supplies by detergent effluent. Soil *must* allow effluent to percolate away readily. Impervious soils, therefore, are not suitable for septic tank systems (see Section 3.12 for soil tests).

3.11 Septic Tanks (Figures 3.6, 7)

a The most important function of the septic tank is to separate out solids that would otherwise quickly clog the soil in the disposal field. The rate of flow of sewage discharging from the house is reduced upon entering the tank, thereby allowing solids to sink into the sludge at the bottom or rise to the scum of the top. The clarified effluent is discharged to the disposal field.

Excavation. Sanitary Systems

b Solids and liquids in the tank are attacked and decomposed essentially by anaerobic bacteria. The treated effluent sewage may be more highly septic and malodorous than the raw sewage (hence the name septic tank), but it causes less clogging than untreated wastes.

c No matter how efficient the septic tank may be at decomposing and digesting wastes, residual solids remain compacted at the bottom and scum containing solids and grease floats partially submerged at the top. The septic tank, therefore, must provide storage space for sludge and scum between periodic cleanings.

d Septic tanks should be at least 5' away from any building, should not be in swampy ground, and should be situated far enough away both vertically and horizontally from wells or other sources of potable water to avoid contamination. Usually, this means that the tank and disposal field should be downhill and at least 50' away from water sources. The depth and direction of flow of underground water sources should be considered when selecting the site for a septic tank system.

e The general arrangement of a typical septic tank system is shown in Figure 3.6. Household wastes discharge through a liquid-tight sewer line into the liquid-tight septic tank. The effluent discharges from the tank into a distribution box and is there directed into the disposal field consisting of drain tile with open joints laid in gravel or crushed stone.

f Tanks must be large enough to handle the expected wastes. Table 3.2 gives recommended tank capacities for normally-expected uses, including garbage grinder, automatic washer, and other household appliances. Materials of construction should not be corroded or otherwise attacked by the contents or surrounding soil. Concrete, properly coated metal, vitrified clay, concrete blocks, and hard-burned bricks have been found suitable. Job-built masonry tanks should have all joints well filled and concrete

Table 3.2 Liquid Capacity of Septic Tank (In Gallons; Provides for Use of Garbage-Grinder, Automatic Washer, and Other Household Appliances)

	Recommended Minimum Tank Capacity	Equivalent Capacity per Bedroom
2 or less	750	375
3	900	300
4[a]	1,000	250

[a]For each additional bedroom, add 250 gallons

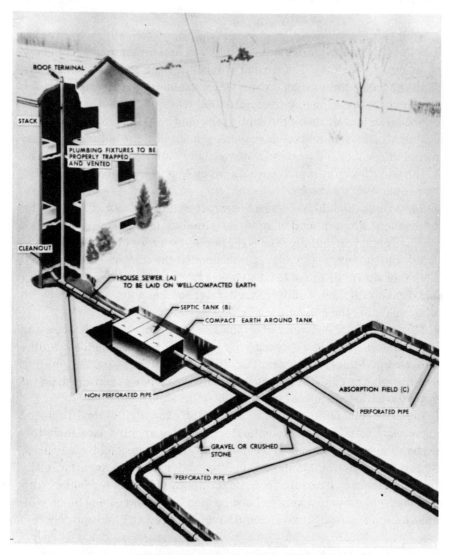

Figure 3.6 Arrangement of House Sewer, Septic Tank, and Absorption Field.

Excavation. Sanitary Systems

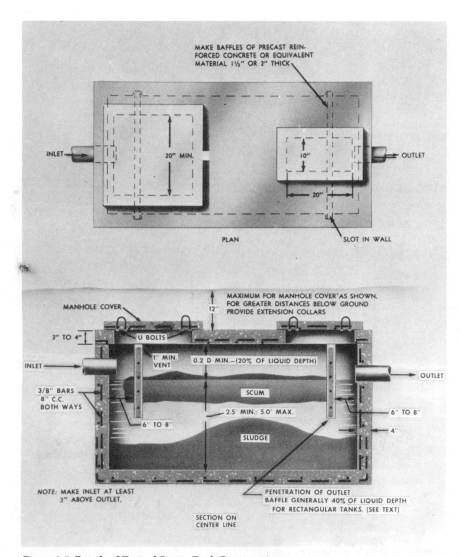

Figure 3.7 Details of Typical Septic Tank Construction.

Dwelling House Construction 30

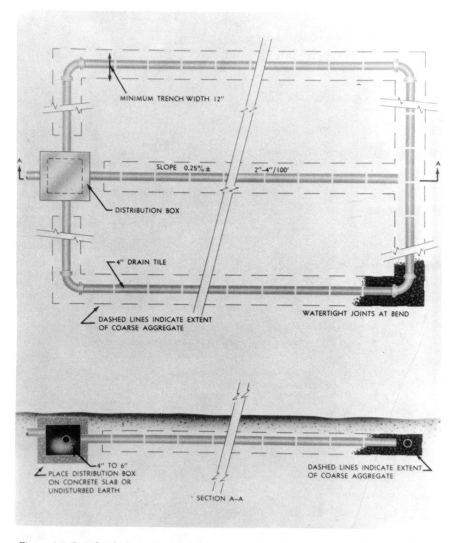

Figure 3.8 Details of Absorption Field.

Excavation. Sanitary Systems

blocks should be plastered on the inside with two $\frac{1}{4}''$ thick coats of portland cement and sand mortag. Prefabricated concrete tanks should be reinforced for handling and should have walls at least $2\frac{1}{2}''$ thick. Concrete access covers should be at least $3''$ thick.

g Adequate access for cleaning and inspection (Figure 3.7) must be provided at both the inlet and outlet ends, and in each compartment if there are several.

h A typical single-compartment tank is shown in Figure 3.7. The liquid level rises to the bottom of the outlet. The inlet is at least $1''$ and preferably $3''$ above the outlet to allow for temporary rises in liquid level during discharges from the house sewer. Sludge settles to the bottom and scum rises to the top. A baffle provided at each end prevents undue penetration of scum into the inlet and outlet, especially the outlet. Vents above the baffles allow gases to escape, usually back along the house sewer and up through the house vent.

i The proportions of the tank can be varied considerably, but in plan the smallest dimension should not be less than $2'$. Liquid depth may range from $30''$ to $60''$. Space must be allowed for scum and gas above the liquid line; about 30 percent of the scum usually rises above the liquid line. For tanks with straight vertical sides, the space above the liquid line should be about 20 percent as high as the depth of the liquid. In horizontal cylindrical tanks about 15 percent of the cross-section of the circle should be above the liquid level, i.e., the liquid depth should be between 75 and 80 percent of the diameter of the tank.

3.12 Disposal Field (Figure 3.8)

a The first step in the design of a septic tank system is to determine whether the soil is suitable for percolation of the effluent. If it is not, the system is not feasible. The soil must have an acceptable percolation ratio without interference by ground water, which should be at least $4'$ below the surface. Rock formations or other impervious strata should be at least $4'$ below the surface.

b A percolation test is first run to determine the characteristics of the soil and to establish the design of the disposal field. A procedure found satisfactory by experience is as follows:

1. Make six or more tests spaced uniformly over the site.
2. Dig or bore a hole $4''$ to $12''$ in diameter to the depth of the proposed absorption tunnel. Scratch the bottoms and sides of the hole carefully to

remove any smeared spots. Remove all loose material and add 2″ of coarse sand or fine gravel to protect the bottom of the hole from scouring.

3. Presoak the soil by filling the hole with water to at least 12″ above the gravel and keep it at that depth for at least 4 hours, or preferably overnight. Determine the percolation rate 24 hours after water is first added to the hole. (If the soil is sandy and has little or no clay, the presoaking is unnecessary and one preliminary filling is enough.)
4. Run the percolation test by bringing the depth of the water in the hole to 6″ over the gravel. From a fixed reference point measure the drop in water level.
    i. If water remains in the hole overnight, measure the drop in a 30-minute period.
    ii. If water does not remain in the hole overnight, bring the depth to 6″ over the gravel, measure the drop at approximately 30-minute intervals (adding water if necessary over a 4-hour period), and use the drop in the last 30 minutes to calculate percolation rate.
    iii. In sandy and other rapidly percolating soils in which 6″ of water seeps away in less than 30 minutes, time intervals between measurements are 10 minutes and the test is run for 1 hour. The drop during the final 10 minutes is used to calculate percolation rate.

c Table 3.3 gives the required absorption area in square feet per bedroom for various percolation rates.

d Two types of soil-absorption systems or disposal fields are generally used: standard trenches and seepage pits. Either should be kept a safe distance from water supply and dwellings. Local conditions dictate the details, but in general, absorption systems should be kept at least 100′ from any water supply well, 50′ from any stream or watercourse, and 10′ from dwellings or property lines. Seepage pits should be avoided where wells are shallow, or where underground channels such as in limestone formations may connect with water sources.

e Standard trenches are most commonly used. They consist of 12″ lengths of 4″ agricultural drain tile, 2′ to 3′ lengths of clay sewer pipe, or perforated nonmetallic pipe laid with open joints to insure reasonable flow into the soil. Individual laterals are preferably less than 60′ and in any event not more than 100′ long. Tiles and trench bottom should have a grade of 2″ to 4″ per 100′; never to exceed 6″ in 100′. A typical layout is shown in Figure 3.8. Many different designs can be used.

Excavation. Sanitary Systems

Table 3.3 Absorption-Area Requirements for Private Residences (Provides for Garbage-Grinder and Automatic-Sequence Washing Machine)

Percolation Rate (Time Required for Water to Fall 1″, min)	Required absorption area, in sq ft per Bedroom,[a] Standard Trench[b] and Seepage Pits[c]
1 or less	70
2	85
3	100
4	115
5	125
10	165
15	190
30[d]	250
45[d]	300
60[d,e]	330

[a] In every case, sufficient area should be provided for at least 2 bedrooms.
[b] Absorption area for standard trenches is figured as trench-bottom area.
[c] Absorption area for seepage pits is figured as effective side-wall area beneath the inlet.
[d] Unsuitable for seepage pits if over 30.
[e] Unsuitable for leaching systems if over 60.

Table 3.4 Distances between Trenches

Trench Width, in.	Minimum Distance between Centerlines of Trenches, ft
12 to 18	6
18 to 24	6.5
24 to 30	7.0
30 to 36	7.5

f Depth should be at most 18″. Freezing rarely occurs in a carefully constructed system kept in continuous operation. Tiles must be well surrounded by gravel. Pipes under surfaces usually cleared of snow, such as driveways, should be insulated.

g Trenches vary from 18″ to 24″ in width; not less than 12″ in any event, with tile laid in at least 6″ of gravel. Alignment boards may be used under the tiles to keep them on a straight uniform grade, but these are optional. The total length of trenches may be calculated from the percolation tests (Table 3.3) and the width of the trench. If, for example, the percolation rate is 10, and there are four bedrooms, the total absorption area required is 4 x 165 = 660 sq ft. If trenches are 2′ wide, the total length required is 330′. If there are 6 laterals, each is 55′ long.

h Minimum spacing center to center of tile lines varies between 6′ and 7.5′ (Table 3.4), depending on the width of the trenches. In the above example, 6.5′ should be used. The total area of the field, therefore, becomes 6 x 6.5 x 55 or 2,145 sq ft.

4 Foundations

4.1 General

a In many respects, foundations are the most important part of any structure. Once built, little can be done to alter them. If they are adequate, the building remains stable, level, and plumb; if not, differential settlement causes cracks and leaks in the foundation, sloping floors, binding windows and doors, cracked plaster, and general racking of the superstructure.

b Foundations normally consist of two principal parts—footings and walls. Materials generally employed are cast concrete or concrete block, although brick and rubble masonry are found. When no basement is desired, the substructure may consist merely of piers resting on footings.

c Many basementless houses are built on slabs resting on soil at grade level. A peripheral footing integral with the slab is employed, and similar integral footings support bearing partitions, chimneys, and other local loads.

d An important aspect of all foundation design and construction in cold areas is to make sure that expansion ("heaving") of soil because of frost action will not cause distortion of the structure. This means the provision of good drainage to carry away soil water which might cause trouble, or extending the bottoms of foundations or footings below the frost line, the greatest depth to which frost penetrates.

FOOTINGS

4.2 General

a Footings may be required under foundation walls, piers, and posts, to spread the load to such an extent that settlement either is negligible or is uniform under all portions of the structure. In firm soils, such as compact sand and gravel, footings can often be omitted if the foundation wall is cast concrete, especially if it is reinforced. Walls of concrete blocks, brick, and stone should have footings, no matter what the soil.

4.3 Footing Design (Figure 4.1)

a Footings for dwelling house foundation walls are seldom designed but are merely built by rule of thumb. Generally speaking, the depth of the footing should be the same as the thickness of the wall above it. Its width is 3″ to 4″ greater on each side than the wall above. As a matter of design, the footings ought to be proportioned to the loads transmitted to them from

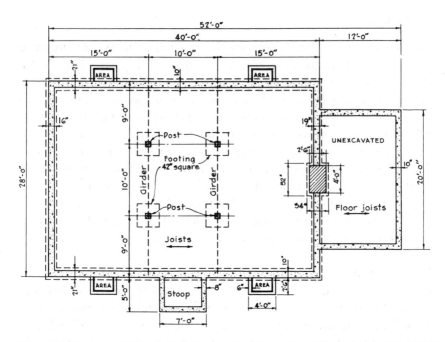

Figure 4.1 Footing and Foundation Plan. Footings Proportioned for Approximately Equal Load Distribution.

the building, in order to spread the loads uniformly on the soil and, thereby, prevent differential settlement. Dwelling-house loads are generally so light, however, that wider footings than those indicated are seldom necessary. Nevertheless, where unusually soft soils are encountered, the footings should be carefully proportioned to spread loads evenly and over sufficient area to avoid overloading the soil beneath.

b In Figure 4.1 is shown a typical house foundation plan. Part is excavated, part is not; a chimney is placed at one point, and there are several areaways, as well as a stoop. Two girders running the width of the house are supported at their ends by the foundations and at two intermediate points by posts. The foundation walls support the exterior house walls, the floors, and the roof. Floor and ceiling of the ell are supported by the long outer wall and by the right-hand wall of the main foundation. The roof of the ell is supported by its two short foundation walls.

c If this house were to be built in yielding soil, it would be well to investigate the probable loads on the various portions of the footing and to proportion them accordingly. The following computations are based upon

Foundations

customary design assumptions respecting the weights of cast concrete and masonry, live load plus dead load on floors and roofs, and weights of walls and partitions. Computations are based upon a 1' length of wall, upon the total weight of chimney, and upon the total tributary load on a post.

d Examination of the computations in Table 4.1 shows that minimum loads per foot of length are found under the ell walls. At 1,200 to 1,360 lb per ft under 10"-thick walls, the bearing on the soil is approximately 1,400 to 1,600 lb per sq ft. This is not excessive for moderately dense clay, and footings could be omitted under the ell walls. With loads of 1,600 lb per sq ft, the sizes of footings shown in the computations are found to be necessary for the other walls, chimney, and posts, if approximately uniform load intensities are to be maintained on the soil. With other allowable bearing loads on soil, other footing sizes are needed.

Table 4.1 Footing Sizes for Approximately Equal Load Distribution (See Figure 4.1)

		Footing (Width)
Main portion, load per running foot		
Long walls		
Foundation, 8' x $^{10}/_{12}$ x 140 lb/cu ft	940 lb/ft	
Walls, 2 stories high, 18' x 20 lb/sq ft	360	
Ends of girders (see posts) spread over 10' of footing, 4½' x 1985 lb/ft = 8,900 lb ÷ 10 =	890	
Roof, 14' x 30 lb/sq ft	420	
	2,610 lb/ft	19"
Short walls		
Left		
Foundation, 8' x $^{10}/_{12}$ x 140 lb/cu ft	940 lb/ft	
Floors, first and second, 2 x 7½' x 50 lb/sq ft	750	
Wall, 2 stories plus gable, 23' x 20 lb/sq ft	460	
	2,150 lb/ft	16"
Right		
Foundation (omitting chimney)	940 lb/ft	
Floors, main house	750	
Ell, 6' x 50 lb/sq ft	300	
Ceiling, ell, 6' x 10 lb/sq ft	60	
Wall	460	
	2,510 lb/ft	19"

Table 4.1 (cont.)

		Footing (Width)
Ell, load per running foot		
Short walls		
Foundation, 6' x $^{10}/_{12}$ x 140 lb/cu ft	700 lb/ft	
Wall, 1 story, 10' x 20 lb/sq ft	200	
Roof, 10' x 30 lb/sq ft	300	
	1,200 lb/ft	none
Right wall		
Foundation, 6' x $^{10}/_{12}$ x 140 lb/cu ft	700 lb/ft	
Wall (part gable), 12' x 20 lb/sq ft	240	
Floor, 6' x 50 lb/sq ft	300	
Ceiling, 6' x 20 lb/sq ft	120	
	1,360 lb/ft	none
Posts		
Load per foot of girder		
First and second floors, 2 x 12½' x 50 lb/sq ft	1,250 lb/ft	
Bearing partitions above, first and second floors. 18' x 20 lb/sq ft	360	
Attic floor, 12½' x 30 lb/sq ft	375	
	1,985 lb/ft	
Loads per post, 9½' x 1985 lb/ft =	18,900 lb	42" x 42"
Chimney block, approximate total weight		
Average solid masonry 38' x 3' x 1.33' x 120 lb/cu ft	18,300 lb	34" x 52"

e With only two posts per girder, the load per post becomes large and the footings correspondingly large. Unless they are reinforced, the footings must be approximately 20″ deep and, therefore, require a considerable amount of concrete, as well as adding over 200 lb per sq ft to the load on the soil. It would be preferable to use three posts, thereby cutting down the load per post as well as reducing the size of the girder.

f It should be noted that floor and roof loads are "live," i.e., intermittent, whereas the chimney load is permanent. Where such a condition exists and the soil is soft, it is a good plan to tie the footings together by embedding short lengths of reinforcing steel in the concrete at the juncture of the footings. In soft soils, this is also advisable at corners and other spots where breaks in the direction or width of footing occur. Such spots are otherwise likely to develop cracks because of load concentrations and changes.

4.4 Forms (Figure 4.2)

a Footing forms are often omitted when the soil is firm enough to stand as a vertical wall for a short time. A shallow trench the size of the footings is dug and concrete is immediately deposited in it. While this is the easiest way to form the footings, it often leaves them uneven on top and may consequently cause some trouble with the subsequent erection of wall forms. Tops of such footings should, therefore, be made as level as possible by lining up with level and cord.

b When the top of the footing projects above the level of the basement excavation, or when the soil will not stand as a vertical cut, it is necessary to build some kind of footing form. This is most conveniently done by placing on edge in the trenches 2″ planks of the proper width, and holding them in position by stout stakes driven into the ground at intervals of 5′ to 6′. The planks are lightly nailed to the stakes after their top edges are levelled at the proper height. Concrete deposited in such forms can easily be brought to a level and straight line.

4.5 Sizes

a Unreinforced solid concrete foundation walls for full-depth basements are usually 10″ to 12″ thick, although 8″ often meets requirements. Very thin walls and piers must be used with considerable caution, for while they may meet load-bearing requirements, they have a tendency to fail in shear or to buckle because of earth and water pressure. It is a good rule, therefore, whenever walls or piers less than 10″ thick are used, to reinforce

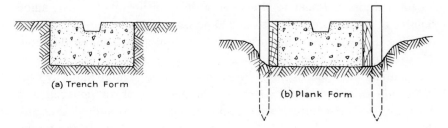

Figure 4.2 Footing Forms.

them. Another useful rule of thumb is to limit the height of unreinforced walls to approximately ten times their thickness; if the ratio is greater, reinforcement should be used.

b Frequently, the thickness of the wall is determined by the superstructure, rather than by the loads it is to carry. The thickness of a frame wall is usually 5" to 6". An 8" wall will do in this case. A brick veneer wall is at least 9" thick and at least 9" or 10" of foundation wall must be provided. For stone face and backing, 16" to 18" may be necessary (Chapters 5 and 10).

c The nature of the soil has its influence. Heavy impervious clay generally creates a water condition outside the walls; and if the basement is to be kept dry, the walls may have to be heavier, water-proofed, and provided with special drainage or a combination of all three.

4.6 Forms (Figures 4.3, 4)

FORM LUMBER

a. Specification Clause All concrete walls shall be formed with sound material properly joined, braced, spaced, and otherwise supported to insure tightness and rigidity. Proper forms shall be provided for all openings in the walls, including boxes or sleeves for all pipes entering or leaving the building.

b Dwelling house foundation forms may be built of lumber which is used subsequently in the superstructure. This consists of studding and rough flooring, sheathing, or plywood, the studs usually 2" x 4" and the boards 1" x 6" to 1" x 10" square-edged or matched material (Chapter 5). As this lumber is to be used again, it should be cut as little as possible. For instance, some stud heights of the house may be 7'-8" and others 8'-3", but studs are not usually to be had in these lengths. Therefore, 16' material is purchased and cut into two pieces. Such irregular lengths may be used for form studs, as these need not be uniform in height. Further, since lengths of boards are multiples of 2' and plywood is usually 4' wide, it is ordinarily best to space the form studs 2' on centers so that as little waste as possible results from cutting.

c Sheathing, if of boards, must be sound and free of large knots or knot holes; otherwise it does not withstand the pressure of wet concrete and does not provide a smooth wall. Studding must be examined, and any defective pieces laid aside to avoid any danger of failure in the forms. The same holds true in even larger degree for the rangers.

Foundations

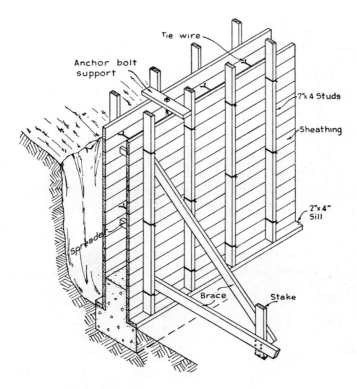

Figure 4.3 Simple Stud Form.

4.7 Types of Forms

a. Simple Stud The simple form shown in Figure 4.3 consists of vertical studs faced with boards or plywood. This is simple but not very robust and should be used, if at all, only for minor walls. Fresh concrete weighs approximately 150 lb per cu ft, and when soft exerts a hydrostatic pressure equal to 150 lb times the depth in feet. Consequently, if such a form is filled too rapidly, too great pressure is exerted against it and it either bulges or collapses.

b The studs are set opposite one another and are tied together at several points in their height. Since these ties take all the stress, there must be enough of them to withstand the pressure of wet concrete.

c Braces are required at frequent intervals with this type of form because there is no lateral stiffness except that provided by the sheathing. Furthermore, there is nothing to keep the form in alignment except the braces.

Dwelling House Construction

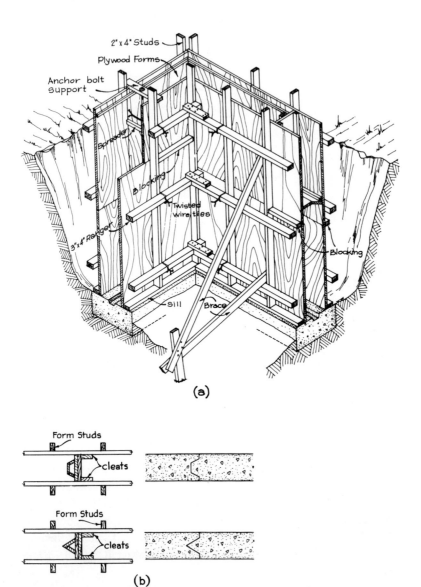

Figure 4.4 Stud and Ranger Form. (a) Section of Form. (b) Method of Framing Vertical Joints in Concrete Walls. (c) Prefabricated Plywood Box Forms.

Foundations

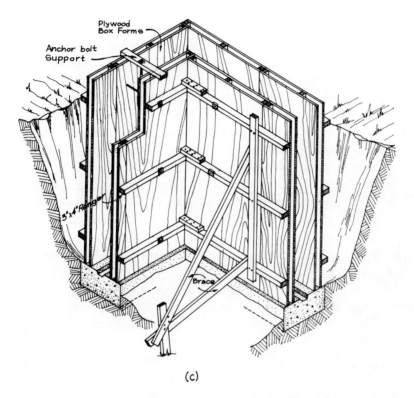

(c)

d. Stud and Ranger Because of their lack of rigidity, simple stud forms are reinforced, when greater strength is needed, by rangers running horizontally, as shown in Figure 4.4. Rangers are placed on both sides of the form and are tied together. With this construction only enough braces are needed to keep the form from overturning. With studs, rangers, and braces properly spaced, it is possible to deposit concrete to the full height of the wall without danger of distortion.

e Rangers should always be used if it is important to have a perfectly straight wall, free of undulations and true to line. Moreover, if the rangers instead of the studs are tied together, studs on opposite sides of the wall need not be in pairs. The rangers stiffen the entire form so that fewer braces are necessary. Any breaks or weaknesses in the form are held in check by the rangers.

f Rangers are commonly 3″ x 4″ or two 2″ x 4″'s.

Dwelling House Construction

g Standard practice is to space studs 2' on centers. The strength of a 1″ board (Chapter 5) or of plywood in bending controls this distance, and the load on the board, in turn, is governed by the hydrostatic head or height of soft concrete above it. For heights up to 5', deflection is not noticeable in such a board if studs are 2' on centers. For heights more than 5', studs must be closer together, unless the concrete is deposited slowly enough to allow the initial set to take place at the bottom, thus reducing the hydrostatic pressure. Forms are generally built completely around the wall, and the concrete is deposited in layers.

h Rangers are spaced varying distances apart vertically, but are closer together at the bottom to resist the increased pressure. The lowest ranger is ordinarily 1' from the bottom, the second is 2' to 3' above that, and the third about 1' down from the top. In dwelling house work, three rangers are usually enough, but they should not be more than $3\frac{1}{2}$' to 4' apart vertically if casting is continuous.

i Rangers must be especially straight and should be at least 16' long. Their ends should be butted and the joints spliced with pieces of board at least 3' long. If rangers are doubled 2″ x 4″'s, splices are made by staggering the joints.

4.8 Prefabricated Reusable Forms

a Prefabricated reusable sectional forms may be employed. These are commonly boxes made of studs with waterproof plywood on each side. The sections are set adjacent to each other and braced (Figure 4.4c). Such sectional forms are especially useful in tract housing where many identical foundations are built. Prefabricated forms may also be metal.

4.9 Spreaders and Wiring

a In order to insure uniform thickness in the completed wall, the forms must be kept a uniform distance apart and be held there while the concrete is setting. This is accomplished by spreaders and wire ties or other devices, the spreaders giving the proper thickness and the ties preventing the forms from yielding (Figure 4.5). The principle is illustrated here with spreaders and ties, but a number of proprietary devices are commonly used to achieve the same effect.

b Spreaders are short metal bars or wood sticks (Figure 4.5a), usually about 1″ x 2″ in cross-section and of length equal to the thickness of the

Foundations

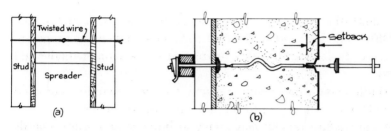

Figure 4.5 (a) Spreaders and Wires. (b) Metal Bar Spreader-Tie With Snap-Off Ends.

wall. They are inserted between the sides of the form and are held in place by the pressure exerted against their ends by the ties, which should be fairly heavy, soft, black, annealed iron wire because it must withstand both tension and severe twisting stresses without breaking. Number 8 wire is best; not only is it strong enough, but it twists rapidly and does not cut into the wood as badly as does smaller wire. However, No. 10 or 12 is suitable for simple stud forms, which are usually small.

c Wires are inserted as the forms go up. In the simple stud form, the wires may be placed in the joints between the boards. The ranger type of form requires $\frac{3}{8}''$ to $\frac{1}{2}''$ holes to be bored in the sheathing to allow wires to pass around rangers on both sides. Wet concrete readily chokes these holes.

d Many proprietary devices are available to perform the combined functions of spreader and tie wire. These are usually easily and quickly put in place in the forms and stay in place in the concrete when the forms are removed. Some provide a device which can be snapped off or otherwise removed at the surface of the concrete, leaving a depression which can be filled with mortar or caulking material. One type is illustrated in Figure 4.5b. This device holds forms the proper distance apart, but the ends of the tie are snapped off when forms are dismantled, leaving part of the bar in place, and leaving a small depression at the surface that can be left exposed or mortared over.

4.10 Erection

a Forms may be erected one side at a time or both sides may be thrown up simultaneously. In the former procedure, the outer side is built first and ties may be inserted loosely in the sheathing as it goes up. If the form is a large one, the studs are erected in place, are nailed to a lower horizontal $2'' \times 4''$ member called a shoe, and are temporarily braced. The sheathing is then nailed to the studs. (Small forms can be built advantageously on the flat and lifted into place when finished.) The inner side is then erected. As it goes up, the ties and spreaders are placed between the two sides and tightened.

b When both sides are built simultaneously, both sets of studs are erected at once and the sheathing goes up on each set at the same time.

c In either procedure, rangers may or may not be placed as the form boards go on, but usually it is best to place them at once so that the form is straight from the very start.

d Braces are set more or less simultaneously with the studs and boards. Enough must be placed at the beginning to hold the form in alignment, and when sheathing is in place, sufficient more must be placed to hold the form rigid. The simple stud type may require braces at every other stud, the ranger type seldom any oftener than every fourth or fifth stud.

4.11 Runways

a When concrete is to be deposited by wheelbarrow or buggy, runways may be built which are partially supported by the forms. For easy dumping, the top of the runway should be at the level of the top form board and should be wide enough to permit two vehicles to pass.

b With the ready-mix concrete commonly employed, it is usually possible to bring the ready-mix truck into position and to deposit the concrete directly by chute into the forms, thus eliminating the runways.

4.12 Top of Form (Figure 4.4)

a The importance of levelling the top of the concrete wall cannot be overemphasized. In order to stop the concrete at the proper height, a guide must be provided which indicates the top of the wall. This may consist of nails driven 2' to 3' apart in the top form board along a straight level line established at the proper elevation. A more accurate method is to nail a wood strip along this line and cast the concrete to the underside of the strip. Some form builders cut the top of the sheathing along this line so that the wall cannot possibly be carried too high and a good level surface can readily be achieved.

4.13 Openings and Sleeves (Figure 4.4)

a Provision must be made in the form for openings such as doors and windows. Finished wood frames should not be concreted in place because

Foundations

they are likely to swell and become distorted. Openings should be boxed and the frames set later. In the boxing should be placed provision for later anchoring the wood frames to the concrete. These may be bolts, clip angles, or other similar devices. When steel window frames are used, the frames are built into the boxes with the edges of the frames protruding. Concrete is cast around the frame, thereby firmly anchoring it in place.

b Pipes for water, gas, oil, and electricity are brought through the foundation, and openings for these must be left in the wall. Openings are most conveniently provided by placing in the forms pieces of pipe or metal sleeves whose length equals the thickness of the wall and whose inner diameter is slightly larger than the pipes that are to pass through them.

4.14 Anchor Bolts (Figures 4.3, 4)

a If the superstructure is to be frame, it should have a sill, generally bolted to the foundation walls. These bolts, commonly ½" in diameter, provided with 2" washers or hooked at the lower end, and usually spaced 4' to 8' apart, are suspended in the forms at the proper height and concreted in (Figures 4.3, 4.4). Later the sill can be slipped over them and fastened down by nut and washer (Chapter 5). Other anchoring devices include metal straps and a variety of proprietary devices. Powder-actuated studs driven (by explosive cartridges) through the sill into the concrete are also used.

4.15 Material

a. Composition and Mixes Concrete is a mixture of stone (called "aggregate"), sand, cement (almost always portland cement), and water. When first mixed, it is plastic, but it hardens when the cement hydrates and "sets."

b The ingredients are mixed in various proportions to obtain strength, density, or economy, and in large structures these proportions are carefully worked out. In dwellings and other smaller structures, simple easy-to-measure ratios are employed. Of these, 1–2–4, 1–2½–5, and 1–3–6 are most common, the ratios meaning 1 cubic foot of cement to 2 cubic feet of sand to 4 cubic feet of stone, and so forth. (One sack of cement is 1 cubic foot; 4 sacks are a barrel.) The higher the cement content (1–2–4, for example), the greater the strength, *provided the water to cement ratio is constant,* but the greater the cost since cement is the most expensive ingredient. Strength is seldom a great factor in small structures, so the approximate 1–2½–5 mix is most commonly employed, with 1–3–6 frequently used for relatively large masses requiring no great strength or hardness.

c. Mixing Concrete may be mixed at the job by hand or machine, or it may be purchased already mixed. The latter is becoming more and more common, especially in larger urban centers, and has many advantages. Hand mixing is used only for small batches.

d. Machine Mixing The usual concrete mixer consists of a rotating drum containing oblique blades. The dry ingredients—sand, aggregate, and cement—are placed in the drum and mixed, after which water is added and the mass turned over long enough to achieve thorough mixing. The drum is then tilted downward and the soft mix is deposited in whatever vehicles are employed to transport the concrete to the forms.

e. Workability Hard to define, "workability" is one of the most important attributes of good concrete. Plasticity is another term. It means the property of flowing easily (not running like water) into and filling all the parts of a form completely without segregation of the ingredients of which the concrete is composed. Good workable concrete does not show free water at the surface; when poured out of a bucket it forms a pancake curling under at the edges, and it quakes when shaken or prodded. Workability is a matter of good proportions, size of aggregate and sand particles, and, above all, the proper amount of water. *Small variations in water content make large differences in workability, strength, and density.*

f. Depositing Concrete must not be dropped from a great height—best not over 3' to 4'. For greater height, a trough or chute should be used. Dropping causes sand, water-cement paste, and aggregate to separate. Individual loads should be spread, not merely dumped. Spreading remixes the ingredients which may have started to separate and mixes the entire mass in the forms more uniformly.

g. Working Once in the forms, the concrete must be worked into all corners and any entrapped air must be removed. This is done by spading and rodding or by means of mechanical vibrators. Rodding and spading are done most vigorously in the center and less so at the form faces. Concrete must be carefully worked. If it is neglected, "honeycombs" or pockets form in which there is no concrete. Too much working tends to separate the ingredients, the lighter material rising to the top as a milky scum called "laitence." It consists of a very fine, over-watered or "drowned" cement which is worthless, soon disintegrates, and causes much of the trouble

Foundations

encountered with concrete. Too much working brings this to the top or to the face of the wall where it subsequently forms a film that continually dusts off, or leaches when water strikes it.

h. Filling Forms may be filled in horizontal layers all around the structure. or in full-height sections adjacent to each other. In small structures, the horizontal layer method is commonly employed, because it is easier and permits lighter forms. It does require that forms be built completely around the wall. Sometimes the forms cannot be filled in one day's operation by this method; consequently, horizontal joints occur between successive days' work.

i The vertical section method, more commonly employed on large structures, calls for heavier forms and for bulkheads to be built inside the forms between successive sections (Figure 4.4b). However, forms need not be built entirely around the structure, since sectional forms can be re-used. Their chief advantages are the elimination of horizontal joints and the regular spacing of vertical joints in the wall. Vertical joints must be provided in long walls because shrinkage otherwise causes irregular vertical cracks. In small foundation walls, such as in dwelling houses, this is not a particularly serious matter and is seldom taken into account.

j Where watertight construction is required (as in cisterns or other liquid-containing structures), concrete must be cast continuously to completion because joints of any kind are almost sure to leak under hydrostatic pressure.

k. Bonding Bonding new to old concrete is one of the most important and difficult aspects of all concrete work. In continuous pouring, the bond is automatic because successive wet masses are merged into each other. The merging process can be carried on anytime within two hours or so, but after that, the concrete enters into its first hardening stage, called the "initial set," and should not be disturbed. New concrete will not merge satisfactorily with the old after this time. After about 8 hours, the concrete enters its second hardening stage, called the "final set," and new concrete will not merge with the old at all.

l Fresh concrete can be bonded to hardened material only by adhesion. For a good bond:

1. Laitence must be removed. Laitence forms a plane of cleavage, does not adhere to either old or new concrete, and gradually crumbles or leaches away. It should be thoroughly chipped, scraped, or brushed off.

Dwelling House Construction

2. Any dirt, dust, or other foreign matter must be cleaned off thoroughly by wire brushing and by flushing the surface with a hose. Water must be allowed to drain away before new concrete is deposited.
3. The surface should be roughened to increase the area for adhesion. This can be done by chipping when laitence is removed. Sometimes, to form projecting keys for the new concrete, fairly large pieces of aggregate are embedded in the surface of the old concrete while it is still fluid.
4. The new concrete must be carefully deposited, spread, and thoroughly rammed and rodded so that no air remains at the surface of the old. A grout with low water-cement ratio may be placed first and thoroughly worked into the new concrete.
5. Mixes of sand, cement, and a variety of latex materials based on plastics and other polymers are available to provide strong bonds to old concrete. These may be used with new concrete to join the new to the old.

m. Curing Concrete should not be permitted to become too hot, too cold, or too dry. The best concrete is obtained by curing at $50°$ to $70°F$. Above that range, the speed of set is accelerated, but strength and hardness are impaired, largely because of the too rapid evaporation of water. High temperatures are, therefore, best offset by constant sprinkling.

n Low temperatures may have some adverse effects on concrete—they retard the set and stop it completely below freezing temperature because the water congeals and no longer reacts with the cement. Frozen concrete is dormant, and after it has thawed, continues to set, probably with some loss of strength. Alternate freezing and thawing, however, has a disrupting influence and may ruin the concrete. Therefore, concrete is best kept from freezing, and if it has once frozen, should be kept heated until it has passed well beyond the final set. Hard concrete rings when struck with a hammer; frozen concrete gives a dull thud and shows a wet spot where the hammer strikes.

p Concrete permitted to dry before it has completely set soon disintegrates because hydration of the cement is incomplete, and the unhydrated portion acts as if it were dust mixed into the mass. During cure, therefore, concrete should be kept as wet as possible. Several days to a week of wetting down with a hose, and covering with wet burlap or with an impervious plastic film is good treatment.

Foundations

4.16 Stripping Forms

a After the concrete has attained its final set, it can stand unsupported and forms may be removed. If just enough nails have been used to hold the forms together, they come apart easily and with little broken material. Inasmuch as studs and boards are to be re-used, the stripping should be done carefully so as to waste as little material as possible. Boards and studs should be cleaned of nails and any adhering concrete should be scraped off. They should then be piled neatly and straight with spacers so that they may dry quickly without warping or twisting.

b Protruding wires are cut off flush with the face of the wall and patched over with mortar, or proprietary spreader-ties are removed as directed. Any open spots (which should not occur in a well-rodded and spaded wall) are filled with a 1–3 cement-sand mortar to which has been added about 15 percent of lime.

4.17 Finishing

a Exterior faces of concrete walls exposed above grade may be treated in some manner to remove the marks of form boards and to improve their appearance, but the rough wall is frequently left as it comes from the forms.

b Several finishing methods are employed, of which the following are the most common:

1. *Coating.* Cement plaster, consisting of sand, cement, and water, is trowelled or spatter-dashed onto the wall. This is the easiest and by far most common method, but frost action may eventually dislodge the coating.
2. *Rubbing.* As soon as possible, the form boards are removed and the surface rubbed down at once, e.g., with a carborundum brick and plain water or a paste (grout) of water and cement.
3. *Brushing.* Forms are removed as soon as the concrete will stand and the still-soft surface is wire brushed.
4. *Hammering.* The hardened surface is gone over with a bush hammer, a hammer with a serrated face.

CONCRETE BLOCK

4.18 Units (Figure 4.6)

a Instead of solid concrete, cast in place, walls are frequently built of concrete block small enough to be handled by one man, and laid up much the same as brick (Figure 4.6).

Dwelling House Construction

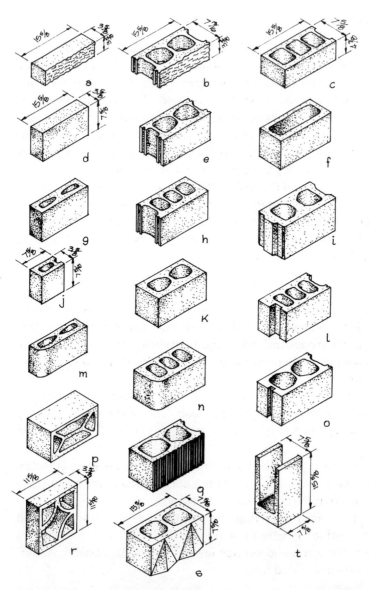

Figure 4.6 Types of Concrete Block. (a) 4″ x 4″ Split. (b) 4″ x 8″ Split. (c) 4⅞″ Starter. (d) 4″ x 8″ Solid. (e) 8″ Stretcher. (f) 8″ Conduit. (g) 4″ Stretcher. (h) 8″ Stretcher-Corner. (i) 8″ Control Joint. (j) 4″ Half-Stretcher. (k) 8″ Double Corner. (l) 8″ Wood Sash. (m) 4″ Single Bull Nose. (n) 8″ Single Bull Nose. (o) 8″ Steel Sash. (p) 8″ Screen. (q) 8 ″Striated. (r) 4″ x 12″ x 12″ Screen. (s) 8″ Sculptured. (t) 8″ x 16″ Lintel.

Foundations

b Although at one time there was a great multiplicity of sizes and shapes, concrete blocks have now been more or less standardized and those units which are most frequently used are $7\frac{5}{8}'' \times 15\frac{5}{8}''$ on the vertical face, with thicknesses varying from $3\frac{5}{8}''$ to $11\frac{5}{8}''$ in $2''$ multiples. The $7\frac{5}{8}''$ thickness is most common. With $\frac{3}{8}''$ mortar joints, all of these blocks are $8'' \times 16''$ on the face. To decrease weight and to save material, vertical open spaces are left in the interiors of the blocks. For corners, special blocks with smooth end faces supplement the ordinary blocks with hollowed ends. Many special blocks, including concrete brick, are made and these call for various special interlocking and cross-tying systems. Special sculptured or otherwise treated surfaces are available. Coatings such as those based on various polymeric materials (Chapter 16) may be applied. Surfaces may be ground smooth.
c Blocks should be well cured by autoclaving or aging, to reduce subsequent shrinkage and cracking in the wall.

4.19 Use

a Concrete blocks are ordinarily used for foundations only, but they are frequently built into upper walls, often faced with stucco, brick, or with integral surfaces (see above). Blocks with a finished surface or split blocks are employed with no further facing. In such instances, they are in the same category as cut stone, and may be made especially to conform to particular wall openings, or the openings must be arranged to meet the stock sizes of such blocks.

4.20 Laying up

a. Coursing (Figure 4.7) Blocks are laid end to end "stretcher" fashion completely around the wall; the first course is brought to a true and level line on top of the footing so that subsequent courses will also be true. The next course is laid on top of the first but with the vertical joints offset or "broken" half the length of a block. Therefore, joints in every other course are directly above one another. Blocks lap over each other at the corners and provide a good bond (Figure 4.7). If walls are to be concealed below ground, standard blocks may be employed at the corners in place of the special smooth-ended corner blocks.

b. Wall Height Since blocks are normally $8''$ high, including a $\frac{3}{8}''$ mortar joint, any wall height which is not a multiple of $8''$ has to be attained by (1) courses of brick, (2) half blocks laid on the flat, (3) thickening all horizontal "bed" joints, or (4) combinations of these methods.

Dwelling House Construction

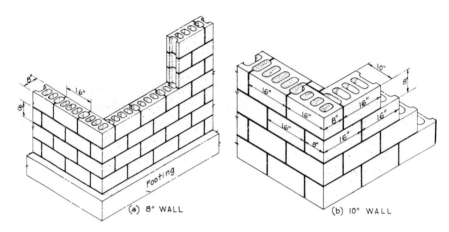

Figure 4.7 Laying Concrete Block Walls.

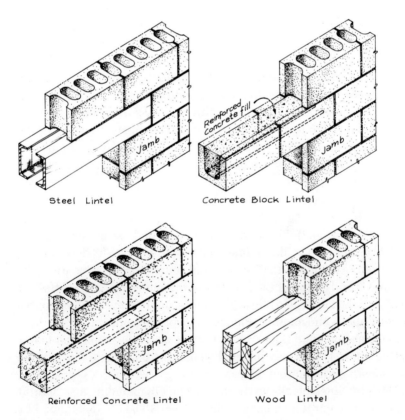

Figure 4.8 Lintels.

Foundations

c. Openings in Walls (Figure 4.8) Wherever doors, windows, or other openings occur, frames frequently are erected before the wall is started so that the wall may be built tightly against the frames. Anchorage blocks are built into the wall directly adjacent to the frames and frames are subsequently fastened securely to these blocks for anchorage. Preferred construction calls for openings to be built into the wall and frames inserted later. Offset jamb blocks are employed to provide a recess into which the frames are fitted. Nailing blocks, strap anchors, or other fastening devices are built into the wall to hold the frames in place.

d Concrete blocks can be carried across openings on lintels. Several types are shown in Figure 4.8. They include wood, steel, and one-piece cast-in-place or precast reinforced concrete lintels. Another type is formed by laying special U-shaped lintel blocks side by side, placing reinforcing steel in the channel thus formed, and filling the channel with concrete. This is usually done in place.

e A commonly-used deflection limitation for hard brittle materials is $\frac{1}{360}$th part of the span. For example, the maximum deflection permitted on a 10' span (120") is $\frac{1}{3}$". For concrete block, with their greater tendency to crack, especially at the joints, it is safer to limit the deflection still more; $\frac{1}{480}$th of the span is recommended by engineers.

f. Mortar Although high-cement mortars have long been commonly used for concrete block, there is a decided tendency toward higher-lime mortars because of their greater workability and the more intimate contact they give between block and mortar. Therefore, a mortar composed of 1 part cement, 1 part lime putty, and 5 to 6 parts sand is generally recommended. Sufficient water is added to obtain good workability. Special mortars formulated for blocks are available.

g. Returns, Pilasters, and Piers (Figures 4.7, 9) Where changes in direction or "returns" in walls occur, care must be exercised to build the corner straight and plumb, and to make sure that no long vertical joints occur. Such vertical joints are avoided by carrying the succeeding courses of block into the corner alternately from one side and the other, much as logs in a log cabin are laid on top of each other at the corner.

h Pilasters are short sections of the wall which are increased in thickness throughout its height, usually to carry concentrated loads such as girders. Since there is a return at each side of the pilaster, the same precautions

Dwelling House Construction

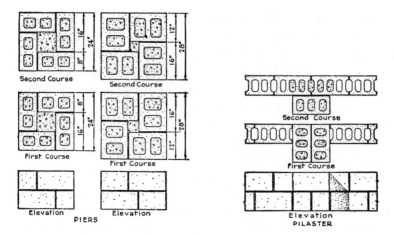

Figure 4.9 Piers and Pilasters of Concrete Block.

respecting long vertical joints hold as for corners. The pilaster must be built as an integral part of the wall by bonding wall and pilaster block together. To obtain odd dimensions, brick and half-block must often be employed.

i Piers are free-standing posts, usually square in cross-section, which support concentrated loads. Again, by lapping block in succeeding courses, long vertical joints are avoided. Frequently, voids in concrete block piers and pilasters are filled with concrete to obtain completely solid structures.

4.21 Anchor Bolts

a Anchor bolts may be needed to prevent displacement of the superstructure by wind, earthquake, or other forces. In the case of concrete block and other masonry units, this is not accomplished as easily as in cast concrete, but it can be done by carrying the bolts down some depth into the wall. In concrete block, this should be equal to at least two courses of block. For firm anchorage, the lower ends of the bolts are provided with steel plate washers which are embedded in the horizontal mortar joint. The bolts extend upward through the openings in the block and are fixed in place by filling the openings with mortar. Steel strap anchors are commonly bent at the bottom and embedded in a mortar joint.

Foundations

4.22 Waterproofing (Figure 4.10)

a In wet locations concrete block walls are apt to leak, so extra precautions must be taken to insure dry walls. Drain tile around the outsides of footings are essential to prevent a head of water from being built up in some soils. Drains lead into dry wells or into the municipal sewerage system (Chapter 3). Even in relatively dry locations it is desirable to "parge" or plaster the outside of the wall with two layers of mortar, each $\frac{1}{4}''$ thick, trowelled and pressed on firmly. This helps seal the fairly porous surface (Figure 4.10). A foaming agent admixture in the mortar is helpful in promoting workability, adhesion, and water-tightness.

b. Asphalt Coating and Membrane Waterproofing (Figure 4.10) Both of these are essentially the same as the waterproofing and built-up roofing found in larger structures. Asphalt coatings are asphaltic mixtures applied either hot or cold to the wall. Their effectiveness depends upon the care and thoroughness with which they are applied and upon the wall not cracking. Membrane waterproofing consists of alternate layers of hot pitch and builders' felt. This is a much more dependable method of waterproofing.

4.23 Advantages and Disadvantages

a Advantages of block construction are the elimination of forms and speeding of construction. Block walls are ready for the superstructure

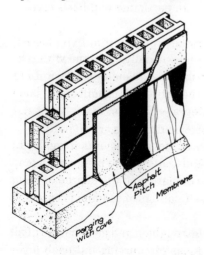

Figure 4.10 Parging and Waterproofing Concrete Block Foundation Wall.

almost as soon as they are finished, and need be allowed to stand only long enough for the mortar to set. Solid concrete walls must be allowed to harden and then forms must be removed.

b Disadvantages of block walls are possible poor anchorage, greater likelihood of leakage, and greater likelihood of cracks developing. A concrete block wall has not the strength of a solid wall, and any settlement is more apt to cause cracks than in solid concrete.

4.24 Reinforced Block Masonry (Figure 4.11)

If lateral forces, such as may be caused by earth pressure or earthquakes, are expected, block walls may be reinforced in a variety of ways, as shown in Figure 4.11. The footing is reinforced against cracking, and vertical rods, embedded in the footing, are carried upward at corners, openings, and intersections, through the holes in the cores of the blocks. These holes are filled with mortar or concrete. Wire reinforcing is laid in the horizontal joints. At the top of the wall a continuous reinforced concrete beam is formed by lintel blocks, reinforcing steel, and concrete.

4.25 Types of Rubble Masonry

a Foundation walls for dwellings as well as for other types of buildings are at times built of stone. Since the house may be built entirely of stone, this discussion will not be confined to the foundation walls in particular, but will consider the subject as a whole, with particular emphasis on random rubble.

b In general, stone masonry is divided rather roughly into two classes, rubble and ashlar. Ashlar is a facing of stone—usually cut stone with regular squared joints—backed with other material, usually brick. Rubble masonry is usually homogeneous, that is, the entire wall is built of the same stone, takes all of the load, and is an integral structure. Many large and famous buildings such as ancient churches and castles are so built.

c Rubble masonry is composed of stones or stone fragments irregular in any or all of the elements of size, shape, and jointing. The irregularity may be somewhat ordered in any of these particulars, and the degree of such ordering has given rise to the broad classifications of random rubble and coursed rubble.

d. Random Rubble (Figure 4.12) This is the roughest and most casual of all stonework. Little attention is given to ideas of coursing but each layer

Foundations

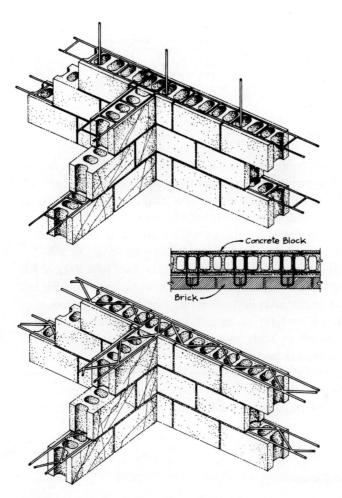

Figure 4.11 Reinforced Block. (Top) Vertical. (Bottom) Horizontal. (Center) Brick to Block Tie.

Figure 4.12 Random Rubble.

Figure 4.13 Coursed Rubble.

should contain stones bonding through the wall in sufficient number to produce a closely knit structure. The remaining interstices are filled with smaller stones of convenient size and shape. Such walls may be used for both foundations and superstructure. The most attractive random rubble is composed of good-sized stones, with a minimum of small chips as fillers. It should never be pointed with full mortar pointing to hide the natural shapes of the stones. The pointing should be kept well back of the outer surface. "Builds" or "vertical" joints need not be truly vertical but "beds" should be approximately horizontal for stability and appearance.

e. Coursed Rubble (Figure 4.13) As its name implies, this is assembled of roughly squared stones in such a manner as to produce, at intervals, approximately continuous horizontal bed joints separated by variable vertical distances from similar neighboring joints. It is necessary to have only fairly level beds and approximately vertical builds and faces. Such masonry may either be irregularly coursed or squared as the specifications may direct.

Floors

4.26 Specification Clause

CONCRETE FLOOR

a Lay a concrete paving 3″ thick over the entire basement floor, all to be placed upon a properly levelled and tamped gravel filling, and at correct grades. Concrete shall be composed of 1 part cement, 2 parts sand, and 3 parts gravel. Finish with a wood float so as to produce a smooth surface free from cracks or imperfections.

b The basement floor is not placed at the same time that foundation walls are built. It must wait until the house is framed and all the necessary underground service lines have been placed under the basement, but it should

Foundations

be cast as soon as possible in order to allow it to dry thoroughly before the finish lumber arrives. Otherwise, dampness originating in the basement may cause swelling and warping in the kiln-dried millwork used for interior finish.

c Unless the subsoil is itself dry, sandy, and gravelly, it is wise to put down a layer of gravel or crushed stone, from 4″ to 6″ thick, before the concrete is cast (Figure 4.14). To avoid future settlement or cracks, the fill must be well tamped, preferably dampened at the same time to make it compact better. Too much water merely soaks it and tends to float the finer particles to the top.

d Concrete for basement floors should be quite stiff, to reduce the amount of water which must be evaporated, to increase the density, and to provide a harder, more wear-resistant surface than would be found if the concrete were soft.

e If floor drains are provided, the floor should be gradually but uniformly sloped toward the drains from all directions. To make sure that the slope is correct, stakes are set and the tops of the stakes are brought to the proper pitch with a level and straight edge. Concrete is cast and levelled to the tops of the stakes, which are removed as the concreting progresses. After the concrete is deposited and brought to the proper elevation, it is surfaced by working the face with a wooden float (wood block provided with a handle) or a steel trowel. Floated finishes are slightly rough and pebbly; trowelled surfaces are smooth.

f Although concrete basement slabs are commonly cast directly against foundation walls, it is wise to leave a space to be filled subsequently with bitumen or other sealant. This is shown in Figure 4.14. Otherwise, shrinkage of the slab is likely to open a peripheral crack which may let in water.

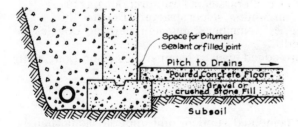

Figure 4.14 Concrete Floor.

5 Framing

5.1 General

a The frame is the structural skeleton of the building, and in light construction such as dwelling houses it consists of relatively small pieces of timber ranging from 2" to 6" or 8" in thickness and from 2" to 12" in width. The portions of the frame are assembled on the foundations and to it are fastened windows, doors, roof covering, exterior covering, interior covering, and floor.

b The species of wood used in framing varies with the locality, and in any one place is that species which is most available and economical. By and large, spruce, hemlock, yellow pine, and Douglas fir are most commonly used. Since appearance is no factor (because the frame is subsequently covered) those grades are employed which are most economical but still possess the requisite strength and rigidity.

5.2 Specification Clauses

ROUGH LUMBER

a All framing lumber shall be of representative quality of the grade specified. The contractor has the right to use good second hand joists approved by the architect, but shall not use other second hand lumber.

LUMBER SIZES AND GRADES

b The sizes and spacing of, and kinds of lumber, unless otherwise shown on the plans, or required by the building authorities, shall be in accordance with the following schedule:
Studs and plates for bearing partitions and walls, 2" x 4", not more than 16" on centers, select structural hemlock.
Studs and plates for non-bearing partitions, 2" x 4" or 2" x 3", not more than 16" on centers, Number 1 hemlock.
Floor joists, 2" x 10", 16" on centers, select structural Douglas fir.
Main rafters, 2" x 8", 16" on centers, select structural Douglas fir.
Roof and wall sheathing, $\frac{3}{4}$", Number 2 common matched Douglas fir boards not over 8" in width.
Underflooring, $\frac{3}{4}$" square-edged boards, Number 2 common Douglas fir not over 8" in width.

PLYWOOD

c Plywood single-layer combined subfloor and underlayment shall be $\frac{1}{2}$" C-C plugged exterior grade Group 1 nailed with 6d deformed-shank nails 6" on centers along edges and 12" at intermediate points. Wall sheathing shall be $\frac{3}{8}$" Standard C-D interior grade. Roof decking shall be $\frac{1}{2}$" C-D interior grade.

Framing

5.3 Lumber

a "Softwood" lumber, cut from coniferous ("evergreen") trees, is used almost exclusively for framing. Softwood is classified as *seasoned, or dry if its moisture content is 19 percent or less, and unseasoned, or green* if its moisture content is greater than 19 percent.

b Softwood is classified according to use as:
1. *Yard lumber.* Grades, sizes, and patterns generally intended for ordinary construction and general building; mostly used for house construction.
2. *Structural.* Lumber more than 2″ in nominal thickness, used where large stress-graded engineered lumber is required.
3. *Factory and shop.* Used primarily for remanufacturing, e.g., into sash and doors.

c Softwood is classified according to extent of manufacture as:
1. *Rough.* Sawed, edged, and trimmed but not dressed (surfaced).
2. *Dressed (surfaced).* Planed to obtain smoothness and uniformity of size, on one side (S1S), two sides (S2S), one edge (S1E), two edges (S2E), or a combination (S1S1E, S1S2E, S2S1E, S4S). S4S is the most common in dwelling house construction.
3. *Worked.* Not only dressed, but matched, shiplapped, or patterned (see Figure 5.10).
    i. *Matched.* Worked with a tongue on one edge and a groove on the other.
    ii. *Shiplapped.* Rabbetted on both edges.
    iii. *Patterned.* Shaped to a patterned or molded form.

d Softwood is classified as to nominal size, i.e., the size originally cut from the tree. Actual sizes (Table 5.1) are less because of drying shrinkage and dressing. Actual sizes are different for seasoned, or dry, lumber (19 percent or less) and unseasoned, or green, lumber (more than 19 percent).
1. *Boards.* Less than 2″ nominal thickness, 2″ or more nominal width. Boards less than 6″ wide may be called strips.
2. *Dimension.* Nominal thickness 2″ to less than 5″, nominal width 2″ or more. Classified as framing, joists, rafters, studs, small timbers, etc.
3. *Timbers.* Least nominal dimension 5″ or more. Classified as beams, stringers, posts, caps, sills, girders, purlins, etc.

e Yard lumber is graded on the basis of quality as:

1. *Select.* Good appearance and finishing qualities.
    i. Natural (transparent) finishes:
        a. Practically clear (A Select).
        b. Generally clear, high quality (B Select).
        Often these grades are combined into "B and Better."
    ii. Paint finishes:
        a. High-quality paint finish (C Select).
        b. Intermediate between high-finishing and common grade (D Select).
2. *Common.* Suitable for general construction and utility where appearance is of secondary importance.
    i. Standard construction:
        No. 1 Common. Better type construction.
        No. 2 Common. Good standard construction.
        No. 3 Common. Low-cost temporary construction.
    ii. For less exacting purposes:
        No. 4 Common. Low quality.
        No. 5 Common. Lowest recognized usable grade.
3. Other terms are often used in the trade. The select boards may be called, for example, Supreme (B and Better), Choice (C), and Quality (D). The Nos. 1 to 5 Common boards may be called, e.g., Colonial, Sterling, Standard, Utility, and Industrial. Light Framing Common dimension lumber may be called Construction, Standard, Utility, and Economy, whereas Structural Light Framing Common dimension may be called Select Structural, No. 1, No. 2, No. 3, and Economy. The latter terms are also used for structural joists and planks.

f Structural lumber, not often used for dwelling house construction, but used for engineered structures, has assigned values for modulus of elasticity and working stresses in bending, compression parallel and perpendicular to the grain, and horizontal shear.

g Factory and shop lumber is graded according to the sizes and percentages of pieces clear on one or both sides that can be cut from it for such things as sash and door parts and other general cutup uses.

h. Plywood This consists of thin sheets or piles of wood, called veneers, glued together with the grain direction of adjacent plies at right angles. An odd number of plies (from a minimum of three) is used, and thicknesses ordinarily used in dwelling house construction range from $5/16''$ to $1\frac{1}{4}''$, with 3-ply, 5-ply, and 7-ply being the most common. Two grades, exterior

Framing 65

and interior, are made. Exterior grades must have completely waterproof glue lines, but the interior grades must themselves be highly moisture-resistant. Somewhat better-quality veneers are used in exterior-grade plywood than in interior.

i Some 50 species of wood are used in making plywood, and these are classified into 5 groups. The most common species are Douglas fir, the yellow pines, other pines, redwood, hemlock, and spruces. A group number is assigned to a given piece of plywood depending upon the surface plies, the lower-group face ply determining the group number (Table 5.2).

j Veneer grades are:

N. Special order, natural finish, select heartwood or sapwood.
A. Smooth and paintable; neat repairs permitted.
B. Solid surface veneer; circular repair plugs and tight knots.
C. Knot holes to 1", some to 1½" within specified limits; some splits; minimum veneer for exterior.
C. plugged: improved C veneer, splits limited to ⅛", knot and borer holes limited to ¼" x ½".

Table 5.1 Nominal and Actual Sizes of Board and Dimension Lumber.

	Actual (in.)	
Nominal (in.)	Unseasoned	Seasoned
Board		
1 x 2	25/32 x 1 9/16	¾ x 1½
1 x 3	25/32 x 2 9/16	¾ x 2½
1 x 4	25/32 x 3 9/16	¾ x 3½
1 x 6	25/32 x 5 ⅝	¾ x 5½
1 x 8	25/32 x 7½	¾ x 7¼
1 x 10	25/32 x 9½	¾ x 9¼
1 x 12	25/32 x 11½	¾ x 11¼
Dimension[a]		
2 x 2	1 9/16 x 1 9/16	1½ x 1½
2 x 3	1 9/16 x 2 9/16	1½ x 2½
2 x 4	1 9/16 x 3 9/16	1½ x 3½
2 x 6	1 9/16 x 5 ⅝	1½ x 5½
2 x 8	1 9/16 x 7½	1½ x 7¼
2 x 10	1 9/16 x 9½	1½ x 9¼
2 x 12	1 9/16 x 11½	1½ x 11¼

[a] Thicknesses of seasoned nominal 3" and 4" lumber are 2½" and 3½", of unseasoned lumber 2 9/16" and 3 9/16". Widths are the same as given above.

Table 5.2 Plywood Groups and Plywood Identification Index

Groups

Group 1	Group 2	Group 3	Group 4	Group 5
Birch	Cedar, Port Orford	Alder, Red	Aspen	Fir, Balsam
Sweet	Douglas Fir 2[b]	Cedar, Alaska	Bigtooth	Poplar, Balsam
Yellow	Fir	Pine	Quaking	
Douglas Fir 1[a]	California Red	Jack	Birch, Paper	
Larch, Western	Grand	Lodgepole	Cedar	
Maple, Sugar	Noble	Ponderosa	Incense	
Pine, Caribbean	Pacific Silver	Spruce	Western Red	
Pine, Southern	White	Redwood	Fir, Subalpine	
Loblolly	Hemlock, Western	Spruce	Hemlock, Eastern	
Longleaf	Lauan	Black	Pine	
Shortleaf	Almon	Red	Eastern White	
Slash	Bagtikan	White	Sugar	
Tanoak	Red Lauan		Poplar, Western[c]	
	Tangile		Spruce, Engelmann	
	White lauan			
	Maple, Black			
	Mengkulang			
	Meranti			
	Pine			
	Pond			
	Red			
	Western White			
	Spruce, Sitka			
	Sweetgum			
	Tamarack			

Identification Index

Thickness (in.)	Standard (C-D) INT C-C EXT			Structural I (3) C-D INT Str. I C-C EXT	Structural II (3) C-D INT	
	Group 1	Group 2 or 3 (4)	Group 4 (5)	Group 1 only	Group 1	Group 2 or 3 (4)
5/16	20/0	16/0	12/0	20/0	20/0	16/0
3/8	24/0	20/0	16/0	24/0	24/0	20/0
1/2	32/16	24/0	24/0	32/16	32/16	24/0
5/8	42/20	32/16	30/12	42/20	42/20	32/16
3/4	48/24	42/20	36/16	48/24	48/24	42/20
7/8	--------	48/24	42/20	--------	--------	48/24

[a] Douglas fir 1—Washington, Oregon, California, Idaho, Montana, Wyoming, British Columbia, Alberta.
[b] Douglas fir 2—Nevada, Utah, Colorado, Arizona, New Mexico.
[c] Black cottonwood.
Source: American Plywood Association.

Framing

D. Knots and knot holes to 2½", some larger; limited splits.
k Principal interior types of plywood are:
Standard C-D. Unsanded, for wall and roof sheathing and subflooring.
Structural IC-D, IIC-D. Unsanded structural grades where strength is of maximum importance.
Underlayment. Underlayment or combination subfloor-underlayment under resilient flooring.
C-D plugged. Utility built-ins, backing for wall and ceiling tile, etc.
2-4-1. Combination subfloor underlayment.
l Principal exterior types are:
C-C. Unsanded, waterproof glue, exposed and severe service.
Underlayment C-C plugged, and C-C plugged. Underlayment and combination underlayment-subfloor under resilient flooring in moisture conditions, also tile backing, refrigerator rooms, etc.
Structural I C-C. Engineered applications requiring maximum exposure resistance and Group I species.
B-B plyform, Class I, II; concrete forms, high re-use.
m The various groups are recommended for floors, roofs, and wall sheathing requiring different spans between supports and various loading conditions, as indicated in the section of this chapter on framing.

5.4 Nails

a Nails are discussed in detail in Chapter 14. For framing, common nails are employed. In general, a nail should be about three times as long as the material through which it is driven. For example, to nail ¾" sheathing to studs, the nail should be 2¼" long. An 8d nail is 2½" long and, therefore, meets the requirements. For fastening dimension lumber, 16d and 20d nails are generally employed for nominal 2" and 40d to 60d for nominal 3" stock. The 16d size just penetrates two 1½" thicknesses whereas 20d has sufficient length to permit the protruding pointed end to be bent over

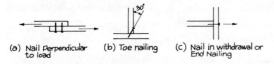

(a) Nail Perpendicular to load   (b) Toe nailing   (c) Nail in withdrawal or End Nailing

Figure 5.1 Nailing. (a) Perpendicular to Load. (b) Diagonal or Toe Nailing. (c) End Nailing, Parallel to Load.

Table 5.3 Recommended Nailing Schedule

Framing, Using Common Nails	
Joist to sill or girder, toe nail	3 – 8d
Bridging to joist, toe nail each end	2 – 8d
Ledger strip	3 –16d at each joist
1″ x 6″ subfloor or less to each joist, face nail	2 – 8d
Over 1″ x 6″ subfloor to each joist, face nail	3 – 8d
2″ subfloor to joist or girder, blind and face nail	2 –16d
Sole plate to joist or blocking, face nail	16d @ 16″ o.c.
Top plate to stud, end nail	2 –16d
Stud to sole plate, toe nail	4 – 8d
Doubled studs, face nail	16d @ 24″ o.c.
Doubled top plates, face nail	16d @ 16″ o.c.
Top plates, laps and intersections, face nail	2 –16d
Continuous header, two pieces	16d @ 16″ o.c. along each edge
Ceiling joists to plate, toe nail	3 – 8d
Continuous header to stud, toe nail	4 – 8d
Ceiling joists, laps over partitions, face nail	3 –16d
Ceiling joists to parallel rafters, face nail	3 –16d
Rafter to plate, toe nail	3 – 8d
1-inch brace to each stud and plate, face nail	2 – 8d
1″ x 8″ sheathing or less to each bearing, face nail	2 – 8d
Over 1″ x 8″ sheathing to each bearing, face nail	3 – 8d
Built-up corner studs	16d @ 24″ o.c.
Built-up girders and beams	20d @ 32″ o.c. along each edge

Source: *Manual for House Framing*, American Forest Products Association. For additional nailing schedules, see Tables 5.4, 6–8, and Chapters 8, 10, 12.

(clinched), and therefore provides superior gripping power. Studs are commonly toe-nailed with 8d, 10d, or 16d nails.

b Nails may be driven (Figure 5.1) perpendicular to the direction of load, parallel or in withdrawal (end nailing), or at an angle to the face of the piece being nailed (toe nailing). The best is perpendicular loading and poorest is withdrawal, especially if the nail is driven parallel to the grain of the wood. Toe nailing is employed, for example, to fasten studs to sills.

c A recommended nailing schedule for framing is given in Table 5.3. Recommended nailing is also discussed in connection with specific construction details in this and other chapters. See also Chapter 14.

Framing

OUTSIDE WALL FRAMES

5.5 General

a In the United States, three chief types of frames are found in light construction; the type most commonly used depends upon the locality. Historically, the eastern or braced frame, also known as the "barn frame" and "old-fashioned frame," is the oldest and is used in the older portions of the country, particularly in New England. Although strong and rigid, it is fairly complex and, consequently, quite expensive. Two other types of frames have evolved which are much simpler and hence more economical to construct. These are the western or platform, and the balloon. The platform frame is the most common of all.

PLATFORM FRAME

5.6 General (Figure 5.2)

a The platform frame has the following parts:
Sill
Posts
Headers or Bands
Soles or Sole Plates
Studs
Plates
Sheathing

b The distinctive feature of this frame is that each floor forms a platform upon which are erected the studs for that story; each story is, therefore, an entirely separate entity and, theoretically at least, the construction could be repeated as often as desired to form a building of any height. The only exception is that the corner posts may be continuous, instead of only one story in height.

5.7 Sill (Figures 5.1, 4, 6)

a The sill's function is to transfer the loads and wind stresses from the frame to the foundation wall, and to form a base upon which to erect the balance of the frame. In order to perform these functions properly, certain features must be incorporated into its construction.

Dwelling House Construction

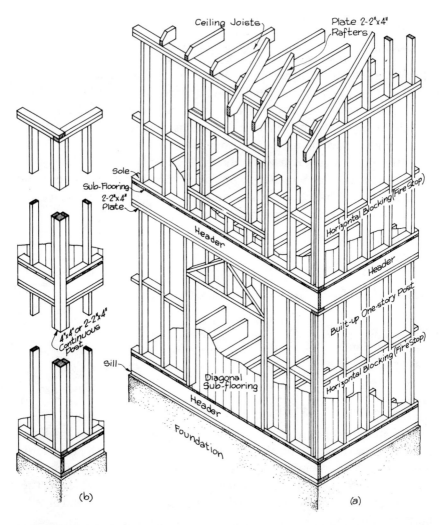

Figure 5.2 Platform Frame. (a) Standard. (b) Continuous Corner Post.

Framing

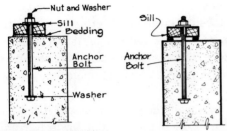

Figure 5.3 Anchor Bolt. (Left) Bedding. (Right) Caulking Beads.

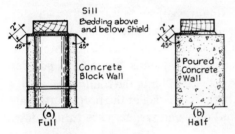

Figure 5.4 Termite Shields.

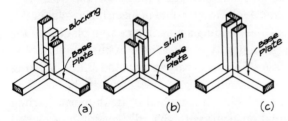

Figure 5.5 Built-Up Corner Posts. (a) Studs Plus Blocking. (b) Offset Plus Shim. (c) L Plus Stud.

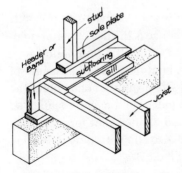

Figure 5.6 Sill Detail, Platform Frame.

b Sills are customarily 2″-thick materials, 4″, 6″, or 8″ wide. If good continuous bearing on level cast-in-place concrete or concrete blocks with filled holes is available, the sill can be 1″-thick material, or it can be eliminated altogether, provided that there is some means of holding floor joists in position. The top of the foundation must be level to avoid unevenness in the floor.

c. Anchors These have generally been set in the foundation wall (Chapter 4). They serve to hold the sill down firmly to the foundation, to prevent any sidewise slipping and to prevent tendency to overturn. They also insure a tight fit between sill and foundation wall.

d Anchors are commonly ½″ bolts (Figure 5.3) set in the wall as already described (Chapter 4). Other types of anchors include various steel straps fastened to the foundation by embedment or mechanical fastening, and to the sill by nails, screws, bolts, and other means. Sometimes powder-actuated steel studs are driven through the sill into the foundation.

e If bolts are employed, holes are bored in the sill at the proper intervals, the sill is slipped over the bolts, and nuts and washers are turned down tight.

f. Bedding The sill must be level and straight if the rest of the frame is to be plumb and the floor joists level. One method of assuring a level bed is to employ special compressible bedding strips that can be placed on the foundation and the sill pressed tightly against them by the anchor bolts. Nuts and washers are tightened as necessary to level the sill. A caulking bead along each edge of the sill similarly provides a watertight bed. The classical method of bedding sills has been a bed of fresh mortar strung along the top of the foundation wall, with the sill immediately placed on it, levelled, and held in place by the anchors such as nuts and washers on anchor bolts (Figure 5.3).

g. Termite Shields (Figure 5.4) Where termites are known to be present, one way of protecting the wood parts of the house is the use of metal (noncorroding) shields. These extend across the top of the foundation wall continuously around its periphery, and are bent down 2″ at an angle of 45° both inside and out. Where the shield fits around anchor bolts the joint is sealed. In solid concrete walls, the half-shield may be employed provided it is solidly bedded and there is no chance that termites may work up the unprotected side. Termites are also combatted by poisoning the soil around the foundation as it is backfilled.

Framing

h. Decay-Resistant Treatment Where sills are in danger of decay because of damp surroundings, it is advisable to use treated lumber, e.g., lumber that has been impregnated with pentachlorphenol or a salt such as zinc chloride. In such instances, it is extremely desirable to have the individual pieces cut and bored before being treated; moreover, the treatment must be by pressure tank if it is to be effective. Whenever it is necessary to cut the pieces after treatment, the exposed surfaces must be thoroughly soaked with the preservative.

5.8 Posts (Figures 5.2, 5)

a Posts are almost always built-up, as shown in Figure 5.5. This arrangement provides a return for nailing interior wall surfaces such as wallboard and lath. As already noted, posts may run the full height of the building, but frequently run only one story. When they are discontinuous in this manner, the corners are apt to form points of weakness and are difficult to keep absolutely straight (Figure 5.2a).

b Figure 5.2 shows an alternative construction in which the corner post is either a solid 4" x 4" or two 2" x 4"'s nailed together. It runs full height and is erected first. Spiked to the post are the end wall studs which support the ends of the plates and at the same time afford nailing for lath or wallboard.

5.9 Headers or Bands (Figures 5.2, 6)

a Headers, also called bands, are on-edge members of the same size as the joists, which rest directly on the sill and run around the periphery of the building. The joists are framed against the headers in the manner shown in Figures 5.2, 5.6. This type of sill and header combination is known as a box sill and is a distinctive feature of the platform frame.

b Before any of the superstructure is erected, the first floor subflooring is put down to form a platform upon which the carpenters can work, frame the material for walls and partitions, and assemble their members. Walls and partitions can be quickly assembled and nailed together flat on the floor and then tilted up into position. This feature of the platform frame is extremely convenient and is a distinct advantage over other types, although the Tee sill (Section 5.19, Figure 5.15) also provides a working platform.

5.10 Studs (Figures 5.2, 6–11)

a Wall studs are almost universally 2″ × 4″ although studs 2″ × 6″ or 2″ × 4″ widened with 2″ × 2″ strips are sometimes required to provide a "wet wall" thick enough to permit the passage of plumbing lines. For one-story construction, 2″ × 3″ studs are often employed.

b. Spacing Because wood lath was originally cut 4′ long or some other multiple of 16″, it has become almost universal practice to set studs and joists 16″ on centers (see Figure 5.18 and Section 5.23, a3.) although joists are frequently set at different spacings. There is no structural reason for the 16″ spacing; it is merely another instance of traditional usage (but see Section 5.52). Because of this standard stud spacing, wallboards and plaster bases are customarily made in stock sizes which are multiples of 16″, commonly 48″.

c. Openings Wherever an opening such as a window or door occurs, such opening must be framed in the studs to receive whatever member is to be placed in it. Framing consists of horizontal members called headers, vertical members called trimmers, and short studs called cripples.

1. *Windows.* Window openings must be framed large enough horizontally to admit the window frame and provide adjustment space between frame and trimmer. Vertically, the opening must be large enough to admit the frame with adjustment. Since window sizes are specified by the size of the glass panes (called "lights" in builders' language), the horizontal opening dimension is made enough larger to allow for two sash stiles, two frame thicknesses, and adjustment and blocking. The vertical dimension is similarly increased for three sash rails, the window sill, and the head of frame. The foregoing dimensions are for double-hung windows. Dimensions of openings for casement and other windows are derived similarly.

2. *Doors.* Door sizes are specified by the overall dimensions of the door; consequently, the opening must be framed enough larger to admit the frame plus sufficient additional room to allow for adjustment, blocking, and levelling.

3. *Framing.* Figures 5.2, 7) Framing around openings must perform two functions: (1) it must be sturdy enough to carry the superimposed wall loads around the opening without sagging or distortion; and (2) it must provide ample support for the frames of windows and doors and provide nailing for interior wall covering, sheathing, grounds, and exterior finish where it adjoins the opening. For both these reasons, the headers and

Framing

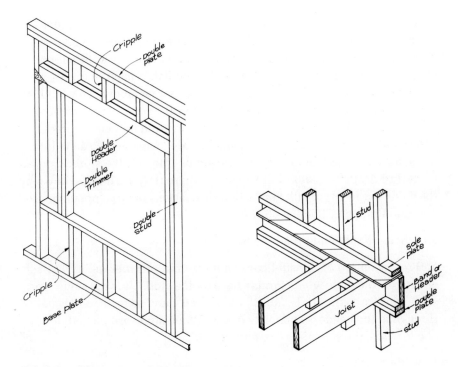

Figure 5.7 Framing Around Window Opening.
Figure 5.8 Floor Framing At Wall, Second Floor, Platform Frame.

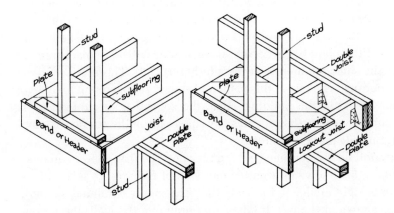

Figure 5.9 Framing Overhang. Left, Cantilevered Joists. Right, Projecting Lookouts.

trimmers should at least be double 2" x 4" members. When openings are large horizontally (more than 4', for instance) the upper header must be made considerably stronger than usual because it has to support a greater load over a longer span. This is accomplished most simply by making the header a pair of 2" x 6", 2" x 8", or larger members. This method, while simple, has the disadvantage that it introduces a large amount of wood which may shrink and swell across the grain and cause cracks. A better way is to insert a pair of diagonal 2" x 4" members so as to form a truss, as shown in Figure 5.2. The diagonals transmit the superimposed loads directly to the sides of the opening where they are carried down by the studs to the girt or sill. For further discussions of large openings, see "Special Framing," Sections 5.56ff.

5.11 Sole or Sole Plate (Figures 5.2, 5–8)

a Directly on top of the subfloor is a horizontal member called the sole or base or sole plate which forms a base for the studs. It is the same cross-sectional size as the studs, i.e., 2" x 4", 2" x 6", and so forth, and is nailed through to the headers and joists below. When studs are placed directly above joists, the sole plate can be 1" thick.

5.12 Plates (Figures 5.2, 7, 8)

a Plates occur at every floor level above the first, and at the eaves. Usually they are doubled 2" x 4"'s. The lower 2" x 4" is nailed through or toe-nailed into the upper ends of the studs, and the upper 2" x 4" is nailed to the lower one. Joints in the plates should be "broken" or staggered at least one foot.
b On top of the plate is the second-floor header with the joists framing into it in the same manner as at the sill (Figure 5.8). Therefore, the box construction occurs again at this point and at every subsequent floor level which may exist in the building. Also, the second floor subflooring is carried out to the edge, the second story sole rests on the subflooring and studs on the sole plate, as before.
c The second floor sometimes extends beyond the first to form an overhang. Figure 5.9 shows alternative methods of framing. If second-floor joists run at right angles to the wall, they are simply extended as cantilevers, as shown at the left. If joists are parallel to the wall, the second or third joist back is doubled, and short joists or "lookouts" are framed as shown at the right.

Framing

5.13 Sheathing (Figures 5.2, 10, 11)

a Traditionally, boards called sheathing boards have been employed to cover the outside of the frame, but these have been challenged by various wallboards, particularly plywood, because these wallboards come in large sheets that are quickly applied. (See also Chapter 12.)

b Plywood (Section 5.3) consists of thin sheets of wood or veneers, called plies, glued together with adjacent plies at right angles. Most plywood used for sheathing is 3-ply, $3/8''$ thick, although greater or lesser thicknesses such as $1/2''$ and $5/16''$ may be employed. Appearance is not important, and blemishes such as occasional knot holes and splits are permitted, but the plywood must be structurally sound, and capable of withstanding stresses that may be caused by, e.g., wind loads.

c Because plywood is applied in large sheets (commonly $4' \times 8'$) it can impart a great deal of rigidity to the wall when nailed as recommended in Table 5.4. So nailed, the plywood is comparable to diagonal sheathing boards (see below) in both stiffness and strength, and much superior to unbraced horizontal sheathing boards. If the plywood is glued, stiffness is greatly increased.

d Fibrous wallboards, made of processed wood, cane, newsprint, and other materials, have become increasingly common sheathing materials, as has gypsum (see also Chapters 10–12). They are sufficiently rigid to withstand handling as well as to provide moderate nail-holding power. Half-inch-thick boards of this type are frequently employed, and provide approximately the same insulating value as wood sheathing. Because the boards are used in large sheets (commonly $4' \times 8'$ to $12'$) they impart a high degree of stiffness to the wall when well nailed (Table 5.4) in spite of the low bearing value per nail. Static tests indicate that the stiffness of a panel sheathed with well-nailed wallboard is greater than unbraced horizontal wood sheathing but less than diagonal sheathing (see Par. j below). When these wallboards are nailed as recommended in Table 5.4, let-in bracing may be omitted (Par. k below).

e The wood boards that are used for sheathing may also be employed for roof boards or roofers, and for subflooring.

f Boards may be square-edge, shiplap, or tongue and groove (Figure 5.10). The joints must be made on studs and should be "broken," that is, joints in successive tiers of boards should not be made on the same stud.

Table 5.4 Nailing Schedule for Plywood Wall Sheathing[a] (Plywood Continuous over Two or More Spans), Fiber Board Sheathing, and Gypsum Sheathing

Panel Identification Index[b]	Panel Thickness (in.)	Max Stud Spacing (in.) for Exterior Covering Nailed to:	
		Stud	Sheathing
Plywood:			
12/0, 16/0, 20/0	5/16	16	16[d]
16/0, 20/0, 24/0	3/8	24	16 24[d]
24/0, 30/12, 32/16	1/2, 5/8	24	24
Fiber Board:	1/2		
	25/32		

[a] When plywood sheathing is used, building paper and diagonal wall bracing can be omitted.
[b] See Table 5.2
[c] Common smooth, annular, spiral-thread, or galvanized box, or T-nails of the same diameter as common nails (0.113" dia. for 6d) may be used. Staples also permittted at reduced spacing.
[d] When sidings such as shingles are nailed only to the plywood sheathing, apply plywood with face grain across studs.
Sources: Plywood schedule, American Plywood Association, Tacoma, Wash.; fiber board and gypsum, *BOCA Basic Building Code, 1970* (Building Officials and Code Administrators)

Framing

Nail Size[c] and Type	Nail Spacing (in.)	
	Panel Edges (when over Framing)	Intermediate (Each Stud)
Plywood:		
6d	6	12
6d	6	12
6d	6	12
Fiber Board:		
1½" galv. roofing or 6d common, or 16-ga. staple 1⅛" long, minimum crown of ⁷⁄₁₆"	3	6
1¾" galv. roofing or 8d common, or 16-ga. staple 1½" long, minimum crown of ⁷⁄₁₆"	3	6
Gypsum:		
12-ga. 1¼", large head, corrosion resistant	4	8

g Each board should be nailed with at least two nails at every stud, and with three if the boards are more than 8" wide (Table 5.3). Such tight nailing makes for stiffness and helps to prevent warping and twisting if the boards become wet.

h Because there are no braces in the platform frame itself, the lateral rigidity of the house must be provided by the sheathing or other bracing; otherwise the racking stresses are thrown into the interior wall covering, and cracks are an almost certain consequence.

i If each sheathing board of horizontal sheathing were nailed with just one nail at each stud there would be practically no resistance to racking in the frame because nothing would prevent rotation of the studs. When two nails are used in each board at each stud, the resistance to racking is equal to the sum of the bearing of each nail against the board times half its distance from its partner in the same board. Considerable strength is achieved, but the total strength depends upon the bearing of nails against wood, and continual racking back and forth may eventually decrease the strength. A third nail adds no strength because the center nail merely acts as a pivot. Increasing the number of nails to four slightly increases

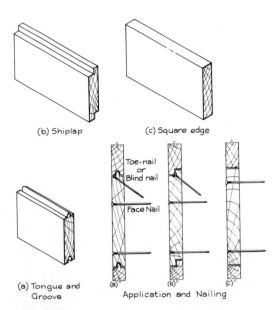

Figure 5.10 Sheathing Boards. (a, b, c) Tongue and Groove, Shiplap, Square-Edge. Toe-Nailing and Face-Nailing.

Framing

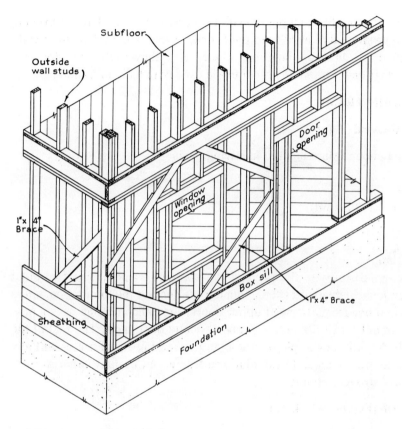

Figure 5.11 Let-In Bracing, Platform Frame.

the strength because the center nails contribute some resistance to rotation. In wide boards, three or four nails are needed to prevent warping and buckling, irrespective of any resistance to racking contributed by the additional nails.

j With diagonal sheathing, racking stresses are transferred directly down the sheathing boards to the sill and thence to the foundation. Diagonal sheathing is, therefore, a stiffer, more enduring type of construction and is best employed on any frame which is not braced in itself. At diagonally opposite corners of the house, sheathing should run down from post to sill, and at the other two corners it runs up from post to plate.

k Let-in bracing is often employed in the platform frame. It consists of boards, usually 1″ x 4″, let into the studs at the corners of the house and run diagonally from post to sill. Properly done, this kind of bracing adds a great deal of rigidity to the frame; its use is sufficient to obviate the need for diagonal sheathing. Figure 5.11 shows let-in bracing in a platform frame. The same procedure is used with balloon framing.

BALLOON FRAME

5.14 General (Figure 5.12)

a The balloon frame has the following parts:
Sill
Posts
Ribbon (or Ribband)
Plate
Studs
Sheathing

b A comparison with the parts of the platform frame shows two chief differences. Instead of a sill and header platform at the foundation, and a plate and header platform at upper floors, the balloon frame uses a one-member sill, and a ribband to support second floor joists. The rest of the members—sill, posts, plate, studs, and sheathing—perform essentially the same functions as in the platform frame, but certain important differences are to be noted.

5.15 Sill (Figures 5.3, 4, 12)

a Sills for balloon frames are similar to those for platform frames. What has already been said about sills—anchor bolts, bedding, termite shields, and decay-resistant treatment—holds for this sill as well (see Section 5.7). There is no header or band.

5.16 Posts (Figures 5.5, 12)

a Posts for the balloon frame are built up similarly to those for the platform frame, but they run full height from sill to roof plate, as shown in Figure 5.12.

5.17 Ribband (Figures 5.12, 13)

a This member is more commonly called the "ribbon" and forms the support for upper-floor joists. It is a square-edged board, 4" to 8" wide, which

Framing

is let ("housed") into the studs so that the top of the ribband is at the elevation of the bottoms of the joists. The joists are set on the ribband and nailed to the studs. The ribband occurs only in those walls which run perpendicular to the direction of the joists; it would have no function in the other walls.

5.18 Plate (Figure 5.12, 14)

a The same observations hold true for the plate in the balloon frame as in the platform (Section 5.12). As shown in Figures 5.12, 14, principal studs on the gable wall (left side of figure) are carried to a plate, and gable studs extend from there to marginal rafters. This is necessary because studs would not readily be procurable in sufficient length to run from sill to rafter, nor would it be feasible to frame and handle such long studs. The same detail is used in platform and braced framing.

5.19 Studs (Figure 5.12)

a As far as sizes and spacing are concerned, the same rules hold true for all frames (Section 5.10). There are, however, certain important differences in construction in balloon frames which are to be carefully noted:

1. *Length*. The balloon frame is characterized by studs that run full length from sill to plate. This is the balloon frame's distinguishing feature.

2. *Loads*. Loads are all carried by the studs and each stud carries its own load from plate to sill without any lateral distribution by girts or headers such as occurs to some extent in the other two types of frames. Because of this and because of the great length of the studs, it is most important in the balloon frame that studs be sound and straight.

3. *Openings*. Framing around openings, and sizes of openings are the same as in the platform frame (Section 5.10) except that large openings are much more commonly spanned by large headers rather than by trusses. Trussing an opening in a balloon frame is a more difficult task than in either of the other types, because there is no girt or header at the second floor line against which the diagonals can bear. Some expedient such as a secondary header above the diagonals (as shown in Figure 5.12) must therefore be employed. Since this makes the trussed opening even more complicated, and since the chief advantage of the balloon frame is economy and simplicity, this kind of construction is seldom found. It is, however, better than the simple large header with its tendency to shrink and swell.

b. Firestopping The vertical spaces between studs are likely to act as flues to transmit flames if fire breaks out. These potential flues must be

Dwelling House Construction 84

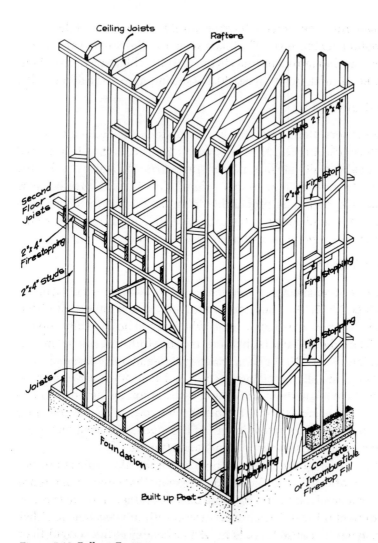

Figure 5.12 Balloon Frame.

Framing

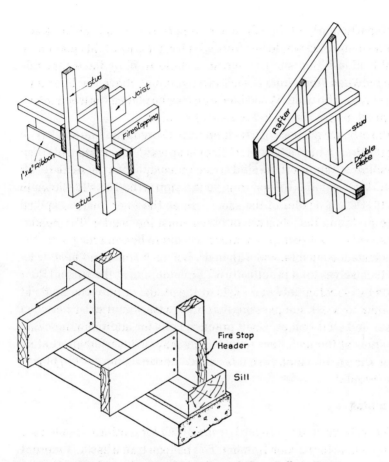

Figure 5.13 Ribbon and Firestopping At Second Floor, Balloon Frame.
Figure 5.14 Roof Plate and End Rafter for Gable Roof.
Figure 5.15 Tee Sill Construction

closed off, particularly at the sill, where otherwise an unbroken sweep from basement upward would be afforded. Firestopping at this point may be incombustible filling such as incombustible wool or flake that fills the space between studs from the sill to a point just above the floor line. Concrete may also be used. Wood blocking may be placed between joists. Types of firestops are shown in Figures 5.12, 13.

c Since the primary purpose of firestopping is to break the flue and hence the potential draft which might carry flames upward, any system of framing at the sill which stops off the stud spaces accomplishes the same result as brickwork. One method of securing such a stop is the Tee sill, shown in Figure 5.15. Here a header of the same size as the floor joists is spiked across the studs and the joists are butted against the header. The header in this case acts as a firestop. To obtain adequate bearing for joists, the sill in this instance must be wider than 6″, because studs and header together in themselves total practically 6″. Firestopping at the second floor is obtained by blocking between joists at the studs, or between studs at the floor line, to break the passage that otherwise would exist between joist spaces and stud spaces. Good practice calls for additional blocking between studs at the half-story height. This may be inserted alternately sloping, as shown, in which case it is called "herringbone," or the blocks may be horizontal.

5.20 Sheathing

a The observations made in regard to sheathing for platform framing are equally applicable for balloon framing. The balloon frame has no inherent lateral stiffness, and this must be supplied by sheathing, let-in bracing, or both.

BRACED FRAME

5.21 General (Figure 5.16)

a As indicated above, the braced frame, also known as the eastern frame, is used primarily in the Northeast, mostly in New England. It is a descendant of the heavy frames employed in the houses built during the Colonial period. It is a sturdy but fairly complex frame, even in its contemporary simplified version.

b The braced frame has the following parts:

Framing

1. Sill, resting on the foundation wall and supporting the balance of the frame.
2. Posts, comparatively heavy pieces placed at all corners of the frame and at intermediate points as well.
3. Girts, horizontal members at the second floor. One supports the ends of the floor joists and is called the dropped girt; the other, or raised girt, is parallel with the floor joists.
4. Plate, a horizontal member at the top of the frame, generally supporting the roof rafters.
5. Braces, diagonal pieces set in the angles between posts and sill and girts and plate to stiffen the frame.
6. Studs, set between the sill and the girts and the girts and the plate.
7. Sheathing; boards, plywood, and wallboard covering the outside of the frame.

c In most braced frames, the sills, posts, and girts are solid timber and are usually 4″ x 6″ in cross-section. Sometimes these members are fabricated of 2″ thick stock nailed together, but this is not customary in braced-frame construction.

5.22 Sill (Figures 5.16, 17)

a Unlike the platform and balloon frames, the sill for the braced frame is almost always 4″ thick; the usual size is 4″ x 6″. It is anchored and bedded in the same way as the other sills, usually with bolts. Half-lap joints (Figure 5.17) are employed at corners, and half-lapped splices at intermediate joints. Preferably, an anchor occurs at such a splice. A built-up sill is similarly lapped at corners and intermediate points.

5.23 Posts (Figures 5.16, 18, 19)

a Posts are usually 4″ x 6″ but occasionally 4″ x 8″ is specified or required by building codes. Posts are placed at all corners of the exterior walls and at intermediate points where important partitions intersect the exterior walls or where it is necessary to make a joint in the girt. Important features are:

   1. *Length.* Posts run full length from sill to plate.
   2. *Joints.* In older construction, posts, girts, and sills were mortised and tenoned. This is seldom done now.
   3. *Nailing for Interior Wall Finish.* At all corners, it is necessary to provide two surfaces at right angles to each other to which lath or other interior

finish material may be fastened. A 4″ x 6″ or 4″ x 8″ post provides only one surface at the corner; consequently, it is necessary to nail an additional 2″ x 2″ piece to the face. The inner edge of this piece is in line with the inner edges of the studs as shown in Figure 5.18. With built-up posts such as are employed for platform and balloon frames, the members can be offset to provide this support, as shown in Figure 5.18.

5.24 Girts (Figures 5.16, 19)

a Originally, the girts were strong enough to carry the entire floor loads, but today they depend in considerable measure upon support from the studs below. Their chief functions are to act as horizontal ties holding the frame together laterally, and to afford reactions for the corner braces. Construction features to watch are:

1. *Joints.* Girts are never spliced, but run continuously from post to post. In a long wall, this calls for a number of intermediate posts set at important cross-partitions, such as floor-bearing partitions. The mortise and tenon joint shown in Figure 5.19 is often omitted, since it is expensive to fabricate.

2. *Raised and Dropped Girts* (Figures 5.16, 19) Girts which run parallel to the direction of the floor joists are set with their top edges at the same elevation as the tops of the joists. This permits floor boards to be run out and nailed to the girts, a procedure which helps tie floor and wall together. Girts which run at right angles to the joists are set with their top edges at the elevation of joist bottoms, and joists rest directly on these girts.

5.25 Plate (Figure 5.16)

a Occasionally the plate is a solid 4″ x 4″ but more commonly and preferably, two 2″ x 4″s are used as in the platform and balloon frames.

5.26 Braces (Figure 5.16)

a End corners of the outside walls and important intermediate corners are braced to withstand the lateral racking caused by wind and nonuniform loads. The braces are the most distinguishing characteristic of this type of frame and give it its name.

b. Single Diagonal Brace (Figure 5.16) This type of brace is generally a 2″ x 4″ or 3″ x 4″ strut. In the first story, it runs diagonally from the intersection of post and girt to a point on the sill approximately 4′ away (three stud spaces, if possible). In the second story, it runs from intersection of

Framing

post and plate to a similar point on the girt. Bearing is obtained at the post by fitting the lower story braces into the intersection of post and girt, and the upper story braces against a 2″ x 4″ block securely spiked against the upper end of the post, as shown in Figure 5.16. The lower ends of the braces may be fitted against blocks spiked to the tops of girts and sills, although common practice is to fit them against the lower ends of studs. Braces are continuous, with studs cut to fit the braces; not the reverse.

5.27 Studs (Figure 5.16)

a In the braced frame, the studs are supposed to act merely as supports for the interior and exterior covering while the girts and posts carry floor and roof loads. Actually, the studs support a large portion of the load while the girts and posts serve to tie the entire skeleton together. Studs extend (Figure 5.16) from sill to girt in the first story and from girt to plate in the second. They are toe-nailed to sill and girts.

b A variant of the stud to sill construction shown in Figure 5.16 is the Tee sill described above (Figure 5.15).

c Framing around openings is similar to that for the platform frame, and includes headers for relatively small openings and trussed construction for wide openings.

d. Firestopping The firestopping is similar to that described under balloon framing. Masonry or incombustible fill firestopping may be placed over the sill or above the dropped girt, or blocking may be inserted between joists or between studs at the floor line.

e Good practice calls for additional firestopping between sill and girt and between girt and plate. This takes the form of horizontal 2″ x 4″ blocking between studs half way up the story height, as shown in Figure 5.2, 12, 16. The blocking may be inserted alternately sloping as shown ("herringbone" stopping), or the blocks may be horizontal. A belief persists that this blocking, especially the herringbone variety, imparts additional strength to the wall, but such contribution is negligible.

5.28 Sheathing (Figure 5.16)

a All of the types of sheathing described above are employed with the braced frame. This frame, however, has lateral bracing built into it and does not depend upon the sheathing to resist lateral racking, as is true of unbraced platform and balloon frames. Consequently, when wood sheathing boards are used, they can be applied horizontally instead of diagonally, without let-in bracing.

Dwelling House Construction 90

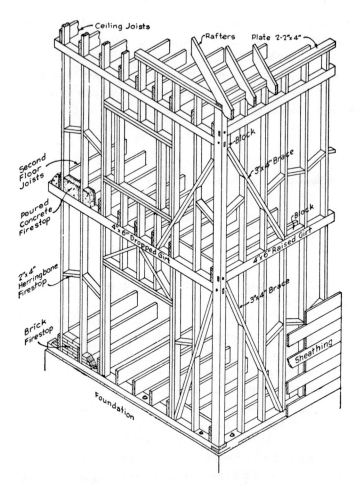

Figure 5.16 Braced Frame.

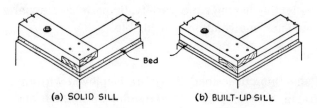

Figure 5.17 Sill for Braced Frame.

Framing

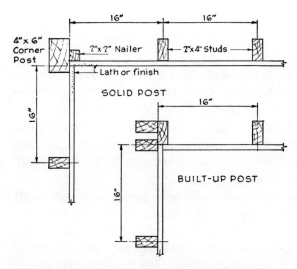

Figure 5.18 Corner Posts and Stud Spacing Details.

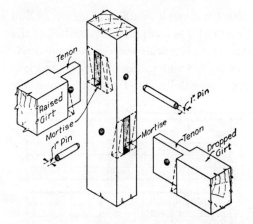

Figure 5.19 Pinned Mortise and Tenon Connections of Girts to Posts, for a Braced Frame.

FLOOR FRAMING

5.29 General

a Floor framing is simpler in its details than wall framing and is essentially the same no matter what kind of wall frame—platform, balloon, or braced—is used. The floor, regardless of the finish which it is to bear, consists of rough or subflooring resting on joists carried by girders, walls, or partitions. Girders in turn are supported by posts or piers which rest on footings. Joists are sometimes known as floor beams, but the term "joist" is much more commonly employed.

b The simplest way to frame a floor would be to let joists run full length from one outside wall to the other, but this would entail excessively long pieces difficult to handle and expensive to purchase. Moreover, if these joists had no intermediate support they would have to be extremely deep to provide the requisite strength and stiffness. For these reasons shorter lengths, supported at their ends on girders, walls, or partitions, are employed.

c Figure 5.20 shows a framing plan for a small dwelling. The left half of the house has joists running perpendicular to front and back walls and the total distance is covered in two spans, the interior ends of the joists resting on a girder which runs parallel to the front and back walls. The right half of the house is framed with joists paralleling the front and rear walls;

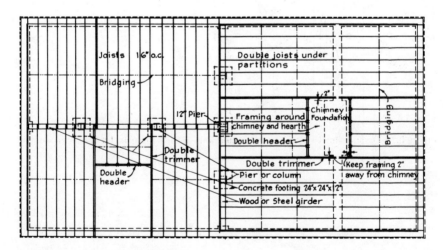

Figure 5.20 Floor Framing Plan.

Framing

two sets of joists again make the total span but their intermediate ends rest on a masonry or concrete wall. In general it is simplest to have all joists run the same direction but this is by no means necessary nor is it always feasible. Generally, joists should be so arranged as to give the shortest spans with the fewest intermediate supports. This makes for economy in labor and material, for the shorter the span the shallower the joists can be and the fewer the intermediate supports the less labor required in framing.

5.30 Posts (Figure 5.21)

a Supports for girders are commonly hollow steel pipe, steel pipe filled with concrete, wood posts, masonry piers (brick or concrete block), and cast concrete piers.

b No matter what the column or pier may be, it must have an adequate footing and a cap of some kind upon which the girder bears. The footing is usually concrete poured in place and is commonly 2' square by 1' thick. For large masonry piers these dimensions must be increased; for small wood or steel columns carrying light loads the dimensions can be reduced somewhat (Chapter 4).

c Brick piers must be at least 12" square unless loads are very light and the piers are short. For large loads and tall piers the dimensions are increased to 16". Concrete block piers (Figure 5.21a) have the same limitations, except that since most blocks are 16" long, the piers are seldom smaller than that dimension in each direction.

d Cast concrete piers should not be less than 10" round or square unless reinforced with laterally-tied vertical rods.

e Wood posts must not be permitted to stand in water or to become permanently damp. Therefore, their lower ends should be raised off the floor several inches so that any water standing on the floor will not cause them to rot. As shown in Figure 5.21b, c, a base of concrete or masonry is built on top of the footing, projects above the floor level, and supports the column. When wood posts are used, the surrounding air must be dry enough to prevent rot; unless the posts have been treated with preservative.

f Steel pipe columns, hollow or filled, are fitted with cap and base plates (Figure 5.21c). Frequently, the base plates are set directly on the footings and the entire foot of the column is concreted in. While this anchors the column firmly, it also permits water to reach and rust the steel parts. Rust-resisting coating should, therefore, be applied liberally to these parts before they are set and concreted. It is preferable to set these plates on concrete or masonry bases, the same as is done for wooden posts, or to set anchor bolts in the concrete and fasten the column and base plate to these bolts, on top of the floor slab.

g Caps for columns of all kinds are commonly flat steel plates from $\frac{1}{4}''$ to $\frac{1}{2}''$ thick. Where wood posts are as wide as the girders above, the cap plate can be omitted.

5.31 Girders (Figure 5.21)

a. Wood Girders in house and other light construction are wood, or steel shapes such as I and wide-flange beams; sometimes reinforced concrete. Wood girders may be either solid or built up and either type is entirely satisfactory. The built-up timber consists of pieces of 2″ or 3″ dimension stock set on edge and nailed or lag-screwed together to form a larger piece. Figure 5.21 shows a girder built up of three 2″ pieces. The solid timber (Figure 5.21) is simplest and gives the most actual timber for a given cross-section, as is shown by the fact that a nominal 8″ wide solid timber is actually $7\frac{1}{4}''$ or $7\frac{1}{2}''$, whereas four 2″ pieces are actually four times $1\frac{1}{2}''$ or $1\frac{9}{16}''$ or 6″ to $6\frac{1}{4}''$ wide. A built-up timber, therefore, would have to be made up of five 2″ pieces in order to contain as much wood as the solid stick. On the other hand, solid pieces run only from support to support whereas built-up girders can be made continuous from one end of the building to the other, thereby tying the entire frame more closely together and adding to the stiffness of the girder. Solid timbers are also apt to shrink and check (split) more badly than built-up timbers. The larger sizes of timber call for an extra item in the lumber order whereas the built-up timbers are made of the same stock that is used for floor joists. As a matter of practical engineering, joints in built-up girders should be well staggered or "broken" and ought to occur at approximately the quarter-points of the spans, not directly over the supports.

b Wood girders may be laminated by gluing smaller pieces together to form larger timbers. Usually nominal 2″ material is dressed smoothly and the necessary number of laminations glued together to form the desired depth. Casein is commonly used for indoor applications, but a waterproof glue is needed for exposure outdoors. After the glue has hardened, the

Framing

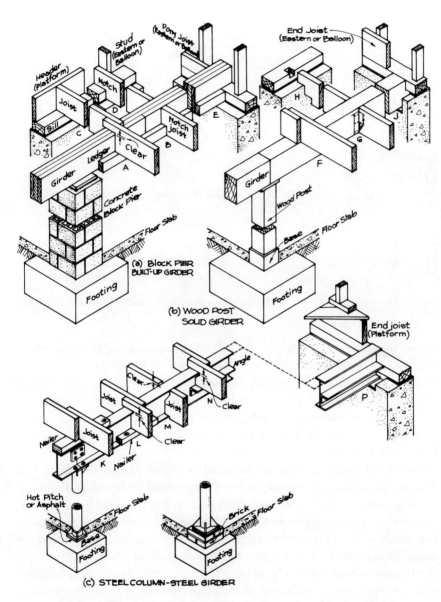

Figure 5.21 Details of Post and Girder Construction for Masonry, Wood, and Steel.

Dwelling House Construction

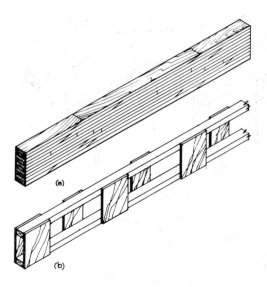

Figure 5.22 (a) Glued Laminated Girder. (b) Joist Beam with Lumber Chords and Plywood Intermittent Web.

timber is again dressed on all four sides to obtain a clean, smooth surface (Figure 5.22). Scarf or fingered joints with slope $\frac{1}{10}$ or shallower are needed in the upper and lower $\frac{1}{4}$ depth to resist bending stresses. Other joints can be butt joints. A built-up joist beam with solid lumber or laminated top and bottom flanges and intermittent plywood web is shown at b. Web members must be designed to resist the shear at various points in the length of the beam.

c. Steel (Figure 5.21c) Steel beams used as girders introduce one or two new problems, but are essentially the same as wood girders in their action. Usually, because of their weight, they run only from support to support instead of being continuous, and are held in place by bolts which engage the cap plates of steel pipe columns. They are held together, in addition, by steel splice plates bolted to the ends of two butted members.

d. Framing at Exterior Walls Where the ends of girders, either wood or steel, rest in exterior walls, recessed ledges are provided and the bottoms are best covered with steel bearing plates or carefully levelled (Figure 5.21b, c). Such openings must be made large and deep enough to provide adequate ventilation around the ends of girders; otherwise dampness causes wood girders to rot and steel girders to rust.

Framing

e. Spacing Since joists are cut in lengths which are 1' or 2' multiples, the best spacing of girders is such as to allow full use of these lengths without waste, and if no other factors govern, girders should be so spaced. Frequently, however, girders are placed under important bearing partitions and these in turn are regulated in their spacing by room sizes. Sometimes, moreover, it is desirable to place girders over basement partitions so that they will not project into the ceiling space in finished basement rooms.

f. Framing into Sills and Girders The simplest and in many ways the best way to frame joists into their supports is merely to set them on top (Figure 5.21, *CFK*). When so supported, joists are easily toe-nailed and spiked to sill and stud, making a good joint; similarly, they are easily lapped at the girders, spiked together and toe-spiked to the girder, again affording a strong joint. However, in-line and cantilever framing (Section 5.52) make it desirable to keep joists in alignment.

g Joists resting on girders pile up a good deal of wood in a direction perpendicular to the grain and make for excessive shrinkage if the wood is not seasoned. Measurements show that, with wide climatic and geographic variations, wood reaches an average final moisture content of roughly 12 percent in dwellings. Consequently, even commercially "dry" wood (seasoned to 19 percent moisture content or less) is likely to show some shrinkage in place in a dwelling. This is especially serious at the girders, less so at the sill. Moreover, dropped girders may interfere with headroom, may require pipes and ducts to be carried very low, and may therefore call for extra depth in the basement.

h Girders should, for the foregoing reasons, be kept as high as possible. This is best accomplished by the cleat or ledger as illustrated in Figure 5.21, *ABE*. The notched joists at *A* are to be carefully noted. Bearing must not in any case be permitted to occur at the notch; it must all occur at the bottom of the joist where it rests on the ledger. Hence the notch must be made high enough to clear the girder by at least ½", so that subsequent shrinkage in the joist will not cause bearing at the notch. Otherwise the joist is almost certain to split at that point. This is a general rule covering all deep notches; bearing must not occur at the notch, otherwise splits are almost certain to occur. This does not apply to shallow notches as illustrated at the sill at point *C*. These merely take up the slight inequalities in depth which often occur in joists, and allow the tops of all joists to be at the same level. When it is desired to have girders and joists flush at the bottom, joists may be notched as shown at B. Notches at this depth are not objectionable on deep joists of long span, carrying fairly light loads. Heavy loads on short spans are apt to cause splits at the notch, as is also true of D unless the bottom of the joist is shimmed to bear on the foundation wall. The more complicated notch shown at H is open to the same general criticism. It was common in older construction, especially in the eastern braced frame, but is seldom used now.

i Specially-formed metal grip plates or framing anchors are available to fasten the ends of joists to the sides of girders by nailing (Figure 5.23). Joists may also be hung from girders with hangers as illustrated in Figure 5.21, G, and Figure 5.23. It is to be noted that the stirrups shown at Figure 5.21, G, allow as much total drop at the top of the joist, due to shrinkage, as if the joist rested directly on top of the girder.

j When steel girders are used (Figure 5.21c) joists may rest directly on top, in which case a 2" x 4" (see K) may be bolted to the top flange to provide nailing for the joists. Joists may also rest directly on the bottom flange of the beam (see M) and here again nailing may be provided by nailers bolted to the beam (see L). Support may also be afforded by angles bolted to the web of the beam (see N). To provide for shrinkage, notches at L and N and tops of joists at M should clear the top flanges of the girder by at least $\frac{1}{2}"$.

k Any of the sill details CDH in Figure 5.21 may be combined with any of the girder details ABFGKLMN, by altering the height of the girder and its wall supports EJ or P. As drawn, E corresponds to A and B; J corresponds to F; and P corresponds to N.

l. Framing around Openings Openings are framed in essentially the same way as in walls, namely, by a combination of trimmers and headers arranged around the opening. The joists (Figures 5.20, 23, 24) which run parallel with two sides of the opening are doubled if necessary to carry the loads and are called trimmers. Across the other two sides of the opening are set two other joists called headers, doubled, if necessary, and into these headers are framed the shortened joists, known as tail beams, tail joists, or header joists.

m Where spans are short and loads are light, this whole assembly can be nailed together, otherwise hangers or framing anchors ("grip plates") are used, particularly to support the headers on the trimmers. Sometimes the tail beams are supported on ledgers by notching, as illustrated in Figure 5.24, but this is subject to the criticism already made regarding bearing on notches.

Framing

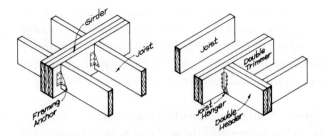

Figure 5.23 Joists and Headers Supported by Framing Anchors and Hangers.

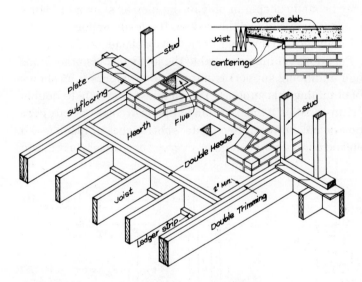

Figure 5.24 Framing Around Chimney and Fireplace.

n Framing around chimneys and fireplaces (Figures 5.20, 24) is commonly held away 2″ and the space between is filled with incombustible insulating material. *Wood members must not be permitted, under any circumstances, to frame into a chimney,* although a pilaster or corbel may be built as part of a chimney to support wood framing members. (See also Chapter 6.)

o. Framing under Partitions Nonbearing partitions resting on the floor and running perpendicular to the direction of the joists need no extra support, but bearing partitions may impose enough load, if they are situated close to the centers of spans, to require additional strength in the supporting joists. These may be made deeper or may be doubled as shown in Figure 5.25 a. When a condition such as this exists, the loads imposed on the joists should be analyzed and the joists designed by elementary engineering procedure. Partitions running parallel to the joists throw undue loads on single joists if additional support is not provided. Figure 5.25 shows two methods of obtaining this support. The simplest is merely to double the joist (b and d) in question if the partition happens to come directly over it. If the partition comes between two joists, short pieces of block (solid bridging) are spiked across between them, every 2′ or so apart, in order to

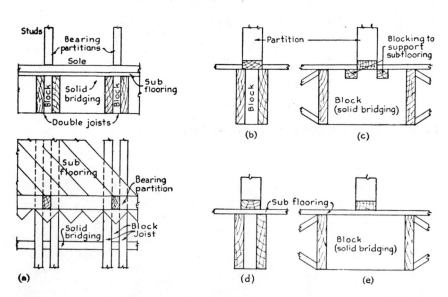

Figure 5.25 Framing Under Partitions. (a) Support for Intermediate Transverse Bearing Partition. (b–e) Support for Non-Bearing Partitions.

Framing

distribute the load to both joists (c and e). Details d and e are the simpler and more common because the subfloor can be run through instead of having to be cut and fitted around the partition sole, as in b and c.

p. Framing Upper Floors Upper floors are framed the same as the first, except that outer ends of joists rest on girts, ribbons, or plates instead of on sills, and inner ends rest on partition caps instead of on interior basement walls or girders. Occasionally, however, girders are used to support second floor joists over large rooms.

5.32 Interior Basement Walls

a Joists supported by interior masonry walls are carried on sills. If there is a chance that lateral forces will be applied to the house (heavy winds, earthquake), these sills should be anchored in the same way as sills on exterior walls.

5.33 Joists (Figures 5.2, 6, 9, 12, 13, 15, 20, 21)

a. Sizes and Spacing Joists must not only be large enough to carry the floor loads, but must also be stiff enough to prevent excessive deflection. The usual rule of thumb is to limit deflection at the center to $\frac{1}{360}$th of the span.

b Table 5.5 lists spans for various sizes of joists at various spacings as limited by the span-divided-by-360 rule. The 40 lb per sq ft loading is a customary building code requirement for the living quarters of a house, with 30 lb per sq ft for sleeping quarters and attics. Under each span as limited above is the extreme fiber stress in bending for that span and loading. For a joist of a given grade and species the modulus of elasticity E and allowable bending stress F_b should be checked to see that the loading is safe.

c Joists are commonly spaced 16" on centers but this spacing can be varied. Joists are generally the same depth throughout the floor. That is, instead of making some spans of the first floor 2" x 8" and some 2" x 12", all spans are 2" x 12". The second floor might be all 2" x 10" or 2" x 8". Quite often, second floor joists are lighter than those on the first floor.

d By varying the spacing it is often possible to save material. For instance, suppose all spans but one in a floor were 12' or less, but this one were 13'. Suppose that, except for this one span, all joists could be 2" x 8". Since deflection for uniform floor loading increases as the fourth power of length, the 2"x 8"'s is to be used on a 13' span would have to be spaced more closely

Table 5.5 Floor Joists[a]

Joist Size	Spacing (in.)	Span and Fiber Stress — Live Load								
		40 lb/sq ft[b]					30 lb/sq ft[c]			
		Modulus of Elasticity E, in 1,000,000 psi								
		1.0	1.2	1.4	1.6	1.8	1.0	1.2	1.4	1.6	1.8
2x6	12.0	9-2 830	9-9 940	10-3 1,040	10-9 1,140	11-2 1,230	10-1 810	10-9 910	11-3 1,010	11-10 1,100	12-3 1,200
	16.0	8-4 920	8-10 1,040	9-4 1,150	9-9 1,250	10-2 1,360	9-2 890	9-9 1,000	10-3 1,110	10-9 1,220	11-2 1,320
	24.0	7-3 1,050	7-9 1,190	8-2 1,310	8-6 1,440	8-10 1,550	8-0 1,020	8-6 1,150	8-11 1,270	9-4 1,390	9-9 1,510
	32.0	6-7 1,150	7-0 1,300	7-5 1,450	7-9 1,590	8-0 1,690	7-3 1,110	7-9 1,270	8-2 1,410	8-6 1,530	8-10 1,650
2x8	12.0	12-1 830	12-10 940	13-6 1,040	14-2 1,140	14-8 1,230	13-4 810	14-2 910	14-11 1,010	15-7 1,100	16-2 1,200
	16.0	11-0 920	11-8 1,040	12-3 1,150	12-10 1,250	13-4 1,360	12-1 890	12-10 1,000	13-6 1,110	14-2 1,220	14-8 1,320
	24.0	9-7 1,050	10-2 1,190	10-9 1,310	11-3 1,440	11-8 1,550	10-7 1,020	11-3 1,150	11-10 1,270	12-4 1,390	12-10 1,510
	32.0	8-9 1,170	9-3 1,300	9-9 1,450	10-2 1,570	10-7 1,700	9-7 1,120	10-2 1,260	10-9 1,410	11-3 1,540	11-8 1,660
2x10	12.0	15-5 830	16-5 940	17-3 1,040	18-0 1,140	18-9 1,230	17-0 810	18-0 910	19-0 1,010	19-10 1,100	20-8 1,200
	16.0	14-0 920	14-11 1,040	15-8 1,150	16-5 1,250	17-0 1,360	15-5 890	16-5 1,000	17-3 1,110	18-0 1,220	18-9 1,320
	24.0	12-3 1,050	13-0 1,190	13-8 1,310	14-4 1,440	14-11 1,550	13-6 1,020	14-4 1,150	15-1 1,270	15-9 1,390	16-5 1,510
	32.0	11-1 1,150	11-10 1,310	12-5 1,440	13-0 1,580	13-6 1,700	12-3 1,120	13-0 1,260	13-8 1,400	14-4 1,540	14-11 1,660
2x12	12.0	18-9 830	19-11 940	21-0 1,040	21-11 1,140	22-10 1,230	20-8 810	21-11 910	23-1 1,010	24-2 1,100	25-1 1,200
	16.0	17-0 920	18-1 1,040	19-1 1,150	19-11 1,250	20-9 1,360	18-9 890	19-11 1,000	21-0 1,110	21-11 1,220	22-10 1,320
	24.0	14-11 1,050	15-10 1,190	16-8 1,310	17-5 1,440	18-1 1,550	16-5 1,020	17-5 1,150	18-4 1,270	19-2 1,390	19-11 1,510
	32.0	13-6 1,150	14-4 1,300	15-2 1,450	15-10 1,580	16-5 1,700	14-11 1,130	15-10 1,270	16-8 1,400	17-5 1,530	18-1 1,650

[a] The first number under each modulus of elasticity value E is the span, in feet and inches, as limited by the allowable deflection (see the following note b), and the second number is the corresponding extreme fiber stress in bending for that span (note b).
[b] Design criteria for deflection under 40 lb/sq ft live load: Deflection is to be limited to span (in.) divided by 360. Strength: The resulting fiber stress value is determined by taking the live load (40 lb/sq ft) plus a dead load of 10 lb/sq ft.
[c] Design criteria, 30 lb/sq ft live load: Deflection limited as given in preceding note b. Strength: As in note b, except that the live load is here 30 lb/sq ft.
Source: National Association of Home Builders Research Foundation, Inc.

Framing

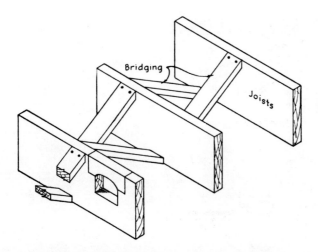

Figure 5.26 Bridging Floors and Cutting Holes for Piping.

to reduce the load per joist. The ratio of spacing would be the ratio $(13)^4$ to $(12)^4$. This is almost exactly 4 to 3. Therefore, the 2 x 8's spaced 12" on centers over this one span would be satisfactory and all joists could be 2" x 8". Similarly, spacing could be increased above 16" in other instances (see also Table 5.5).

5.34 Bridging (Figure 5.26)

a Bridging has traditionally been considered to be needed to support long deep floor joists from tending to buckle sidewise at the bottom. Tests indicate that bridging, except in unusual cases, is not really necessary, but many codes require it.

b As shown in Figure 5.26, bridging consists of short pieces set in crosswise between the joists and nailed top and bottom. Bridging is commonly 1" x 2" or 1" x 3" for joists up to 2" x 10" inclusive, and 2" x 2" or 2" x 3" for deeper joists. Customarily, there is a line of bridging for each 6' to 8' of unsupported length of joist. Steel strapping placed in the same criss-cross manner as wood bridging is also employed. It is run over one joist and under the adjacent one and nailed to their edges.

5.35 Subflooring (Figures 5.2, 6, 8, 9, 12, 16, 27, 28)

a The ¾" square-edged or matched boards used for rough flooring are best laid at 45° to the joists when the finish floor is wood strip flooring

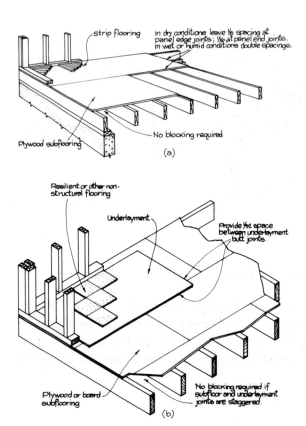

Figure 5.27 (a) Plywood Subflooring with Wood Strip Flooring. (b) Plywood Subfloor and Underlayment for Resilient Flooring.

(Chapter 13). The reason is that such flooring should not run the same direction as the rough floor and, since the finish floor may change direction in various rooms, this can be avoided only by laying the subfloor diagonally. The best board subfloor is 6" or 8" tongue and groove or shiplap driven up tight and nailed with at least two nails at every joist. Unless the flooring is well nailed, it may warp and twist upon wetting and provide a poor nailing surface for the finish floor.

b Plywood is extensively employed for subflooring as single thickness, double thickness with subflooring and underlayment, or underlayment over boards. As a general rule, plywood should be laid with the grain direction of the face plies perpendicular to the direction of the joists. This is particularly true of 3-ply plywood which is much stiffer and stronger in

Framing

this direction than at right angles to it. Thickness depends upon the species group, grade (Section 5.3), and spacing of joists.

c Some recommended grades, thicknesses, spans, and nailing schedules for single-thickness plywood subflooring used under tongue-and-groove wood strip and block flooring (Chapter 13) are given in Table 5.6. Plywood is continuous over two or more joists, and face grain runs across the joists.

d As shown in Figure 5.27, a $\frac{1}{16}''$ space should be left between plywood sheets at end and edge joints respectively. If the finish floor is wood block, edges of plywood should be supported by blocking between joists unless the edges are tongue and groove. End joints in adjacent tiers of plywood are staggered (broken).

e For support of resilient flooring (Chapter 13), plywood underlayment provides a smooth surface over plywood subflooring or over boards. Recommended grades, thicknesses and fastener (nail or staple) spacings are given in Table 5.7. Joints in underlayment are offset from joints in plywood subflooring.

f Plywood subfloor and underlayment may be combined into one thickness to eliminate the double labor of installing two layers. Some recom-

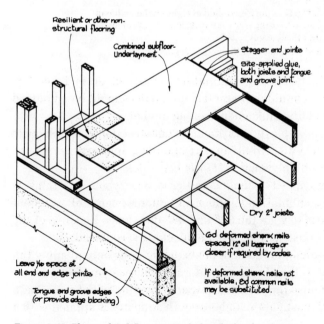

Figure 5.28 Plywood Subflooring Nailed and Glued to Joists.

Table 5.6 Nailing Schedule for Plywood Subflooring [a,c,d] (For Direct Application of Tongue and Groove Wood Strips and Block Flooring; Plywood Continuous over Two or More Spans, Face Grain across Supports)

Panel Identification Index [b]	Plywood Thickness (in.)	Max Span [e] (in.)	Nail Size (Common Nails)	Nail Spacing (in.) Panel Edges	Intermediate
30/12	5/8	12 [f]	8d	6	10
32/16	1/2, 5/8	16 [g]	8d [h]	6	10
36/16	3/4	16 [g]	8d	6	10
42/20	5/8, 3/4, 7/8	20 [g]	8d	6	10
48/24	3/4, 7/8	24	8d	6	10
1 1/8" Groups 1, 2	1 1/8	48	10d	6	6
1 1/4" Groups 3, 4	1 1/4	48	10d	6	6

[a] These values apply for STANDARD C-D INT, STRUCTURAL I and II C-D INT, C-C EXT and STRUCTURAL I C-C EXT grades only.
[b] See Table 5.3. The identification Index appears on all except 1 1/8" and 1 1/4" panels.
[c] In some nonresidential buildings, special conditions may impose heavy concentrated loads and heavy traffic requiring subfloor constructions in excess of these minimums.
[d] Edges shall be tongue and grooved or supported with blocking for square edge wood flooring, unless separate underlayment layer (1/4" minimum thickness) is installed.
[e] Spans limited to values shown because of possible effect of concentrated loads. At indicated maximum spans, floor panels carrying Identification Index numbers will support uniform loads of more than 100 psf.
[f] May be 16" if 25/32" wood strip flooring is installed at right angles to joists.
[g] May be 24" if 25/32" wood strip flooring is installed at right angles to joists.
[h] 6d common nail permitted if plywood is 1/2".
Source: American Plywood Association, Tacoma, Wash.

mended grades, thicknesses, spans, and nailing are given in Table 5.8. Deformed-shank (e.g., annular or spiral shank) nails should be employed to minimize the effect of shrinkage in framing lumber.

g Thicker plywood is employed in systems of construction in which joists or floor beams are spaced 32" to 48" apart. Depending upon species groups and spacing (Table 5.6), plywood is 1 1/8" or 1 1/4" thick

h In one system, called 2-4-1, tongue-and-groove plywood, 1 1/8" thick, is utilized to provide combination subfloor and underlayment of sufficient strength and stiffness to span the 32" and 48" distances. Because of the tongue-and-groove construction, no edge blocking is required.

i When plywood is glued to the floor joists, using nails only to apply the necessary pressure to the glue, a considerable increase in strength and stiffness results because the plywood now acts together with the joist as a Tee beam. A bead of elastomeric high-strength adhesive is applied with a

Table 5.7 Plywood Underlayment For Application of Tile, Carpeting, Linoleum or other Non-Structural Flooring.

Plywood Grades and Species Group	Application	Minimum Plywood Thickness (in.)	Fastener size (approx.) and Type (set nails $\frac{1}{16}''$)	Fastener Spacing (In.)	
				Panel Edges	Intermediate
Groups 1, 2, 3, 4 UNDERLAYMENT INT (with interior, intermediate or exterior glue)	Over plywood subfloor	$\frac{1}{4}$	18-ga. staples or 3d ring-shank nails [a,b]	3	6 each way
UNDERLAYMENT EXT C-C Plugged EXT	Over lumber subfloor or other uneven surfaces	$\frac{3}{8}$	16-ga. staples [a] 3d ring-shank nails [b]	3 6	6 each way 8 each way
Same Grades as above but Group 1 only.	Over lumber floor up to 4" wide. Face grain must be perpendicular to boards.	$\frac{1}{4}$	18-ga. staples or 3d ring-shank nails	3	6 each way

[a] Crown width $\frac{3}{8}''$ for 16-ga., $\frac{3}{16}''$ for 18-ga. staples; length sufficient to penetrate completely through, or at least $\frac{5}{8}''$ into, subflooring.
[b] Use 3d ring-shank nail also for $\frac{1}{2}''$ plywood and 4d ring-shank nail for $\frac{5}{8}''$ or $\frac{3}{4}''$ plywood.
Installation: Apply UNDERLAYMENT just prior to laying finish floor or protect against water or physical damage. Stagger panel end joints with respect to each other and offset all joints with respect to the joints in the subfloor. Space panel ends and edges about $\frac{1}{32}''$. For maximum stiffness, place face grain of panel across supports and end joints over framing. Unless subfloor and joists are thoroughly seasoned and dry, countersink nails $\frac{1}{16}''$ just prior to laying finish floor to avoid nail popping. Countersink staples $\frac{1}{32}''$. Fill any damaged, split or open areas exceeding $\frac{1}{16}''$. Do not fill nail holes. Lightly sand any rough areas, particularly around joints or nail holes.

Table 5-8 Grades, Thicknesses, and Fastening Details for Single-Layer Plywood Flooring

		Maximum Support Spacing [a,b]							
		16" o.c.		20" o.c.		24" o.c.		Nail Spacing (In.)	
Plywood Grade [c]	Plywood Species Group	Panel Thickness	Deformed Shank Nail Size [d]	Panel Thickness	Deformed Shank Nail Size [d]	Panel Thickness	Deformed Shank Nail Size [d]	Panel Edges	Intermediate
C-C Plugged Exterior	1	$\frac{1}{2}''$	6d	$\frac{5}{8}''$	6d	$\frac{3}{4}''$	6d	6	10
Underlayment with EXT glue	2, 3	$\frac{5}{8}''$	6d	$\frac{3}{4}''$	6d	$\frac{7}{8}''$	6d	6	10
Underlayment	4	$\frac{3}{4}''$	6d	$\frac{7}{8}''$	6d	1"	8d	6	10

[a] Edges shall be tongue and grooved, or supported with framing.
[b] In some non-residential buildings, special conditions may impose heavy concentrated loads and heavy traffic requiring subfloor-underlayment constructions in excess of these minimums.
[c] For certain types of flooring such as hardwood strip, wood block, slate, terrazzo, etc., sheathing grades of plywood may be used.
[d] Set nails $\frac{1}{16}''$ and lightly sand subfloor at joints if resilient floor is to be applied.
Source: American Plywood Association, Tacoma, Wash. National Association of Home Builders Research Foundation, Inc., Wash. D.C.

Dwelling House Construction 108

caulking gun to the upper edge of the joist and in the groove of the tongue-and-groove edge of the plywood. As shown in Figure 5.28, end joints are staggered and a $\frac{1}{16}''$ space is left between panels. Nails are spaced 12" on centers along edges and intermediate joists. Table 5.9 gives recommended spans and spacings for selected glued plywood-joist combinations, as compared with Table 5.5 for ordinary joist construction.

j Special preparation is needed for the bathroom floor if it is to be tile (Figure 5.29). In the upper view, the tops of all joists are first chamfered and then 1"x 3" strips are nailed to the sides of the joists, 3"down from the top. On these strips are placed short lengths of rough flooring fitted between joists. The concrete base for tile is then cast on these boards and wire mesh, run continuously over the chamfered tops of joists, is embedded in the final mortar bed upon which the tile are placed. A more desirable method is shown in the lower portion of Figure 5.29. Here the tendency to incipient cracking in the floor above the usual chamfered joists is avoided because the concrete is of uniform thickness. (See also Chapter 13.)

k Joists for bathroom floors must be set with care because they carry especially heavy loads and plumbers are very apt to cut holes and notches in them indiscriminately to get their pipes in place. These joists should therefore be set in consultation with the plumbing foreman and should be cut only under the supervision of the carpenter foreman or the builder

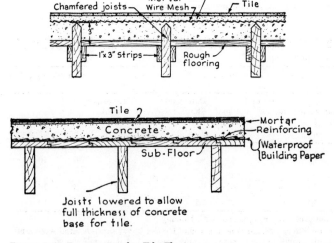

Figure 5.29 Preparation for Tile Floor.

Framing

Table 5-9 Glued Plywood Joist Construction; Allowable clear spans (Partial List)

Species-Grade	Joist Size	5/8"Plywood[a] Joists @ 16"	3/4"Plywood[b,d] Joists @ 16"	3/4"Plywood[c,d] Joists @ 24"
Douglas Fir, Larch, No. 1	2x6	11' 1"	11' 6"	9' 5"
	2x8	14 4	14 9	12 5
	2x10	18 1	18 6	15 10
	2x12	21 9	22 2	19 3
Douglas Fir, Larch, No. 2	2x6	10 6	10 6	8 7
	2x8	13 10	13 10	11 3
	2x10	17 7	17 7	14 5
	2x12	21 5	21 5	17 6
Douglas Fir, Larch, No. 3	2x6	8 0	8 0	6 7
	2x8	10 7	10 7	8 8
	2x10	13 6	13 6	11 0
	2x12	16 5	16 5	13 5
Douglas Fir, South, No. 1	2x6	10 5	10 9	9 1
	2x8	13 5	13 10	12 0
	2x10	16 10	17 3	15 4
	2x12	20 3	20 8	18 8
Hemlock, Fir, No. 1	2x6	10 3	10 3	8 5
	2x8	13 7	13 7	11 1
	2x10	17 2	17 4	14 2
	2x12	20 7	21 1	17 2
Southern Pine KD 15%, No. 1	2x6	11 3	11 8	9 9
	2x8	14 7	15 0	12 11
	2x10	18 4	18 9	16 6
	2x12	22 1	22 6	20 0
Southern Pine, No. 1	2x6	11 1	11 6	9 5
	2x8	14 4	14 9	12 5
	2x10	18 1	18 6	15 10
	2x12	21 9	22 2	19 3
Southern Pine, No. 2	2x6	9 6	9 6	7 9
	2x8	12 7	12 7	10 3
	2x10	16 0	16 0	13 1
	2x12	19 6	19 6	15 11

[a] UNDERLAYMENT INT Group 1, 2 or 3 may be used for combined subfloor underlayment. Underlayment grade may be $19/32$". If separate underlayment or a structural finish floor is installed, STANDARD DFPA $5/8$", $32/16$ or $42/20$ may be used.

[b] UNDERLAYMENT INT Group 1, 2, 3 or 4 may be used for combined subfloor underlayment. Underlayment grade may be $23/32$". If separate underlayment or structural finish floor is installed, STANDARD DFPA $3/4$", $36/16$, $42/20$, or $48/24$ may be used.

[c] UNDERLAYMENT INT Group 1 may be used for combined subfloor underlayment. Underlayment grade may be $23/32$". If separate underlayment or structural finish floor is installed. STANDARD INT-DFPA $3/4$", $48/24$ may be used.

[d] $7/8$" or 1" UNDERLAYMENT of any Group, or STANDARD with appropriate Identification Index numbers, may be substituted for lesser thicknesses if desired.

Source: American Plywood Association, Tacoma, Wash.

himself. Figure 5.26 shows one method of cutting holes for piping which weakens the joist only slightly. Holes may also be strengthened by reinforcing the adjoining areas with wood scabs or steel plates or shapes securely nailed, spiked, screwed, or bolted.

PARTITION FRAMING

5.36 General

a A partition is merely an interior wall and is framed almost exactly the same as the exterior wall of the platform frame, i.e., it possesses a sole plate or sole, studs, and top plate. Studs are usually 16″ on centers.

b Partitions are either bearing or nonbearing—they either help support the joists above or they do not. In bearing-partitions, the studs are 2″ x 4″ and the top plate is a doubled 2″ x 4″, although a single 2″ x 4″ is sufficient if the joists bear on the plate directly above the studs (Figure 5.30a, b). Nonbearing partitions may be of 2″ x 3″ studs or less, and closet partitions are often built with the studs set the 2″ way to save floor space. On the other hand, where plumbing waste pipes come down through partitions they may have to be framed with 2″ x 6″ studs or 2″ x 4″ studs thickened with 2″ x 2″ strips.

c Framing around door openings is the same as in the exterior walls and consists of doubled studs at the sides and a header across the top (Figure 5.31a). For larger openings, the header is made deeper or may be trussed, as shown in Figure 5.31b. Openings for doors are made 1″ to 2″ larger on each side to allow the door frames to be fitted in, squared, plumbed, blocked, and nailed.

d In many ways, the simplest way to build partitions is to put down the subfloor over the entire floor area, then assemble the partitions—sole, studs, plate and all—on the floor and raise them into place. The sole may be put down first and the rest of the partition set on top of it, particularly in the case of nonbearing partitions erected after upper floor joists are in place.

e It may be required that, wherever possible, second-story partitions studs shall rest directly on the plates of first-story partitions and first-story studs directly on girders (Figures 5.30b, 31). This is good practice for bearing partitions in eastern and balloon frames because settlement caused by shrinkage in the outside walls and in the interior bearing partitions is thereby more or less equalized. In platform framing, however, it is

Framing

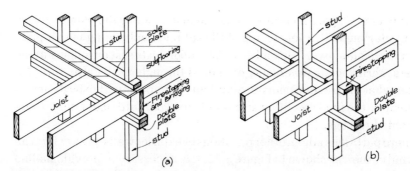

Figure 5.30 Partition Framing. (a) Partition Resting on Subfloor and Joists. (b) Upper Partition Resting Directly on Lower Partition.

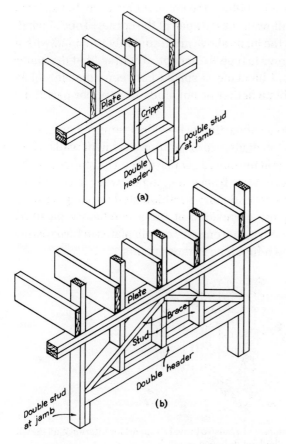

Figure 5.31 Framing Above Openings. (a) Double Header. (b) Trussed or Braced.

Dwelling House Construction   112

better to have all partitions built in the same manner as the outside walls, that is, starting with a sole on top of the subfloor and ending with a plate under the joists above. If, in addition, first-floor joists rest on ledgers at the girders, shrinkage is practically equalized throughout the structure. No matter what the framing, moisture content of lumber ought to be as nearly the same as the final condition as possible to avoid settlement due to subsequent shrinkage.

f Where partitions adjoin other partitions or exterior walls so as to form an internal corner, as shown in Figure 5.32, it is necessary to provide nailing for wallboard or lath, just as it is at the corner posts. The two most commonly used methods are shown, type "a" being prefered. Type "b" is an excellent method provided adequate blocking, securely nailed to the two adjacent wall studs, is provided behind the nailer. Otherwise, the latter is likely to work loose and to allow cracks to open at the corner. Tops of partitions which run parallel to the joists above are frequently provided with a nailer much the same as shown in type "b" except, of course, that the nailer is fastened to the plate and blocking is spiked between adjacent joists above. This detail is feasible whether or not studs from upper partitions rest on this partition.

g Firestopping at the bottoms of partitions is seldom required unless the studs are carried down on top of girders or interior basement walls. Here the same flues are present as at the outside walls and can be stopped in the same way. Similarly, to prevent fire from sweeping between second-floor joists and upward between second-story partition studs resting on first-story partitions, firestopping is provided at the second-story partition bases (Figure 5.30). Partitions resting on soles or sole plates in turn resting on subfloor provide their own firestops at those points.

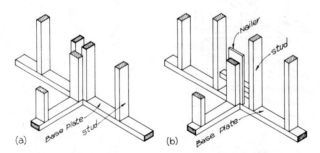

Figure 5.32 Intersection of Partitions and Walls To Provide Support for Interior Wall Finish. (a) Paired Studs. (b) Nailer and Blocking Between Studs.

STAIR FRAMING

5.37 General

a Detailed discussion of stair construction is left for Chapter 13. During the framing of the house it is necessary to allow openings in the floors for the stair wells, and frame for intermediate landings. Rough stairs are commonly built at this time to allow ready access from floor to floor until the finished stairs can be installed.

5.38 Openings

a Ordinary rectangular openings in floors for stair wells are framed with the usual headers and trimmers (Figures 5.23, 24). If the opening contains an angle (Figure 5.33, 34), the corner may be supported in a number of different ways:

1. *Post.* Where header and trimmer intersect, a post is inserted below to carry the load. The post may be free-standing, or may form part of a partition (Figure 5.33a). This is generally the easiest and most satisfactory method.

2. *Cantilever.* If the angle is close to a partition below, and joists from an adjacent span can be run out far enough to frame the angle, cantilevering is possible as shown in Figure 5.33b. A projecting structure such as this tends to deflect more than if it is supported from below. Occasionally a corner is supported by a rod or wood tension member suspended from above.

5.39 Landings (Figure 5.34)

a Landings are small floors intermediate between main floors, and serve to break the continuity of stairs, either in a straight run, or in a turn. Landings are supported by adjacent partitions, usually by spiking but occasionally by resting the ends of landing joists on horizontal intermediate plates provided in the partition framing.

5.40 Stringers (Figure 5.35)

a The structural support for finished stairs is usually provided by rough carriages, also called stringers and horses, running from floor to floor, floor to landing, or landing to landing. Stringers are cut to the profile of the undersides of risers and treads (Chapter 13). and enough are provided to

Dwelling House Construction

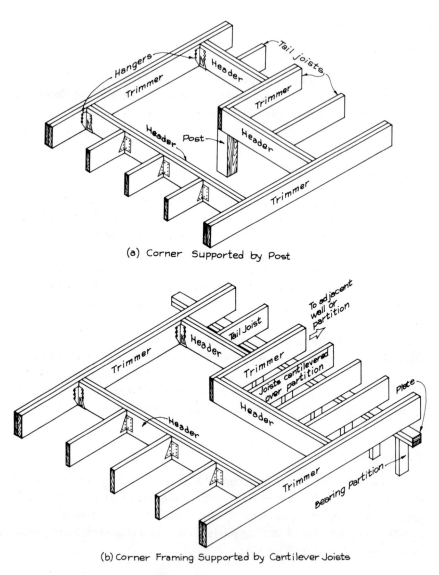

(a) Corner Supported by Post

(b) Corner Framing Supported by Cantilever Joists

Figure 5.33 Framing Around Angular Stair Well Opening.

Framing

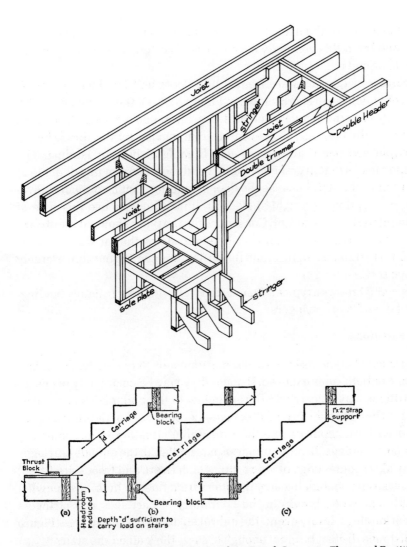

Figure 5.34 Stairway Framing Showing Landing, Rough Supports, Floor, and Partition.
Figure 5.35 Rough Stringers or Carriages for Stairs.

carry the load. For the usual 3-foot wide stair, when built in place rather than shop built, this generally means three stringers, two near the edges of the stair and one at the center.

b Stringers or carriages are framed into floors and landings in various ways, of which several are shown in Figure 5.35. Of these, the detail at *a* is in many ways the simplest and best. The lower end of the carriage bears directly on the floor and is prevented from moving by the thrust block. The upper end bears directly against framing of floor or landing. The objection to it is that in cramped stairs, head room may not be adequate at the bottom of the stair because the floor or landing framing projects downward. In a landing at a right-angle turn, it is difficult to use this type of framing at the interior corner. Consequently, the types shown at *b* and *c* are favored for built-in-place stairs. In shop-built stairs, the carriages are part of the finished stair structure, and the difficulties pointed out above largely disappear (Chapter 13).

c Figure 5.34 shows a typical framed stairwell with floor framing, landing, partition, and stair stringers.

5.41 Partitions

a Stairs may be enclosed between partitions, or they may be open, i.e., with one or both sides exposed. If open, they may be completely exposed, or partitions may be carried up from below to the undersides of the carriages on the open side. In the latter instance, both sides of the stairway are supported throughout its length, a desirable situation in any stairway containing landings. If such support is not available, the outside carriages and stringers (outer edge of stairs into which risers and treads are fitted) must be strong enough to carry the stairs from floor to floor, a somewhat difficult task when breaks in the continuity of carriages and stringers occur at landings. In any event, the unbroken depth of the carriage (Figure 5.35, distance *d*) must be great enough to carry the load on the stair.

ROOF FRAMING

5.42 General

a Although the function of a roof is to protect the house from the elements, it also adds greatly to its appearance. Its shape is therefore dictated both by climatic and by architectural considerations, but with the increasing

Framing

efficiency of roofing materials, the climatic considerations are becoming less important. Traditionally, low flat roofs have been used in warm climates where only water has to be shed, whereas steep roofs have been used in cold climates in order to shed snow as well as water.

5.43 Types

a Roofs may take many shapes, but the most commonly found are the following:
Shed or lean-to
Gable
Hip
Gambrel
Mansard
Deck

5.44 Parts of a Roof (Figure 5.36)
a. Roof Plate Already discussed under wall framing, this is the double 2″ x 4″ member which rests on top of the studs.
b. Ridge The peak of the roof. It is found on all but shed and deck roofs. The ridge board or ridge pole is the member against which the rafters bear. It forms the lateral tie which holds them together at that point. In a hip roof over a square house the ridge dwindles to a point.
c. Eaves The lowest part of the roof. It is the section formed by the rafter ends, plate, and cornice.
d. Cornice The ornamental detail found at juncture of roof and wall. It is made up of a combination of moldings and boards, and usually includes the gutter as a part of its design (Chapter 9).
e. Rafters These are the structural members of the roof. They correspond to the joists in the floors, and carry all the loads, whether dead, snow, or wind. Like the joists, they must be designed to perform this function satisfactorily, but unlike the joists, deflection is usually not a prime consideration. Table 5.9 gives sizes of rafters recommended for the spans, loads, spacings, and species indicated. It corresponds to Table 5.5, which gives sizes of joists. Group I and II roofing correspond to light and heavy (Chapter 8). The live load is an average figure for an average pitch. Building codes commonly specify the loads. For long rafters in which deflection rather than strength may be the controlling consideration. $L/180$ is often

taken as limiting. Codes differ.] Common rafters run from ridge to plate at right angles to the wall. At the intersections (hips) of the roof planes of a hip roof are found hip rafters, running from the exterior corners of the plate to the ends of the ridge. Framing into the hips are shorter rafters, running from plate to hip rafter, called jack rafters. Jack rafters are short rafters of any kind which do not run full length from plate to ridge. They may run from ridge to hip, plate to hip, ridge to valley, valley to plate, or from hip to valley. Jack rafters are, therefore, found on any kind of roof which is broken up in some manner.

f. Hip The intersection of two roofs which meet at an exterior angle of the building. The roofs slope down and away from the hip in each direction.

g. Valley The intersection of two roofs which meet at an interior angle of the building. The roofs slope down toward the valley from each direction.

h. Purlin The intersection of the two slopes of a gambrel roof. The name is also applied to the horizontal member which forms the support for the rafters at that point. More generally, a purlin is a member spanning roof supports, such as a roof beam running from truss to truss, and in turn, supporting rafters or planking.

5.45 Pitch

a The "pitch" of a roof is its slope, and it may be expressed in a variety of ways. The most obvious is to express the slope in degrees, i.e., the size of the angle which the roof makes with the horizontal. This is seldom done, however, because in practice it is cumbersome. More commonly pitch is expressed as a ratio; the units used to express the pitch are the "rise" and the "run." The rise is the vertical projection of the slope (sine of the angle) and the run is the horizontal projection (cosine of the angle). Evidently the pitch can be expressed easily and unmistakably as a ratio of rise to run and this is most commonly done. For convenience, the rise is always expressed as a certain number of inches per foot of run; thus, a rise of 6″ means that the roof rises 6″ for each 12″ that it runs horizontally. The pitch is 6 to 12.

5.46 Analysis and Framing (Figure 5.37)

a. Shed The shed roof, as its name implies, is merely a single inclined plane which may cover the main structure or may cover an ell which adjoins a higher structure, in which case the term lean-to is more appropriate. Many so-called flat roofs are actually shed roofs with a very small slope to shed

Framing

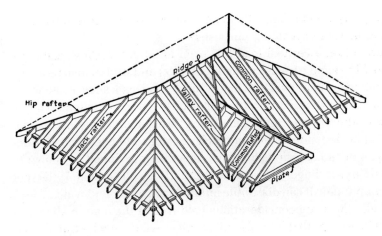

Figure 5.36 Framing Members in Hip Roof.

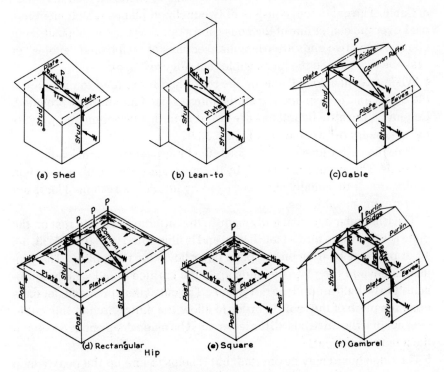

Figure 5.37 Elements of Roofs in Relation to Loads.

water and prevent "ponding." Ponding is to be avoided unless water is wanted on the roof for cooling purposes.

b In Figure 5.37a, a vertical load P is seen to induce vertical reactions in the studs. The simple frame consisting of rafter and studs would be in unstable equilibrium under vertical loads. Any lateral load, such as the wind loads W, would tend to overturn the frame unless the connections of studs to sills and rafters were rigid. Since these are generally merely nailed, they cannot be depended upon for much resistance to overturning. Consequently, a tie, such as a ceiling joist, must be introduced as shown to impart stability, or diagonal braces to the studs may be added. If the roof is a rigid diaphragm, it can distribute lateral loads to the end walls.

c Essentially the same considerations hold true of the lean-to shown in Figure 5.37b, except that lateral loads, such as wind, may be transferred to the main structure. If so, the tie can be omitted.

d Figure 5.38 shows several methods of framing shed and lean-to roofs.

e. Gable The gable roof consists of two inclined planes which meet in a peak over the center line of the house and slope down to two opposite roof plates. At the two ends are triangular sections of wall called "gables" or "gable-ends," hence the name gable roof (Figure 5.37c).

f When a vertical load P acts on the roof, the rafters tend to push out at the eaves and to displace the studs unless some kind of tie is provided. Generally, the attic floor joists are oriented in this direction and provide the necessary tie, as shown in Figure 5.39.

g The stud-rafter-joist combination is unstable against lateral loads, such as winds. These must be resisted by the end walls (braced, plywood, rigid wallboard, or diagonally sheathed) and by interior partitions. Floors act as stiffening membranes.

h The triangular cut commonly made in the rafter to allow it to rest on the plate is known as a bird's mouth joint and is illustrated in Figures 5.38, 39, 40. It provides bearing on the plate and allows for toe-nailing. The projection beyond the plate depends upon the cornice detail which is ultimately to be built around the rafter ends. The cut at the upper end depends upon the pitch of the roof. In order to allow the entire area of this cut to bear against the ridge board, the latter must be made deep enough to equal the length of the cut.

i The ridge board may be omitted, but it helps to line up the peak which otherwise may be crooked. Frequently the board is a 1" piece, but 2" stock is more likely to keep the ridge straight. Two-inch thick ridge boards may or may not be chamfered at the top.

Framing

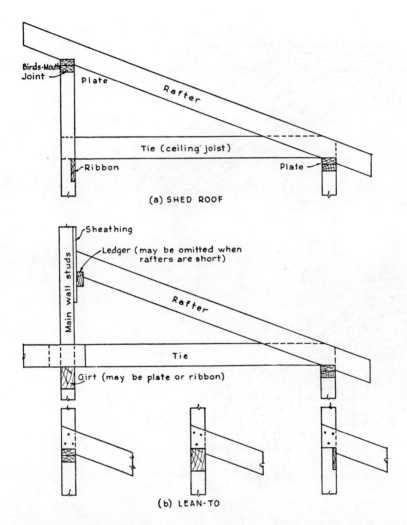

Figure 5.38 Framing for Shed and Lean-To Roofs.

Dwelling House Construction 122

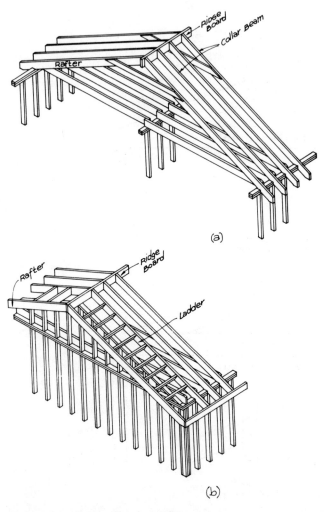

Figure 5.39 Framing for Gable Roof. (a) Rafter, Ridge Beam, and Ceiling Joist Resting on Wall Plate. (b) Framing for Overhanging End.

Framing

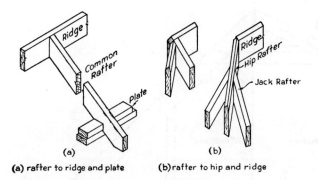

(a) rafter to ridge and plate   **(b)** rafter to hip and ridge

Figure 5.40 Roof Framing Details.

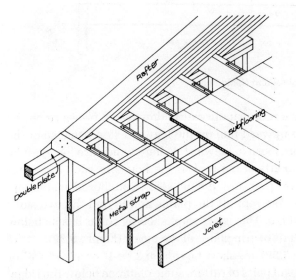

Figure 5.41 Framing to Absorb Outward Thrust of Rafters, Ceiling Joists Parallel to Wall.

Dwelling House Construction

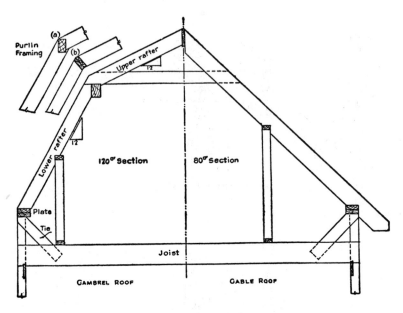

Figure 5.42 Sections of Gambrel and Gable Roofs.

j Wind blowing across a roof usually exerts downward pressure on the windward wide and an upward pull on the leeward side because of the reduced atmospheric pressure to leeward. Where winds are heavy, it is, therefore, not enough to attach the roof to the house in such a manner as to resist the downward pressure, it must also be anchored against uplift, which tends to raise the roof off the plates and to pull the two sides of the roof apart at the ridge. This is accomplished by collar beams below the ridge and various kinds of grip plates at the eaves (Figure 5.39).

k Collar beams (Figure 5.39) are short boards of 2″ x 4″ or 1″ (1″ x 6″ or 1″ x 8″) nailed to the sides of pairs of rafters some distance below the ridge. They tie the pairs of rafters together and prevent them from pulling apart in a high wind; they reduce the clear span of the rafters and thereby reduce the sizes which must be used, and they reduce the outward thrust of the rafters at the eaves.

l If the attic is finished, collar beams are placed at ceiling height and form the ceiling joists. In gambrel roofs, they have the further important function of serving as horizontal ties to resist the outward thrust of the upper rafters at the purlins (Figure 5.42).

m If ceiling joists run parallel with the walls instead of at right angles, some other means of absorbing the outward thrust of the rafters must be provided. One method is shown in Figure 5.41. Short lengths of joists are brought against the sides of the rafters at the plate and butted against the side of a joist. Steel straps tie the two together. Subflooring assists in transmitting the thrust into the floor by acting as a diaphragm.

n A projecting gable may be framed as shown in Figure 5.39b. A "ladder" is built of short "lookout" rafters framing into the side of the end rafter and resting on the gable wall, in turn framed with short studs resting on the wall plate.

o Gable rafters are frequently prefabricated into trussed rafters as shown in Figure 5.43a. These are preassembled in the shop, sent to the site, and are quickly erected, in contrast with the often awkward task of hoisting and assembling pairs of rafters to a plate and ridge. Sizes of rafters (top chords), ceiling joists (bottom chords), and diagonals depend upon the span, but are most commonly all $2'' \times 4''$ for the usually encountered spans, with $2'' \times 6''$ not unusual for top chords. Connections are most commonly made by means of steel plate connectors, of which a few are shown in Figure 5.43c. Many of these have claws or prongs stamped integrally from the sheet. They are usually pressed into the wood by a jacking mechanism in the assembly jig. Glued and nailed plywood gussests are also employed. For heavier loads, split-ring connectors may be employed, as shown in Figure 5.43b.

p. Hip When the roof slopes down in four inclined planes to four plates, it is called a hip roof. The plan may be rectangular or square (Figures 5.36, 37d, e).

q A vertical load applied to the peak of a square hip roof is distributed among the four hips, and the vertical components are transmitted to the corner posts. Any spreading tendency caused by the horizontal components is taken up by the skin of the roof (roof boards) as shown. No lateral ties are theoretically required because the roof is a self-contained unit. Actually, the attic joists do provide a lateral tie in one direction and in the opposite direction on large roofs it may be advisable to employ knee braces or ties on the longest jack rafters. Collar beams add stiffness to the longest jack rafters in both directions.

r End portions of a rectangular hip roof behave in the same way as the square hip just described. The central portion behaves and is framed in the same manner as a gable roof.

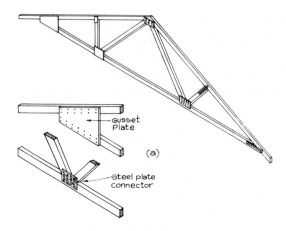

Figure 5.43 Trussed Rafters. (a) Steel Plate Connectors or Glued and Nailed Plywood Gussets.

s Framing the intersection of hip, ridge, and common rafter is shown in Figure 5.40. Jack rafter cuts are also shown.

t Like the ridge board, hip rafters carry no load and need, therefore, not be especially heavy. One-inch boards can be made to suffice but 2″ stock is preferable to keep the hip line straight and to provide nailing. Hip rafters may or may not be chamfered on top.

u. Gambrel Unlike the gable, which is two inclined planes running from peak to plate, the gambrel roof consists of four inclined planes, two on each side of the peak, the upper plane on each side fairly flat and the lower one fairly steep (Figures 5.37f, 42).

v A vertical load P applied as shown in Figure 5.37f induces a tendency to spread at the lower ends of both upper and lower rafters. Ties are therefore required at both points, and usually consist of collar beams (simultaneously acting as ceiling joists) and second-floor joists. So tied, the frame is in equilibrium under vertical loads, but tends to collapse under lateral loads, such as winds, unless vertical braces, such as those shown, are provided.

w From a utilitarian standpoint, the chief advantage of the gambrel roof is the additional space it provides at no increase in the height of the ridge, as compared with the gable. This is especially true of headroom, which occupies a much larger portion of the usable space under a gambrel than under a gable roof. The two sections shown in Figure 5.42 have the same rise and span, but the gambrel has 50 percent more useful space than the gable.

Framing

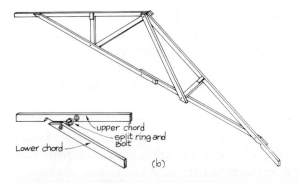
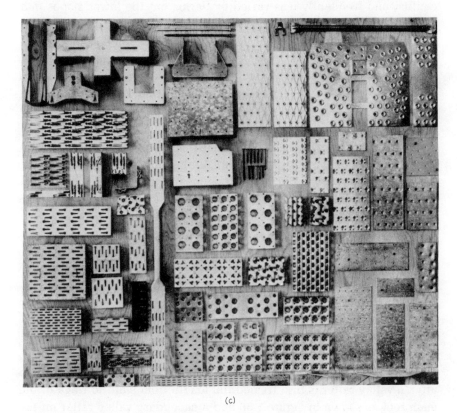

Figure 5.43 (b) Timber Connectors. (c) Types of Steel Plate Connectors.

x Purlins may be framed so as to be completely enclosed by rafter ends, as shown at *a* in Figure 5.42, or may be exposed underneath, as at *c*. The latter is frequently supported by partition studs and may merely be the cap of a partition. The purlin shown at *b* is simpler than either *a* or *c*, may be either solid or built up as shown, and behaves much as if it were a secondary ridge board.

y. Mansard If the gambrel roof is visualized as running up from all four plates, instead of from two, and if the lower plane on each side is made very steep while the upper is very flat, the result is a mansard roof. This type was commonly used on the mansions built in the late nineteenth century and frequently was varied by flaring out the lower slopes in a sharp curve.

z. Deck The deck roof is merely a hip roof with the top sliced off and a flat platform substituted. It was frequently used on large houses with steep roofs where the roof would otherwise have been high. The dotted lines in Figure 5.37e show the relationship of hip and deck roofs.

5.47 Ells (Figure 5.36)

a Roofs over ells are framed in the same manner as the main roof, with ridge board, common and jack rafters, and so forth. Usually, however, the ell is not as wide as the main house; its roof does not rise so high as does the main roof, and its ridge consequently ends against the side of the main roof instead of joining the main ridge.

b Two ways of making the juncture between main and ell roofs are possible. If the ell is small and there is no reason why its attic space should be accessible or continuous with the main attic, the main roof rafters can all be carried down to the plate, the roof boards on the main roof carried beyond the point of intersection of the main and ell roofs, and the ell roof merely rested on top of the main roof by fastening valley members (2″ x 4″ pieces are usually sufficiently large) on top of the main roof and allowing the ell's valley jacks to frame into the valley members. If, on the other hand, the ell is quite large and its attic space is to be continuous with the main attic, a valley rafter is first framed full length from plate to ridge of the main roof as shown in Figure 5.36 and a secondary valley rafter on the other side is brought from main roof plate to the first valley rafter. Jack rafters from both roofs then frame against these valley rafters, the ell roof framing is completed, and roof boards are carried into the valleys from both main and ell roofs.

Framing 129

c Valley rafters, unlike the ridge and hip rafters, must withstand the downward thrust of the two portions of roof which frame into the valley, and must therefore be heavy material. Usually they are made quite deep — 2"x10", or 2"x12" on medium-sized roofs and may be doubled on larger roofs. Where the spans are long they are given additional interior support by posts resting on bearing partitions below.

5.48 Framing around Openings (Figure 5.44)

a. This is essentially the same as framing around openings in walls and floors. The rafters at the two sides of the openings are doubled and are, therefore, trimmer rafters. Across the ends of the opening are placed doubled headers, and the rafters which frame into the headers are called header rafters or tail rafters. If the opening is at the ridge (Figure 5.40c) the two sets of trimmers react against each other at the ridge the same as do pairs of rafters.

5.49 Dormers (Figure 5.45)

a Whenever it is necessary to have a vertical wall rising out of a roof, the construction by which this is accomplished is known as a dormer. Usually dormers are employed to allow windows in the attic space, and such windows are called dormer windows.

b Most dormers are quite small and provision for them is most easily

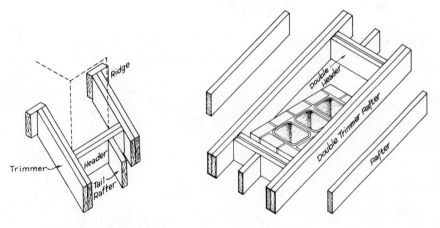

Figure 5.44 Framing around Chimneys.

Dwelling House Construction

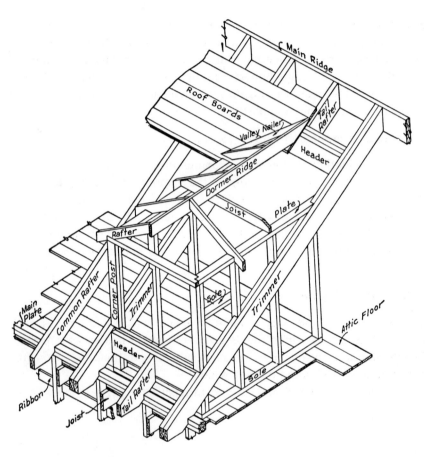

Figure 5.45 Dormer Framing.

Framing

made by framing a rectangular opening in the main roof in the same manner as for chimneys, i.e., by the trimmer-header combination. Within this opening are erected three walls—two sides and one front—which are framed the same as any wall, with sole resting on the attic floor, studs running from sole to plate, and plate which supports the dormer roof. Sometimes, instead of starting at a sole set on the floor, side wall studs are rested directly on the double trimmer rafters, and front wall studs are rested on the double header at the bottom of the opening. Sometimes also, some of the side wall studs start at the floor and others are cut off below the trimmer. This detail depends upon the architectural treatment of the attic space which may or may not be finished. The opening for the window in the front wall is framed the same as any window opening, with double studs at the sides (trimmers) and headers top and bottom. Where space is restricted, the plate across the front may at the same time form the upper header for the window opening.

c The dormer roof consists of the same parts as any other roof and may be either gable or hip. Figure 5.45 is a detail of a gable roof. The rafters rest on the plates, the ridge ends at the main roof, and celing joists span from one side wall plate to the other. This type of construction gives a flat dormer ceiling.

d If a peaked ceiling is desired, the upper header in the original opening in the main roof is framed higher, beyond the dormer ridge board, and two short valley rafters are inserted below it, running from the center of the header to the two side trimmers. The dormer ceiling joists are omitted and the wall finish is carried up under the dormer rafters.

e Large dormers are framed in essentially the same way as small except that instead of using two headers in the original opening in the main roof the upper header is left out and a pair of valley rafters inserted, one running from trimmer to main ridge and the second from trimmer to the first valley rafter. This is the same framing as the second type described for ell roofs (Figure 5.36).

5.50 Roof Boards (Figure 5.45)

a The manner in which roof boards are put on depends upon the kind of roofing which is to be applied. Usually the roof is tight-sheathed, that is, the boards are driven up tightly against each other or plywood is employed. Unlike walls and floors, roof boards are not laid on a diagonal. They are laid with broken (staggered) joints, all joints are made at the rafters, and

Dwelling House Construction   132

they are securely nailed with two 8d or 10d nails per board per rafter. The first board is started at the eaves, is nailed down tight and straight, and the others follow until the ridge is reached, where the top boards from both sides are cut off on a straight line along the ridge.

b Plywood sheets are similarly started at the eaves and carried up to the ridge. Because plywood comes in multiples of 2', most commonly in sheets of 4'x8', rafter spacing is customarily 2'. Thicknesses generally employed are ⅜" and ½". Table 5.10 gives a few values, loads, and spans.

Table 5.10 Plywood Roof Decking[a,b,c] (Plywood Continuous over Two or More Spans; Grain of Face Plys Across Supports)

Panel[d] Identification Index	Plywood Thickness (in.)	Max. Span[e] (in.)	Unsupported Edge— Max. Length (in.)[f]	Allowable Roof Loads (lb/sq ft)									
				(Spacing of Suppogts [in.] Center to Center)									
				12	16	20	24	30	32	36	42	48	60	72
16/0	⁵⁄₁₆, ⅜	16	16	**130** (170)	**55** (75)									
24/0	⅜, ½	24	24		**150** (160)	**75** (100)	**45** (60)							
32/16	½, ⅝	32	28				**90** (105)	**45** (60)	**40** (50)					
48/24	¾, ⅞	48	36							**105** (115)	**75** (90)	**55** (55)	**40** (40)	
2•4•1	1⅛	72	48								**175** (175)	**105** (105)	**80** (80)	**50** (50)	**30** (35)
1⅛" Group 1 and 2	1⅛	72	48								**145** (145)	**85** (85)	**65** (65)	**40** (40)	**30** (30)

[a] These values apply for STANDARD C-D INT, STRUCTURAL I AND II INT, C-C EXT, and STRUCTURAL I C-C EXT grades only.
[b] For application where the roofing is to be guaranteed by a performance bond, recommendations may differ somewhat from these values.
[c] Use 6d common, smooth, ring-shank or spiral-thread nails for ½" thick or less, and 8d common, smooth, ring-shank or spiral-thread for plywood 1" thick or less. Use 8d ring-shank or spiral thread or 10d common smooth shank nails for 2•4•1, 1⅛" and 1¼" panels. Space nails 6" at panel edges and 12" at intermediate supports, except that where spans are 48" or more, nails shall be 6" at all supports.
[d] See Table 5.3.
[e] These spans shall not be exceeded for any load conditions.
[f] Provide adequate blocking, tongue and grooved edges or other suitable edge support such as Plyclips when spans exceed indicated value. Use two Plyclips for 48" or greater spans and one for lesser spans.
[g] Uniform load deflection limitation: $\frac{1}{180}$th of the span under live load plus dead load, $\frac{1}{240}$th under live load only. Allowable live load shown in boldface type and allowable total load shown within parentheses. The allowable live load should in no case exceed the total load less the dead load supported by the plywood.
[h] Allowable roof loads were established by laboratory test and calculations assuming uniformly distributed loads.
Source: American Plywood Association, Tacoma, Wash.

Framing

c The exception to this procedure is a roof which is to be covered with wood shingles. Since wood rots when kept continually damp, and since rain water manages to work under shingles at least to a slight extent, sufficient ventilation should be provided under wood shingles to dry any dampness as quickly as possible. This is most easily done by spacing the roof boards apart several inches. In this case, square-edge strips, usually approximately 3" wide, are nailed to the rafters and are spaced center to center a distance equal to the amount which the shingles are subsequently to be exposed to the weather.

5.51 Roofing Paper

a Roof boards are covered with roofing paper or felt to provide a surface tight against rains. This paper should be a good quality paper, applied so as to give a water-tight roof. (A "square" in roofers' language is 100 sq ft.) This is accomplished by starting at the eaves and applying the paper in layers, each layer overlapping the preceding one by at least 2", and all securely nailed with flat-headed roofers' nails every 6" along the edges. End laps should be at least 3".

b If there is any likelihood that the roof will not be covered with the final roofing for some time and the paper may be subjected to high winds, the nails can be driven through metal discs or caps which hold the paper much more firmly and with less likelihood of tearing than would be true of nails alone. Fiber-reinforced paper, similarly, is more resistant to tearing.

IN-LINE. POST, PLANK, AND BEAM

5.52 In-line Framing

a When large standard-size sheets such as plywood are employed for subflooring, sheathing, and roof boards, it is desirable to frame the house in such a way as to eliminate as much cutting of the wallboard as possible, to save on waste and labor. This makes it desirable to have all framing members—joists, studs, and rafters—in line and at such spacings that ends of wallboard sheets fall on framing members. With plywood of the proper thickness for subflooring, sheathing, and roofboards, framing members are customarily placed 24" on centers.

b The basic platform frame plus trussed rafters or spliced rafter-ceiling joist connections makes such in-line framing possible if joists are also framed in-line instead of lapped, as is customary.

c Figure 5.46 shows such in-line framing. Floor joists are set in line with their ends butted and resting on an intermediate girder as shown at a. Metal or wood splices tie the joists together.

d However, an alternate method of framing shown at b provides better continuity by having the joists cantilevered over the girder alternately from one side and the other, and making the splice at an intermediate point. This is best done at approximately one-third to one-quarter span, because the bending moment approaches zero and only shear must be resisted by the splice. This construction makes the joist continuous over the intermediate girder, and considerably reduces deflection, which is often the controlling factor in selecting joist sizes (Table 5.5). Splices may be metal or plywood, the latter best nail-glued.

5.53 Post, Plank, and Beam (Figure 5.47)

a Instead of the multiplicity of joists and rafters found in standard framing, in post, plank, and beam construction planks, usually of nominal 2″ and 3″ thickness, are supported on beams spaced usually 4′ to 8′ apart. These beams, in turn, are supported on posts or piers. Spaces between posts are filled in with studs or other framing as necessary to provide support for the enclosure. Partitions, likewise, are similar in construction to standard partitions with some exceptions. The supplementary framing in walls provides the lateral stability necessary against lateral loads such as winds. If partitions are tied into the main structure, they also provide such lateral stability in a manner similar to standard construction.

b To take full advantage of the simplicity of plank and beam construction, it is desirable to carry out the design of the house with this type of construction in mind from the very beginning, rather than to try to adapt it after the design has been worked out. Windows and doors, for example, are best located between posts in exterior walls, so as to eliminate headers over the openings. Usually the wide spaces between posts are ample for this purpose. There must, however, be sufficient solid walls to provide the necessary lateral bracing. With a little forethought, it is possible to combine conventional and plank and beam framing.

c The principal advantage of the system is its simplicity, leading to fewer members, which, in turn, can reduce the amount of labor required. A further advantage from the architectural standpoint frequently lies in the exposed plank and beam ceiling, which is often considered aesthetically

Framing

pleasing and also provides higher ceilings than is true of joists and rafters covered with interior finish on their lower edges. The heavier flooring and framing members also lead to increased resistance to fire.

d Limitations associated with plank and beam framing include the lack of concealed spaces for wiring and piping or duct work. Wiring and piping can sometimes be accommodated by making the beams of two pieces spaced apart to accommodate the wires and pipes (Figure 5.48). A soffit board fastened to the bottoms of the two members can be used to conceal that space.

e The planks employed in plank and beam framing are for moderate, distributed loads and not for heavy concentrated loads such as bathtubs, refrigerators, bearing partitions and similar loads. Where these occur, additional framing is needed under the planks to transmit these loads to the beams.

f Where insulation is needed, it can be placed between the beams, but in that case, the lower surfaces of the planks are concealed and insulation is in view. The alternative is to place the insulation above the roof planks, in which case it should be rigid and capable of supporting the roofing material. Furthermore, a vapor barrier between the insulation and the planks is highly desirable.

5.54 Construction Details

a Foundations for plank and beam construction may be continuous as shown in Figure 5.47, or individual piers may be employed to support the posts and the ends of beams.

b Posts must be adequate in size to carry the loads and should, in any event, be at least 4″ x 4″ in cross section or two 2″ x 4″′s securely nailed together. Where the ends of beams abut and rest on a post, there should be at least 6″ of bearing parallel to the beams. This may be obtained by making at least one dimension of the post 6″, if solid, or by fastening at least three 2″ members together, or by using a plate at the top of the post to support the beams (Figure 5.48).

c The sizes of beams depend upon the spans and the loads being carried by them. Many building codes require, in addition to the dead load, a live load of 40 lb per sq ft on floors, and 20 to 40 lb per sq ft on roofs, depending upon the local climate. Beams may be laminated (Section 5.31b, Figure 5.22) or one piece. They must be of sufficient structural quality to carry the loads.

d Fastenings of beams to posts are usually accomplished by means of

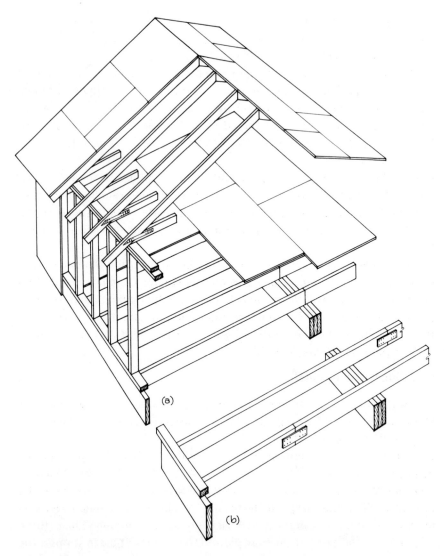

Figure 5.46 In-Line Framing. (a) Joists, Studs, and Roof Elements All in Line. (b) Joists Made Continuous By Cantilevering and Joining with Splice Plates.

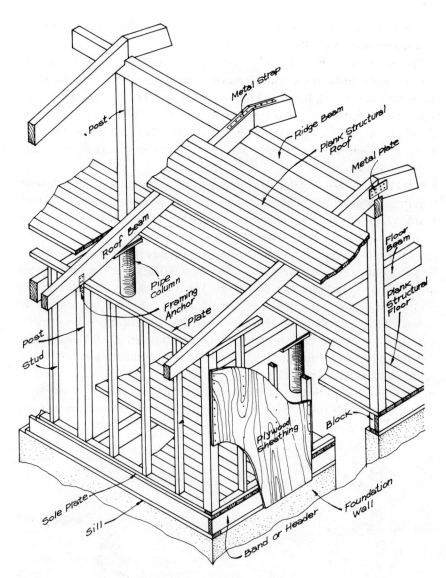

Figure 5.47 Post, Plank, and Beam Construction.

Dwelling House Construction   138

framing anchors or angle clips, nailed, bolted, or lag-screwed to the timbers (Figures 5.47, 49).

e The planks may be square-edged, but usually are matched in some way such as by means of tongue-and-groove, splines, or rabbetted battens, as shown in Figure 5.50. Because planks are frequently exposed, it is highly desirable to have their moisture content at the time of installation as closely as possible matching the moisture content in actual use. This will avoid excessive shrinkage and opening of unsightly cracks at the joints between planks.

f Planking may be laminated by gluing two or more nominal 1″ thick boards together. When three boards are employed to provide nominal 3″ thick planking, the center board or lamination can be shifted slightly sidewise to provide a tongue along one edge and a groove along the other. Furthermore, the center lamination can be shifted longitudinally to provide a tongue at one end and a groove at the other (Figure 5.50). Among the advantages of glued-laminated planking are the fact that knots and other blemishes do not penetrate through the entire thickness, the thin boards before laminating are easily dried to a low moisture content (must be for successful gluing), and better-quality stock can be placed on the outside. The principal disadvantage is the cost of labor and materials as compared with "solid" unglued planking.

g It is highly desirable to have planks continuous over more than one span to increase the stiffness of the floor. The plank that rests only on its ends

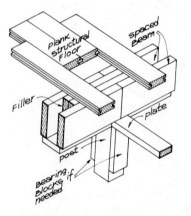
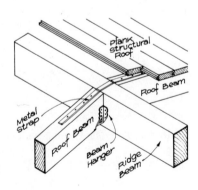

Figure 5.48 Spaced Beam with Post and Plank.
Figure 5.49 Roof Beam Framed to Side of Ridge, Tied with Metal Strap.

Framing 139

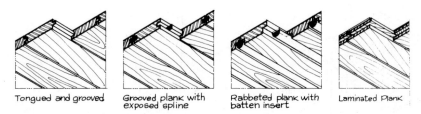

Figure 5.50 Types of Planks.

will deflect approximately 2½ times as much as a plank that is continuous over two spans. From the standpoint of appearance, it is probably best to have all joints between planks made over beams, but random lengths can be employed with joints coming between beams. In this case, it is essential that planks be matched by splines, tongue-and-groove, or similar means. End-matching, i.e., tongue-and-groove at the ends of planks as well as at the sides, is highly desirable.

h Finish flooring should be laid at right angles to the directions of the planks, and if the planks are exposed below, care should be taken that the nailing for the finished floors does not penetrate through the planks.

i Instead of plank, the floor may be heavy plywood, such as the 1⅛" and 1¼" thick plywood described under Subflooring (section 5.35, Table 5.8). If the plywood is left exposed on the underside, a grade and group must be specified that has high-quality surface veneer at least on one side.

j Most partitions in plank and beam construction are non-bearing because loads are supported on beams, in turn resting on posts. If bearing partitions occur, they are best placed directly over beams, which must be large enough to carry the additional load. If bearing partitions cannot be supported on the principal beams, supplementary beams are needed to support them.

k If non-bearing partitions run at right angles to the direction of the planks, no additional support is necessary underneath them. Non-bearing partitions that run parallel to planks, however, should have additional support. This can easily be provided by a pair of 2" x 4"'s laid on edge. If this is not possible because of door openings, the 2" x 4" supporting members can be placed under the plank and, in turn, supported at their ends by framing anchors fastened to principal beams. This type of construction is shown in Figure 5.51.

Dwelling House Construction 140

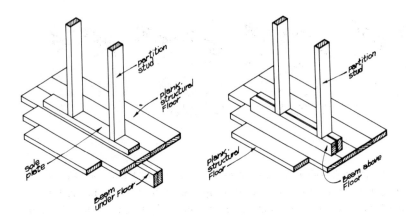

Figure 5.51 Supplemental Support for Partitions. Doubled Sole Placed Under or Over Floor Plank.

l As Figure 5.47 shows, a convenient method of handling roof framing is to utilize sloping beams resting on a ridge beam, in turn supported by posts. Roof beams may rest on top of ridge beams or may be framed into the sides and supported by beam hangers (Figure 5.49). Metal straps across the top are desirable to provide a tie at this point; or metal plates fastened to the sides of the beams may also be employed.

m If a ridge beam cannot be employed and roof beams consequently may provide outward thrusts at the walls, some kind of tie is required as in standard gable construction. One method of handling this is shown in Figure 5.52, where a horizontal beam consisting of two members is framed into the ends of the roof beams and takes the horizontal thrusts at that point. The horizontal beam, in turn, may be spliced at its center or some other intermediate point and rest on a post.

n The design of planking, beams, and posts is controlled by the necessity for adequate strength to carry the superimposed load and, also, to provide sufficient stiffness to avoid excessive deflection under those loads. Building codes commonly specify floor and roof loading as well as wind loading. They may, in addition, specify deflection limitations such as some fraction of the span. These usually range from $L/180$ to $L/360$ to provide sufficient stiffness and a sense of security to the inhabitants. As pointed out above, stiffness in planking can be greatly enhanced by utilizing planking spanning at least two spans and preferably more. For

Framing

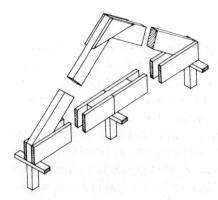

Figure 5.52 Roof Beams and Spaced Ceiling Beam.

example, a 12' long plank will reach across two spans where beams are spaced 6' on centers.

o Framing anchors, angles, plates, and other means of fastening beams to beams and beams to posts must be adequate to transmit the loads involved.

p As is true of any wood construction, it is desirable to detail the bottoms of posts so that they do not rest in moist conditions, and to allow plenty of ventilation around the ends of beams.

SPECIAL FRAMING

5.55 General

a The foregoing sections of this chapter deal with types of framing employed in customary construction. Problems arise, however, particularly in contemporary domestic architecture, involving exceptionally large spans for windows and interiors, openings in corners, wide overhangs, and the like. Usual procedures and rules-of-thumb often do not apply, and special techniques must be developed to meet special cases.

b Frequently, the loads involved and the arrangement of members to handle the loads are such as to require engineering examination and design. No attempt at structural analysis will be made here, but some of the types of framing which have been found successful in meeting some typical design requirements will be described. Sizes of members cannot be given, since actual sizes depend upon the conditions of any particular problem.

5.56 Large Openings (Figure 5.53)

a Small and medium-sized openings in walls and partitions are framed as previously described in this chapter. Large openings require heavier framing to carry the loads and to prevent excessive deflections. Figure 5.53 illustrates several possible methods. Simplest is the heavy single or two-piece girder shown at a. It is quickly erected, strong, and inexpensive. It has the serious drawback that any shrinkage in the girder is immediately made apparent by settling of the superstructure and by cracks in surface finishes. Laminated timbers (Section 5.31) are usually dry when manufactured and can largely overcome this problem. A steel I-beam or pair of channels may be employed, as shown at b. A nailer is necessary on top, and should be provided on the bottom also if finished woodwork is to be applied later. End supports may be steel pipe columns or wood posts. Some kind of tie to the wood framing, such as the angle irons illustrated, must be provided.

b A built-up girder is illustrated at a, with $\frac{1}{2}''$ plywood webs and lumber top and bottom flanges and stiffeners. With nominal $2'' \times 3''$ stock, the total thickness is the same as the normal $2'' \times 4''$ wall. For maximum efficiency, but with more waste, the grain in the surface plies runs in the diagonal direction shown. The large sheets of plywood impart stiffness and rigidity, particularly if they are glued as well as nailed to the frame.

c Figure 5.53d illustrates two methods of supporting joists on built-up girders. Simplest is to rest the joist on top in the usual manner, but the total possible depth of girder is limited. On very long spans, where maximum depth is essential, floor joists may be framed into headers in turn bolted or lag-screwed to the stiffeners of the girder. Girders thus may be any convenient depth—from lower window head to upper window sill, for example.

5.57 Overhangs

a Wide overhangs are frequently required, especially as sunshades over large deep windows. Simple overhangs are easily framed by projecting joists or rafters the required distance beyond the wall. Overhanging second stories are customarily built on the ends of projecting joists. If the overhang runs around the periphery of the building, framing is modified in the manner shown in Figure 5.54. Lookout joists are framed into the side of a doubled regular joist or main header, and smaller lookouts frame into the side of a secondary header. If the overhang is large,

Framing

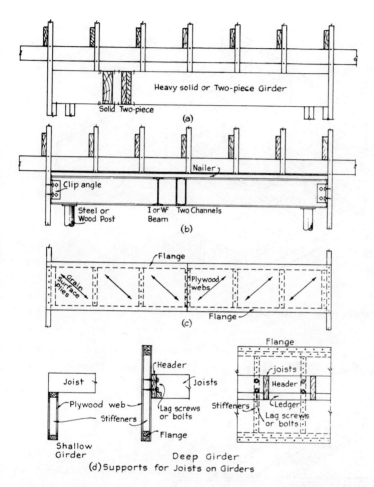

Figure 5.53 Methods of Spanning Large Openings. (a) Timber Girder. (b) Steel Girder. (c) Box Girder with Plywood Web. (d) Joists on Shallow and Deep Girders.

Dwelling House Construction 144

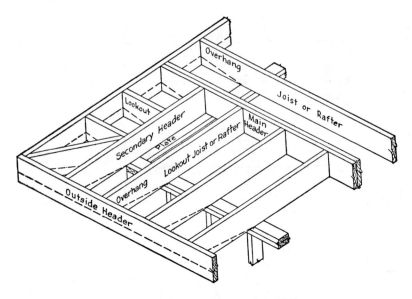

Figure 5.54 Framing for Large Overhang Supported on Wall.

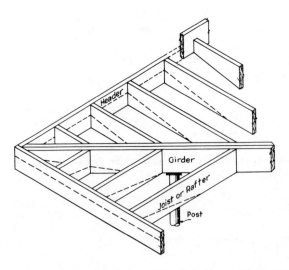

Figure 5.55 Framing for Large Overhang Supported on Post.

Framing

the secondary header may be moved over one or two joist spaces and the lookouts become lookout joists resting on the plate in the same manner as the lookout joists shown. Lower edges of the overhanging portions of all members may be cut on a bevel (dotted lines in Figure 5.54), particularly if the superimposed load is small. Bevelled cuts of this kind increase the angle at which the sun's rays may enter windows unobstructed.

b Details are sometimes found in which large roof areas are supported on occasional posts instead of on continuous walls. One method of handling such details is shown in Figure 5.55. The post, generally kept as slender as possible, supports a projecting girder strong enough to carry the load from its tributary area. Joists or rafters span from girders to outside headers which are supported by the girders. Headers must be strong and stiff enough to support the outer ends of joists or rafters without sagging. This requirement often precludes any bevelling of the lower edges of joists or rafters unless the upper edges of headers are raised as shown in the upper portion of Figure 5.55. The joint at the ends of headers and girder is critical.

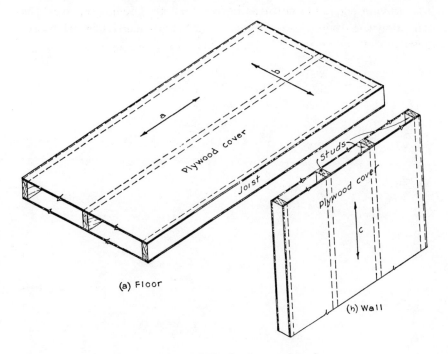

Figure 5.56 "Stressed-Cover" Plywood Glued to Joists and Studs.

5.58 "Stressed Cover" (Figure 5.56)

a A technique which is employed in shop-fabricated panelized construction, but which also finds application in field construction, is the "stressed cover" or "stressed skin." In essence, it consists of a frame to which a continuous skin, usually plywood, is so firmly fastened that the entire structure acts as a unit and a considerable portion of the load is carried by the skin. Strength and rigidity are both markedly increased. Since it is essential that the frame and skin be tightly joined, nailing alone is not enough, and the skin is glued as well as nailed. Nails not only fasten the parts together but apply enough pressure to the glue line to effect a good bond. For floors a common combination is 2″ x 6″ joists, 2′ on centers with ½″ or ⅝″ plywood top and ⅜″ plywood bottom, glued with cold-setting casein or synthetic resin, nailed with 6d or 8d etched or coated common nails 6″ on centers. Walls may be ⅜″ plywood outside and ¼″ inside with 1″ x 2″ studs 12″ on centers. Six-penny finish nails are favored. Although the greatest reinforcement for floor joists is obtained if face grain in the plywood runs in direction *a*, it is often placed in direction *b* to provide maximum stiffness between joists. Otherwise, excessive deflection would occur under concentrated loads placed halfway between joists. In wall panels, face plies run in direction *c*.

6 Chimneys and Fireplaces

6.1 Specification Clauses

CHIMNEYS

a Chimneys shall be built as indicated on plans. All flues throughout shall be lined with hard burned terra cotta flue lining set in full mortar joints smoothed on the inside. Lining for all basement flues shall start 4'-0" below the basement ceiling. In any case where no withes occur between the flues, the joints of the flue lining are to be staggered and a thin filling of mortar grouted between the linings.

b All brickwork shall be laid at least 2" away from all floor framing, and 1" away from all studding or furring. No nails shall be driven into the chimneys. The brick mason shall furnish heavy iron thimbles of sizes shown and shall build them into the flues at the heights directed by the architect or owner.

c Cleanout doors shall be furnished and built in at the base of each basement flue.

d The incinerator chamber shall be built as required, and all equipment furnished by the manufacturer shall be built in in accordance with the directions for setting furnished by him.

e Chimney flashing shall be furnished by the roofer and built in under his direction. All chimney flashing shall be through flashing.

f Two 3"x 3"x ½" steel angles shall be built in above each fireplace opening. The hearths shall be supported by 3" reinforced concrete slabs as shown.

g Chimney shall be topped out and capped as shown.

FIREPLACES

h Backs, jambs, and underfire shall be of fire brick.

6.2 General

a Coincident with the framing of the house comes the construction of chimneys and fireplaces. If the chimney is an outside one, i.e., is on an outside wall, the lower part is built at the same time as the foundation walls and forms an integral part of the wall at that point. Otherwise, it may be begun at any time after the first floor framing is finished. If it follows the framing, openings for it are left in the floor, wall, and roof framing.

b To a large degree, the efficiency of any heating unit depends upon the chimney. At the same time the chimney constitutes a serious fire hazard. Consequently, its design demands careful planning and its construction meticulous attention to detail and the requirements of good workmanship.

c A chimney may be merely a simple flue or it may be a large piece of intricate masonry construction consisting of heater flues, ash pits, incinerators, ash chutes, fireplaces, and fireplace flues, all arranged to fit into the minimum space consistent with maximum efficiency.

d In the following discussion chimney construction is for convenience divided into two parts, the chimney proper and the fireplace, although the two are usually built as one operation. In ornamental fireplaces, however, rough openings may be left in the chimneys at the time of construction, and fireplaces are later built into the rough openings.

CHIMNEY

6.3 Flues and Flue Linings (Figures 6.1, 2)

a The flue is the vertical open shaft through which smoke and hot gases are carried from fire to open air. It must be designed to accommodate the unit to which it is connected. Its capacity is measured by the effective cross-sectional area, which in turn depends upon the path of the smoke and hot gases. When a current of warm air rises through a flue, it ascends in spirals and occupies the greater part of the center of the flue. If a circular flue is lined with smooth terra cotta flue lining, there is little or no drag, and the full area of the flue can be taken as its effective area. If the flue is square, the corners are occupied by cold air and soot, and the upward current of warm air must be considered as occupying only the central circular part of the flue. The effective area of a lined square flue is a circle whose diameter equals the width of the flue. A rectangular lined flue has an effective area equal to the sum of the two semicircles inscribed in the ends of the rectangle, plus the intervening rectangular space. The foregoing principles are illustrated in Figure 6.1.

b The size of flue required for a heating unit is dictated by its draft requirements, and must be determined by the designer of the heating equipment. When it comes to fireplaces, the following rule of thumb has been found reliable: the effective area E of the flue should be from $\frac{1}{10}$ to $\frac{1}{12}$ the area of the fireplace opening. For example, a fireplace opening 42" wide and 30" high has an area of 1,260 sq in. and $\frac{1}{12}$ of that is 105 sq in. A 12" circular flue lining has an actual and effective area of 113 sq in. and would be satisfactory. (See Table 6.1.) A 13" x 13" flue lining, outside measurement, has a gross inside area of 127 sq in., and an effective area of 100 sq in., which would be a trifle small although it would do if smoke chamber,

Chimneys and Fireplaces

Table 6.1 Flues, Areas, and Chimneys With Flue Linings

Nominal Sizes		Actual Inside area (sq in.)	Effective Area E (sq in.)	Minimum Thickness of chimney Wall
Round Flue Lining, Inside Diameter	Rectangular Flue Lining, Outside Dimensions			
...	4½" x 8½"	23.6	21.3	3¾"
6"	...	28.3	28.3	3¾"
...	7½" x 7½"	39.1	30.7	3¾"
...	4½" x 13"	38.2	35.9	3¾
...	8½" x 8½"	52.6	41.3	3¾"
8"	...	50.3	50.3	3¾"
...	8½" x 13"	80.5	70	3¾"
10"	...	78.5	78.5	3¾"
...	8½" x 18"	106	96.5	3¾"
...	13" x 13"	127	100	3¾"
12"	...	113	113	3¾"
...	13" x 18"	177	150	3¾"
15"	...	177	177	3¾"
...	18" x 18"	233	183	3¾"
...	20" x 20"	298	234	8"
18"	...	254	254	8"
...	20" x 24"	357	295	8"
20"	...	314	314	8"
...	24" x 24"	461	346	8"
22"	...	380	380	8"
24"	...	452	452	8"

lip, and length of flue were adequate (see Section 6.11c and Table 6.2).

c A single flue should not be used for more than one heating appliance if maximum draft is to be attained. Although this rule has certain common exceptions, such as the incinerator, it is true that for maximum efficiency there should be only the smoke inlet at the bottom and the outlet at the top.

d It should be apparent from Table 6.1 that square or rectangular flues have no inherent advantage over circular except that they are more easily built into the chimney.

e When the direction of a flue must be changed (Figure 6.2), the angle should in no case be more than 45°, and it is much better to make it 30° or less. The slope of a flue should not be less than 45°, and 60° or more is better. Sharp turns set up eddies which seriously impede the smooth motion of smoke and gases. Soot and dirt collect in sharp corners and on shallow slopes.

Dwelling House Construction

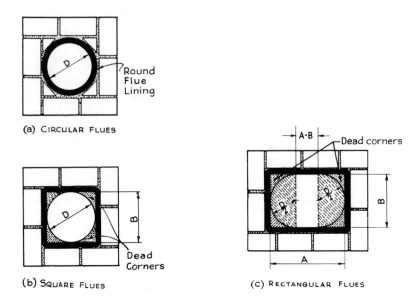

(a) CIRCULAR FLUES

(b) SQUARE FLUES

(c) RECTANGULAR FLUES

Figure 6.1 Total and Effective Areas of Flues.

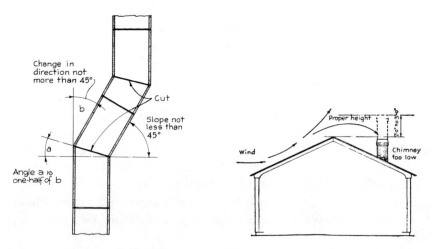

Figure 6.2 Changes in Direction and Slopes of Flues.
Figure 6.3 Height of Chimney Above Roof.

Chimneys and Fireplaces

Table 6.2 Flue Calculations (See Figures 6.7, 8)

$W = 2'6''$ to $5'0''$

$H = 2'6''$ to $4'0''$

but $H < W$ in all cases, and ranges from $2W/3$ to $3W/4$

$D = \dfrac{H}{2}$ to $\dfrac{2H}{3}$

but $\leq 26''$, $\geq 16''$ for coal and $\geq 18''$ for wood

$T \times W = 1.25A$ to $1.5A$

A[a] $\begin{cases} \text{Rectangular Flue} \begin{cases} 20'0'' \text{ above hearth} = \dfrac{W \times H}{10} \\ 30'0'' \text{ above hearth} = \dfrac{W \times H}{12} \end{cases} \\ \text{Round Flue} \begin{cases} 20'0'' \text{ above hearth} = \dfrac{W \times H}{12} \\ 30'0'' \text{ above hearth} = \dfrac{W \times H}{15} \end{cases} \\ \text{Square Flue} \begin{cases} \text{Large enough to inscribe round} \\ \text{flue for similar conditions.} \end{cases} \end{cases}$

[a] Smoke chamber and lip of adequate size.

f Care should be taken to set flue linings close and flush on top of each other, with the joints well filled and carefully struck on the inside to avoid rough spots and lodgment for soot. On the outside the flue lining should be packed tightly with mortar and the brick shoved in close and tight, with all joints in brickwork slushed full, leaving no holes or voids anywhere. This is essential to prevent air leakage through the brickwork into the flue. A section of flue lining can readily be cut with a hammer and small chisel, if it is first set on a solid bed or floor and filled with sand. To change direction, the ends of both flue linings adjacent to the bend must be cut carefully to the proper angle—half the angle of the bend—and fitted tightly (Figure 6.2). In this manner flues may be carried to almost any location.

g Tops of chimneys must be carried high enough to avoid downdrafts caused by turbulence in the wind as it sweeps around nearby obstructions or over sloping roofs. In general, a chimney should project 2' to 3' above the ridge of the house, depending upon its proximity to the ridge (Figure 6.3). The steeper the pitch, the higher the chimney should be carried.

Dwelling House Construction 152

6.4 Chimney Wall Thicknesses; Withes (Figures 6.1, 4, 6, Table 6.1)

a An 8″ brick wall alone is not impervious to wind or weather. However, if we combine flue lining and 8″ of brick, there should be no trouble from this source, and this combination is recommended for the exposed portions of chimneys. Interior chimneys, or the interior portions of chimneys built into outside walls, may generally be only 4″ of brickwork if wood framing is held away from the masonry. For large flues, brickwork should be 8″ in any event (Table 6.1).

b The space between flues, when more than one flue occurs in a chimney, is called the withe. For best construction, this should be 4″, filled in solidly with brick and all joints slushed full of mortar (Figure 6.4a). If space is at a premium the withe can be reduced to 1″, carefully and solidly filled with mortar. In this instance it is absolutely essential that joints in adjacent flue linings be staggered, that individual pieces of lining be sound and uncracked, and that joints between pieces be full, struck, trowelled, smooth, and tight. It is highly desirable, in any event, to avoid having more than two flues adjacent to each other without full 4″ withes (Figures 6.4b, 6).

6.5 Clean-outs (Figure 6.5)

a At the bottoms of all flues except those for fireplaces, there should be a clean-out door, so that soot or other accumulations can easily be removed from the chimney.

b It is unnecessary to carry the bottom of the flue farther below the intake of the smoke pipe than sufficient to install the clean-out door. All clean-out doors should fit tightly in their frames, which in turn should be care-

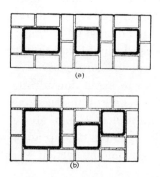

Figure 6.4 Chimney Wall Thicknesses and Withes.

Chimneys and Fireplaces

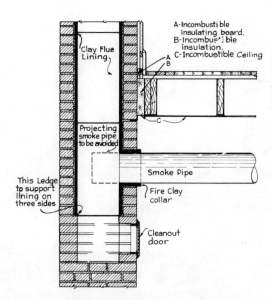

Figure 6.5 Cleanout, Smoke Pipe, and Framing.

fully set in the brickwork and securely anchored to make them proof against air leakage.

6.6 Smoke Pipe (Figure 6.5)

a An improperly located and protected smoke pipe from the furnace, stove, or other heating unit is a dangerous fire hazard in a dwelling house.
b Care should be taken in setting the smoke pipe that it does not extend into the flue, as shown by the dotted line in Figure 6.5. A metal or terra cotta collar should be built into the brickwork and the smoke pipe should be slipped into this collar, flush with the inside of the flue.
c Many building codes require heater-room ceilings to be plastered or protected with incombustible wallboard as an essential part of the protection of the house from fire. In any event, the smoke pipe should be kept away at least 10″ from the floor joists over it even if the ceiling is protected.

6.7 Wood Framing (Figures 6.5, 6, 7, 11, 12)

a Framing members—floor and wall or partition—must be kept free of the chimney and should be insulated from it, especially if only 4″ of brick masonry surrounds the flue lining. The usual header and trimmer construction is employed, with framing members held 2″ away from the

Dwelling House Construction 154

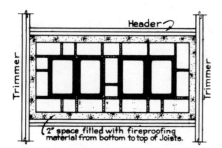

Figure 6.6 Framing around Interior Chimney.

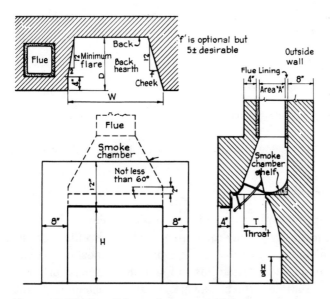

Figure 6.7 Fireplace Relationships (see Table 6.2).

Chimneys and Fireplaces

masonry. See Chapter 5 for a discussion of framing around chimney and fireplace. The space may be filled with incombustible insulating material such as mineral wool.

6.8 Testing

a Before apparatus is connected with any flue, but after the mortar has hardened, it is a good plan to apply a smoke test. One method is to build a smudge fire of paper, straw, wood, or tar paper at the base of the flue and when a dense column of smoke is ascending, tightly block the outlet. Leaks can be located at once by the appearance of smoke, and bad leaks due to carelessness in laying up masonry are often revealed. Such leaks may be into adjoining flues or directly through the walls or between the walls and the lining. A chimney which shows leakage should not be accepted until the defect has been remedied. Since the stopping of a leak after the chimney is completed is usually very difficult, it pays to watch the construction closely as it progresses.

6.9 Footings, Flashing, Cap

a As already noted in Chapter 4, chimneys are generally the heaviest portion of the structure and must have ample footings. Footings should be designed to prevent differential settlement in the building by balancing the various loads on the footings.

b Where the chimney penetrates the roof or comes in contact with the roof as in an outside chimney, it must be flashed to the slope of the roof. Flashing is taken up in detail in Chapter 8, but it may be noted that chimney flashing frequently consists of two parts, cap and base. The base is a bent sheet of metal fitted into the angle between the roof and the chimney. One part extends out on the roof under the shingles and the other lies flat against the side of the chimney. The cap is another bent sheet built into the brickwork and turned down over the base flashing.

c At the back of the chimney a flashed cricket (small gable with ridge at right angles to slope of main roof) is built to divert the water and snow that come down the slope of the roof (Chapter 8).

d Tops of all chimneys must have a good and efficient wash. Frequently, a cap of stone is provided, but in place of stone a thick bed of cement mortar, pitched to all sides to shed water quickly, may be spread over the whole area of the chimney. It is also desirable to let the flues project several inches above the chimney. Water must not be allowed to get into the brickwork at the top of the chimney because it is likely to cause disintegration.

FIREPLACE

6.10 General

a There are six principal items to be watched in the construction of a fireplace:
1. Size and shape of the fireplace opening.
2. Relation of size of fireplace opening to size of flue.
3. Size and shape of the fireplace hearth or "underfire."
4. Smoke shelf.
5. Smoke chamber.
6. Damper.

6.11 Proportions

a Most fireplace openings are rectangular, that is, wider than high, unless the fireplace is very narrow. Average widths for fireplaces are 36″ to 42″, and the height 30″, although the width and height may be greatly extended. A room with 300 sq ft of floor space can well be served by a fireplace 36″ to 42″ wide. Fireplaces 48″, 54″, or 60″ wide are usually constructed in rooms of correspondingly greater dimension.

b The depth of the fireplace should not be too great; 18″ to 20″ suffice for the average fireplace up to 42″ wide, with slight increases in depth for wider openings. The shallower the fireplace, the more heat is reflected into the room.

c The back hearth (underfire) should not be square or contain square corners. The sides of the fireplace should slope inward from front to back approximately 5″ per foot of depth. Experiment has shown this to be the most effective angle, although it can be reduced to 2″ and still be satisfactory. Square corners cause eddies and smoke pockets which may interfere with combustion. In the following formulae (see Figure 6.7) are set forth relationships among the various parts of the fireplace which have been found workable in practice.

d Figures 6.7 and 6.8 show sections which should be carefully noted. The back of the fireplace should be built vertical to $\frac{1}{3}$ height of fireplace opening and then it should be sloped forward, so that at the top it comes to

Chimneys and Fireplaces 157

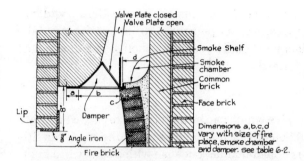

Figure 6.8 Detail of Throat, Showing Damper, Lip, and Smoke Shelf.

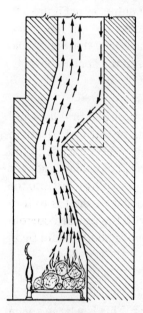

Figure 6.9 Poorly Constructed Fireplace, Lacking Damper and Smoke Shelf. Interference Occurs at Throat.

the back of the damper, or if there is no damper, so as to allow a width of between 8" and 10" for the throat. At a point about 8" above the top of the fireplace opening, the damper should be set and a shelf or ledge called the smoke shelf should be formed.

6.12 Damper

a The damper is a large valve which can be adjusted to regulate the draft. Its position is important. Sometimes it is set directly at the top of the fireplace opening, but this is objectionable, because with a smoky fire in the fireplace, the draft is likely to carry the smoke against the top of the fireplace opening and so force some of it out into the room. A much better position is about 8" above the top of the fireplace opening, as in Figure 6.8. This arrangement allows the draft to carry the smoke up into the space back of the lip at the top of the fireplace opening and below the damper, so that it is more readily drawn through the damper into the space above.

6.13 Smoke Shelf (Figs. 6.7, 8, 9)

a The smoke shelf is one of the most important features of a fireplace and must have close attention. With the central column of warm air rising in the flue, there is a downward current of cool air close to the walls of the flue. If there is no smoke shelf, this column or current of cool air reaches the throat of the fireplace, blocking the smoke and crowding it out into the room, as shown in Figure 6.9. The function of the smoke shelf is to turn this current of cool air upward so that it will join with the upward current of warm air and not impede it. The width of the smoke shelf should be at least 4 inches.

6.14 Smoke Chamber

a The next important feature is the smoke chamber. This is a large space over the damper and smoke shelf (Figure 6.7). The back may go straight up, but the sides, starting at the ends of the damper, are sloped at the rate of 7" in 12" of height, and are carried up, converging, until the distance between the two sloping sides is equal to the inner dimensions of the flue lining, directly on the center line of the fireplace. The front of the smoke chamber may be sloped as needed to meet the flue at the same point. It is bad practice to have the flue off the center line because it causes the draft

Chimneys and Fireplaces

through the fireplace to be uneven. The first section of flue lining should be vertical, after which the flue may be turned or bent as desired.

b Over the top of the fireplace opening, an angle-iron 3″ x 3″ or 3″ x 4″ is generally set to carry the brickwork across the opening (Figure 6.8). Some manufacturers make concrete or metal smoke chambers for each of their dampers. When the top of the fireplace opening is reached, the damper is set, and then over this is placed the metal or concrete smoke chamber of the proper shape and size. Masonry is placed around the prefabricated smoke chamber.

6.15 Hearth

a The hearth consists of two parts: the front or finished hearth and the back hearth or underfire. The front hearth is simply a precaution against flying sparks and, while it must be noncombustible, need not resist intense prolonged heat. It can, consequently, be finished with quarry tile, ordinary brick, stone such as flagstone, concrete, or similar materials. The back hearth must withstand intense heat, as must the cheeks and back of the fireplace. These must be built of heat-resistant materials. Fire brick laid in fire clay is the best combination. As fire bricks are large, they can be laid on edge instead of on the flat. This saves brick and takes up less space. If the color of fire brick is objectionable, hard-burned clay bricks can be employed if it is recognized that these may crack and spall after long exposure to hot fires.

b. Trimmer Arches As shown in Figure 5.24, the floor framing around the space occupied by the front hearth is the familiar header-trimmer combination. The front hearth itself is supported on a concrete trimmer arch as illustrated in Figure 6.10.

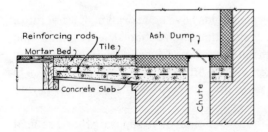

Figure 6.10 Trimmer Arch Support for Front Hearth.

c The framing for the concrete slab is much the same as for a tile floor. Nailing blocks are firmly spiked to the header framing the front hearth, and a brick corbel is built in the chimney at approximately the same height. Short lengths of subflooring are laid on these supports. The concrete slab, at least 3″ thick, is cast on top of this platform and back into the rear of the chimney to form a base for the underfire or back hearth. Wire mesh is embedded in the concrete near the top of the slab and runs from header to back of underfire. If ash dumps are to be provided, box forms are placed in the concrete at the proper place.

6.16 Ash Pit

a The lower portion of a chimney containing one or more fireplaces is quite large and can conveniently be left hollow, both to save masonry materials and to form an ash pit into which ashes from the fireplaces can be chuted. The walls must be at least 8″ thick to support the chimney above, and if the latter is very large the ash pit side and back walls should be 12″ thick. The front wall is usually 8″ thick, but may be 4″ in small chimneys since it does not carry much of the chimney load because this load is transferred to the side walls by the construction over the fireplace openings.

6.17 Construction Time

a The fireplace may be built at the same time as the chimney or later when the house is nearly completed. If the fireplace is to be faced with ornamental material such as tile and face brick which may become soiled or damaged during the construction of the job, it is much better to defer the building of the fireplace until just before the finish flooring is laid.

6.18 Examples

a In Figure 6.11, the foregoing principles have been drawn together into a single large chimney, shown in elevation and section. A main heater flue on the left is kept straight from bottom to top. A secondary flue, third from left, is shown as serving an incinerator. Although it is common practice to have incinerators utilize the main flue, it is better to provide a separate flue.

b Two fireplaces, one offset above the other, have separate flues, each of which is centered over an ample smoke chamber and each of which rises straight for one length of flue lining before bending. No flue is inclined

Chimneys and Fireplaces

less than 60° to the horizontal. Joints in flue linings are staggered in adjacent flues, even though full 4″ withes are provided. Fireplaces have ash dumps and chutes, and are provided with proper dampers, lips, and smoke shelves. The lower trimmer arch is brick, now seldom used but once popular; the upper is concrete. Since the chimney is built into an outside wall, the outer three walls are 8″ thick, but the inner wall from above the lower fireplace to below the roof framing is 4″. At the roof it is corbelled, i.e., built out by projecting successive brick courses beyond the preceding ones (a common procedure) to 8″, so that above the roof all chimney walls are 8″ thick. Flue linings project beyond the generously pitched cap. Cap and base flashing is provided at the roof. The ash pit top is arched, although it could well be corbelled. Its floor is raised above the basement floor to facilitate cleaning. Wood framing is held away from the brickwork at every point.

c Figure 6.11 illustrates traditional fireplace and chimney construction. Figure 6.12 shows a more contemporary design with a raised fireplace hearth and cantilevered front hearth. Although the plan of the back hearth is rectangular rather than splayed, the general principles of lip, damper, smoke shelf, and smoke chamber are the same.

6.19 Prefabricated Chimneys and Fireplaces

a Prefabricated flues typically consist of circular sections designed to fit snugly over each other with sealed joints similar to shiplap or tongue and groove. The inner lining is of a high heat resistant material such as fire clay. Surrounding it are several inches of heat-resistant insulation, in turn covered by a weather-resistant outside circular, square, or rectangular jacket such as rust-resistant metal. Such flues must, of course, be adequate in size, as is true of any flue. Many varieties of prefabricated fireplaces or fireplace linings are to be found. Some are wall-hung, others rest on a base of some kind. Many have metal exteriors with heat-resistant insulating liners such as fire brick. Their sizes, proportions, and flues should be checked to see that they will indeed draw properly without emitting smoke into the room.

Dwelling House Construction

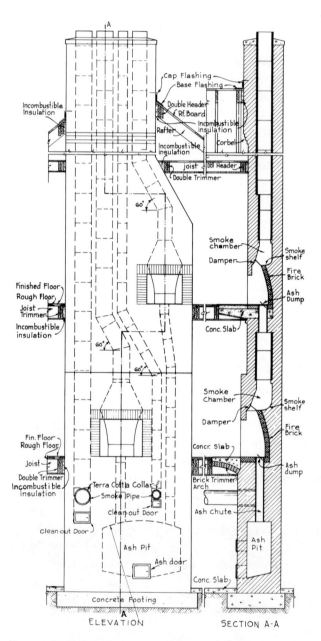

Figure 6.11 Chimney Incorporating Basement Flues, Two Fireplaces, Flue, and Ash Pit.

Chimneys and Fireplaces

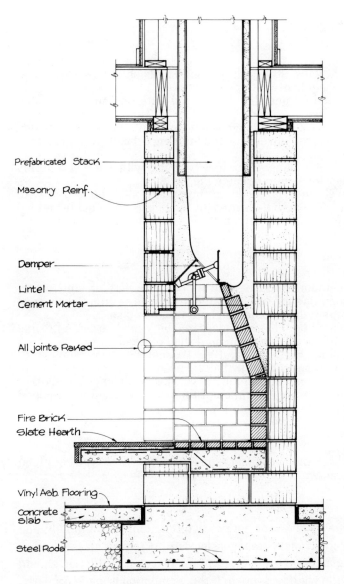

Figure 6.12 Contemporary Fireplace With Raised Hearth, Cantilevered Front Hearth, Damper, Lip, Smoke Shelf, and Smoke Chamber.

7 Windows

7.1 Specification Clauses

WOOD WINDOWS

a. Double Hung Frames Stiles and parting beads shall be yellow pine; the rest of the frame, clear white pine.
b. Priming All wood frames shall be primed with one coat of priming paint before delivery.
c. Sash Sash shall be clear white pine made to detail, $1\frac{3}{8}''$ thick—divided as shown—and shall be delivered to the building primed with one heavy coat of linseed oil.
d. Hanging All double-hung sash shall be hung with approved spring balances of proper tension to balance the sash.
e. Glazing All sash and glazed doors throughout shall be glazed with first quality double-strength flat-drawn or float glass, thoroughly bedded and sealed, using only best compound. No glass shall be set until after sash have been thoroughly oiled.

METAL WINDOWS

f. Materials Aluminum windows shall be 6063-T5 alloy or equivalent. Rigid vinyl components shall be integrated to provide an effective thermal break between aluminum members in both frame and sash.
g. Glazing insulating double-sheet glass $\frac{3}{8}''$ thick shall be set in flexible dry glazing. No putty or glazing compound shall be used.
h. Weatherstripping High-density PVC fix-seal weatherstripping, or silicone-treated wool pile, or both, shall be used.
i. Hardware All mechanical fastening devices, sash balances, and other necessary hardware for the operation of the windows shall be provided as part of the complete window unit.

BASEMENT WINDOWS

j Basement windows shall be steel casements of the lever type equipped with screens as shown.

7.2 General

a A window performs either or both of two functions: to let in light and to provide ventilation. Light transmission is effected by glass or other light-transmitting material such as one of the plastics; ventilation should occur only when the window is open. If the installation is made properly, the window is tight and allows only a small amount of air to leak past when it is closed; if the window is installed carelessly, not only does a great deal of air leak past but rain finds its way through and disfigures the walls adjacent to the windows. Tight windows are a matter of proper construction and proper detailing.

Windows

7.3 Window Types (Figure 7.1)

a Windows might be broadly classified as "fixed" (non-opening) or "ventilating." Ventilating windows generally consist of one or more movable panels called sash, within a fixed frame. Ventilating windows are generally classified by the manner in which the sash open or operate.

b Double-hung windows (two sliding sash) and single-hung windows (one fixed sash, generally above, one sliding sash, generally below) open and close by sliding the movable sash vertically in grooves provided in the sides (jambs or stiles) of the frame, the sliding sash being held in the desired open or closed position by various spring balances concealed in the sash or frame (in jamb or head), by a manually operated or spring-actuated ratchet or peg arrangement between the sash and the frame, or by a friction-type control. Formerly, double-hung sash were customarily counterweighted with sash weights.

c Horizontal "traverse" windows open and close by sliding or rolling the sash horizontally in grooves or tracks provided in the top (head) and bottom (sill) of the frame, the sliding sash generally being limited to the smaller sizes or lighter glazing by their operating friction.

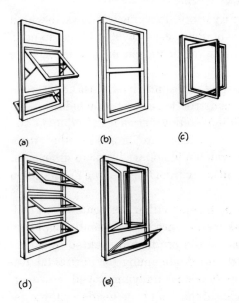

Figure 7.1 Types of Windows. (a) Projected. (b) Double-Hung. (c) Reversible, Pivoted. (d) Awning. (e) Casement (Top) and Hopper (Bottom).

d Casement windows formerly included all those in which the sash were hinged to the frame at one edge (rail or stile) opening inward or outward at the opposite edge, as well as the fixed or fully-encased windows. In current usage, however, the term "casement" generally refers only to those windows in which the sash are supported or "hung" at one side (jamb or stile) and open outward or inward at the opposite side, the outward-opening casements being the more common. Top-hung sash opening outward at the bottom are now generally called awnings, while bottom-hung sash opening inward at the top are referred to as hoppers or hopper-vents. Top-hung in-swinging sash are also available which are said to be easy to clean.

e While sash of the above casement varieties are frequently supported from the frame, or swung, on hinges or butts (Chapter 14) of various types, they are often of the projected type, in the opening of which the retained rail or stile of the sash slides toward the center of the frame on shoes from which the sash pivots as it moves to the open position. In both methods, the sash are held in the open position by extension arms. One end of these arms is attached to a shoe which slides in a track or groove at the edge of the sash, the other end being pivoted on the frame. Projected sash are generally of the awning or hopper vent types.

f Jalousie or louvered windows are a variety of awning windows, in which numerous narrow horizontal lights of glass, each pivoted at its top edge (generally by a metal rail or muntin), operate in unison and overlap slightly in the closed position.

g Various combinations of ventilating sash and fixed glass in a single frame are often used in residential construction as picture windows, window walls, bow windows, etc. Folding windows are a less-frequent combination in which several sash in a single frame are hinged together, turning on pivots supported by shoes which slide on tracks or in grooves in the frame, permitting the sash to fold away from the opening in accordion style.

h Pivoted sash, which open only by turning either horizontally or vertically about the centers of their rails or stiles, are less frequently employed in residential construction because of the problem presented by insect screening. A combination pivoted and double-hung (or horizontal traverse) sash in a reversible window is occasionally employed, however, to simplify the cleaning of both sides of the sash from inside, without removal.

Windows

i Because of the great diversity of window types, materials, and details, only the principal ones are described below, and only in sufficient detail to set forth principles. Wood windows are used to illustrate many of these principles because they were the most common original prototypes, and many later metal and other windows were derived from them.

7.4 Materials

a Windows are generally constructed of either wood or metal, the metals most commonly employed for this purpose being aluminum and mild carbon steel. Windows constructed of stainless steel or bronze are available, but are rarely used in residential work because of the cost. Windows are also constructed entirely or partially of plastics.

b While the great majority of residential windows have traditionally been wood double-hung or outswinging wood casements, increased use is being made of wood and aluminum awning, ventilators, and jalousies, as well as of wood and aluminum horizontal sliding (or rolling) windows. The residential steel projected casements, hopper vents, and awning ventilators are employed principally in foundation walls, brick-veneered frame construction and solid masonry construction.

7.5 Glass

a Window glass is made by the flat-drawn and float processes. Flat-drawing consists of drawing a continuous sheet of glass from a molten mass and slowly cooling or annealing it through a long tunnel. Float glass is made by casting a sheet of glass continuously on a bed of molten tin where it spreads to a uniform thickness and flat surfaces by surface tension forces. The sheet is drawn through and slowly cooled in a long tunnel. These two types of glass provide the bulk of window glass. When still flatter surfaces are wanted to avoid noticeable optical distortion, flat-drawn glass is ground and polished on both sides. This is known as plate glass.

b In building, the thickness of glass is known as its "strength" and the two most commonly used are "single strength" (ss), and "double strength" (ds). Clear-glass designations in current use for residential work are given in Table 7.1. Figured glass (glass with a patterned surface, which transmits light but not clear images) is generally $\frac{1}{8}''$ or $\frac{7}{32}''$.

c Single strength is used only in small lights; wherever there is likelihood of heavy wind pressures or the lights are larger, double strength is needed.

Dwelling House Construction

Table 7.1 Residential Glass

Type	Weight (oz/sq ft)	Max. Area 100 mph (sq ft)
Single strength 3/32"	19.5	14
Double strength 1/8"	24–26	20
3/16" sheet	36–40	32
7/32" sheet	44–45	36
1/4" sheet	50–52	44
1/4" plate	50–52	38

The heavier grades are used for still larger windows such as store fronts and large "picture windows." The latter may be plate glass, because undistorted vision is generally important.

d Maximum glass area depends upon the expected wind pressure. Wind velocities of 100 miles per hour exert pressures of roughly 30 lb per sq ft. Manufacturers recommend that glass areas be no larger than the values given in Table 7.1. With these areas, somewhat less than 1 percent of the sheets, in a large number of sheets, can be expected to break the first time these pressures are reached. For a smaller chance of breakage, smaller areas should be used, as is common in residential work. (See Chapter 16 for discussion of plastic.)

7.6 Sash (Figure 7.2)

a The vertical side members of sash are called stiles, the horizontal top and bottom members are called rails, and the smaller members which divide the glass into smaller panes—called "lights" in building language—are called bars and muntins, or just muntins. Muntins or grilles may be integral as shown in Figure 7.2, lower sash, or snap-in wood or plastic, as shown in the upper sash. The latter are obviously for appearance only, but can be removed to make washing of glass easier. Sash details are many and varied.

b Wood sash and frames are generally constructed of kiln-dried, clear, straight-grained Western or Ponderosa pine, and are factory-treated with a toxic, water-repellant preservative. Stock wood sash thicknesses vary from 1 3/8" to 2 1/4", depending upon the size of the opening and the thickness of the glazing units (single, or double "insulating" glass may be used).

Windows

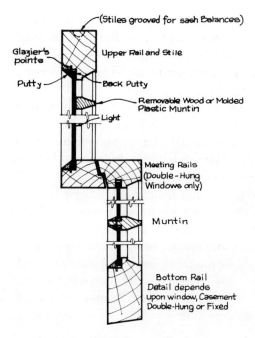

Figure 7.2 Parts of Window Sash.

c Sash members are rabbeted on the outside to receive the lights and putty and are finished with a molding on the inside, often an ogee (cyma), but also any other detail the architect or manufacturer may desire (Figure 7.2).

7.7 Glazing (Figure 7.2)

a The process of inserting lights into sash is called glazing. It must be done carefully or air and water may find their way around the edges of the glass.

b Lights are cut approximately $\frac{1}{16}''$ to $\frac{1}{8}''$ smaller than the sash opening to provide enough clearance and to permit slight distortions of the sash without cracking the glass. Before the light is inserted, the sash rabbet is painted with a drying oil such as linseed, and then a thin layer of putty is spread over the back of the rabbet or struck on to the edges of the glass. The glass is next pressed firmly against the bed of putty. This is called back-puttying and is essential to a tight joint inasmuch as direct glass to wood contact would not be absolutely continuous. Flat, triangular, or diamond-shaped zinc "points" are next laid on the glass and forced into the wood so that $\frac{1}{8}''$ to $\frac{3}{16}''$ is left projecting to hold the glass in place. Various other clips and holding devices are employed, especially with metal sash. Finally, the rabbet is filled with putty sloped to the outer edge of the sash.

c A good putty is made of precipitated whiting (calcium carbonate) or marble dust ground in drying oil. This makes a hard durable putty which adheres firmly to wood and glass. In order to make any putty adhere satisfactorily to wood, the latter must first be painted with drying oil, otherwise the raw dry wood absorbs too much of the oil from the putty.

d Although these putties have had a long history and are particularly suitable for small lights in wood sash, many other glazing compounds and sealants have been devised, especially for metal windows and large lights. Some of these are semi-hardening and remain fairly soft. They are compounded for applications where differential movement of glass and sash, caused by temperature changes, may be expected. For large lights, sealants derived from synthetic polymeric materials (Chapter 16) are employed. Gaskets based on rubber polymers are also used (see under the several metal windows, below).

7.8 Casement Windows, General

a Casement windows are side-hung and swing in or out; outswinging being the most common. The wood casement windows described below are used to illustrate general principles, but there are many variants.

7.9 Frame (Figures 7.3–5)

a Window frames have three chief parts, the top or "head," the sides or "jambs," and the bottom or "sill." Although "jamb" is derived from the word for "leg," the term "jamb" is also used to denote the head; thus, "side jamb" and "head jamb." Wood casement frames are almost always "plank" frames, i.e., they are made of nominal 2″ stock cut and shaped to receive the sash. Shaping the head and jambs consists of cutting rebates or "rabbets" $\frac{1}{2}''$ deep and $1\frac{1}{2}''$ or more wide to accommodate $1\frac{3}{8}''$ or thicker sash. A second rabbet is usually added to accommodate screens and storm windows.

b Figures 7.3 and 7.4 show details of outswinging and inswinging casement windows in frame walls. Head and jambs are practically identical.

Windows

They are wide enough to extend through the wall from outside of sheathing to inside of wall surface and they clear the trimmer and header studs by only a small amount. Sheathing and building paper are carried in as close to the frame as is practicable.

c The sill generally extends through the wall to the interior surface line. The outer edge, unlike head and jambs, projects beyond the sheathing line far enough to provide a base upon which the outside casing rests, and form a projecting shelf which protects the joint between window frame and siding or other exterior finish. The lower side of the sill should be plowed (grooved) to allow the siding to fit in snugly and prevent rain from being swept through under the sill. A separate projecting subsill may be employed, thus allowing the main sill to be the same width as jambs and head. This is a manufacturing convenience.

d The sill slopes down and outward at the rate of about $1''$ in $5''$ so as to shed water readily. Its upper surface ("wash") is usually rabbetted (Figure 7.4) to form a snug watertight fit for the window sash or for screens and storm sash. Although wood screens are shown in Figure 7.3, metal screens are common. They may be placed in separate fixtures for interchangeability with storm sash.

7.10 Inswinging Sash (Figure 7.4)

a Lower rails of inswinging sash require special treatment to prevent rain from penetrating between rail and sill. Generally, the upper interior edge of the sill is provided with an upstanding lip, and the lower edge of the lower rail with a corresponding rabbet as shown. For further protection, a drip molding is fastened to the lower rail to prevent rain from being driven against the lower edge of the rail.

7.11 Mullions and Meeting Stiles (Figures 7.4, 5a, b)

a If several casement windows are in a row in one frame, they may be separated by a vertical member called the mullion (which is simply a two-sided jamb), or they may close against each other in pairs. A typical mullion detail is shown in Figure 7.5a. Sections at the meeting of two casements are shown in Figures 7.4 and 7.5b.

7.12 Casing (Figures 7.2, 3)

a Casing, both exterior and interior, finishes and closes the joint between frame and wall. Since it seals the joint between wall and frame and keeps

Dwelling House Construction

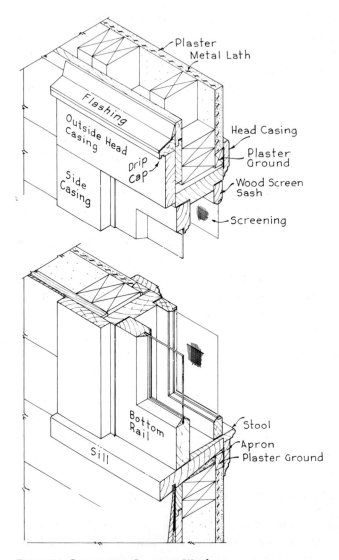

Figure 7.3 Outswinging Casement Window.

Windows

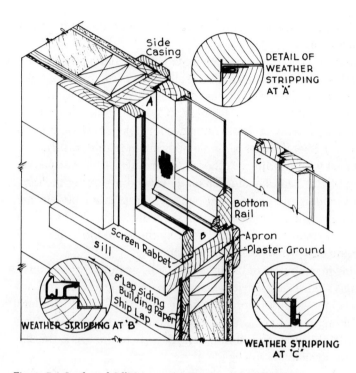

Figure 7.4 Jamb and Sill, Inswinging Casement Window.

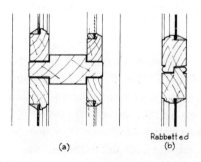

Figure 7.5 (a) Solid Mullion. (b) Meeting Stiles.

wind and rain from penetrating at that point, it is a building detail of considerable importance.

b Exterior casing runs up the two sides (side casing) and across the top (head casing). It is attached firmly to the frame while being preassembled in the shop, and should be "back-primed", i.e., the back and cut ends should be painted at the time of assembly. It is also nailed against the sheathing or provided with anchoring devices for field attachment to the sheathing or frame. Sheathing paper must be brought in under the casing so as to form a seal where siding or other exterior finish abuts the casing. Across the top of the head casing is usually placed a sloping molding, provided with a small vertical fillet at the back, which is called the drip cap. Flashing covers this and is carried up under the exterior finish. This detail can be simplified by merely sloping and flashing the top of the head casing but the projecting cap provides a positive drip.

c Interior casing forms a part of the interior "trim" or finish of the house.

d At the bottom (Figure 7.3), starting at or on top of the wood sill and projecting into the room, is a horizontal flat piece called the "stool." Under it and covering the space between stool and wall finish is the "apron." Starting from on top of the stool and running up the side to the top is the side casing, and across the top is the head casing.

e The stool may be omitted with inswinging casements (Figure 7.4). If so, the side casing is carried down to meet the apron in the same manner that it meets the head casing. The apron, in effect, becomes the bottom casing.

7.13 Plaster Grounds (Figures 7.3, 4)

a If the interior is plastered, the plasterer needs a guide of some kind or he may leave slight waves in the surface of the plaster. For the casing and the apron to fit snugly against the plaster, strips of wood or metal called grounds (the thickness of the plaster) are applied around the opening, and the plaster is finished flush with these strips. Moreover, the casings are slightly hollowed in back so that they can pass over slight irregularities. (See Chapter 12.)

7.14 Metal Windows (Figure 7.6, 7)

a Metal window frames and sash are frequently employed in dwelling construction. The most common metals are aluminum and steel.

b Residential aluminum sash and frames are generally constructed of extruded 6063-T5 tempered aluminum alloy, while the higher strength

Windows

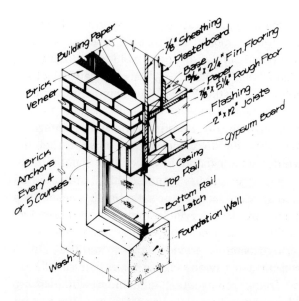

Figure 7.6 Steel Basement Window, Cast Concrete Foundation Walls.

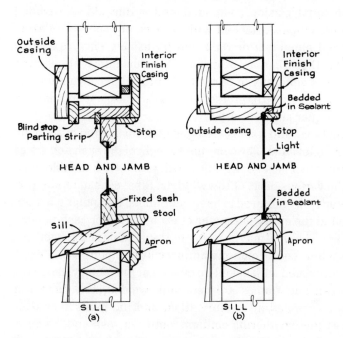

Figure 7.7 Fixed Lights. (a) Separate Sash. (b) Glass Set Directly in Frame.

6063-T6 alloy is used by some manufacturers. The extruded aluminum members are frequently tubular (closed hollow box) in cross section, occasionally semi-tubular (semi-hollow or open), seldom actually flat. Minimum wall thicknesses of 0.055″ to 0.062″ ($\frac{1}{16}$″) are typically specified for tubular (sash) sections, 0.062″ to 0.093″ ($\frac{3}{32}$″) for semi-hollow (frame) sections, and 0.093″ to 0.125″ ($\frac{1}{8}$″) for open or "flat" (sill) sections. Corner joints are generally mortise and tenon or mitered and internally reinforced, and are held by aluminum rivets, stainless steel screws, or welds (helium-arc or electronic flash welds), which should be milled to a smooth finish. Typical factory finish is "as extruded" or "standard mill," protected by two dip coats of clear methacrylate lacquer, or satin finish, caused by a slight caustic etch, protected by clear acrylic lacquer. A more corrosion-resistant finish is available at extra cost by factory anodizing of the aluminum members after extrusion, to provide either a "natural" aluminum or a uniformly colored surface, a wide range of colors being available. For installation in wood frame construction, aluminum windows are provided with integral aluminum nailing flanges or fins, pre-punched for nailing to the wood frame. The glazing of aluminum windows may be accomplished with butyl glazing compound and spring clips, extruded aluminum-glazing channels attached with screws, or "snap-in" glazing moldings of formed aluminum or rigid plastic, usually vinyl (extruded polyvinyl chloride). Direct contact of unprotected aluminum with concrete or mortar masonry is generally to be avoided. In such construction, the aluminum should be anodized or protected with zinc chromate primer and bitumastic or a lacquer such as acrylic.

c Residential steel sash and frames are typically constructed of solid hot rolled low-carbon billet steel. The combined weight of the sash and frame in pounds per linear foot is generally specified, rather than the thickness of the metal of the various parts of the window. Steel casement and projected sash are generally required to have continuous two-point weathering contacts around the entire perimeter of each ventilator sash to minimize air infiltration. The steel may be protected by pickling and galvanizing with a hot-dip "spelter" coat of molten zinc, or it may be electro-galvanized, phosphatized, and primed with baked-on enamel at the factory. Since the joints of steel sash and frames are generally welded, the protective coatings should be applied after fabrication, and the joints should be ground smooth at the weathering contacts between sash and frame to

Windows

provide a continuous seal. Steel frames are generally provided with anchors or flanges (pre-punched) attached, to engage the masonry or wood frame for installation purposes. Glazing is usually accomplished with non-hardening glazing compound and spring clips or steel glazing beads attached with screws.

d Metal windows must be handled with great care prior to installation, to avoid bending or warping, since a vent sash in a frame which is out of square or out of "wind" (twisted), is difficult to operate and to close tightly.

e Basement windows are commonly metal (Figure 7.6). If the foundation wall is cast concrete, the window frames may be built into the forms (Chapter 4) so that when concrete is placed the projecting legs of the frame are anchored directly and tightly into the concrete. Frames may also be inserted later, as shown in Figure 7.6. Similarly, when foundations are of masonry, such as concrete block or brick, the frame may be built directly into the wall or may be inserted later. In any event, the wash provided at the sill should be steep—approximately 1 to 3 is sufficient—in order to shed water quickly. Water should not be permitted to stand in contact with frames; otherwise rapid corrosion may occur.

7.15 Fixed Windows (Figure 7.7)

a Frequently it is desired to have windows stationary. Glass in such windows may be set in sash, in much the usual way, or may be set directly into the frame without any sash. Fixed lights, especially when large, are more easily installed and removed when set in a separate sash, but a larger clear area is obtained if sash are omitted.

b Fixed lights in separate sash can be casement windows minus the hinges, or they may be single sash in double-hung frames (Section 7.16, ff) as shown in Figure 7.7a. Evidently, no provision need be made for screens, but storm sash may be desired.

c Figure 7.7b shows one method of installing lights directly in the frames. The critical point is at the sill, because the joint must be tight against rain outside and condensate inside. Back-puttying is essential.

7.16 Double-Hung Windows, Wood

a In double-hung windows, two counterbalanced sash slide up and down in the frame. Formerly, counterbalances were commonly cast iron weights attached to one end of a cord fastened at the opposite end to the sash and carried over pulleys set in the upper ends of the sides of the frame. Weights slid up and down in "weight pockets" between window frame and studs. Figure 7.9b shows this arrangement. Today, a variety of spring or otherwise activated balances are used to offset the weight of the sash (Figures 7.8, 9).

7.17 Sash (Figures 7.2, 8, 9)

a Double-hung window sash are in pairs with the upper and lower sash slightly different. The lower rail of the upper sash and upper rail of lower sash are especially shaped with a double bevel to fit snugly against each other when closed, but to move apart without friction when opened. These two rails are called the meeting rails. Each projects beyond the parting strip (Section 7.18b) whereas the rest of the rails and stiles do not.

b With sash weights and certain types of spring balances, the stiles are grooved at the edges adjacent to the jambs so that the sash cords or balances can be carried down past the edge of the sash.

7.18 Frames and Casing

a A double-hung frame contains more parts than a casement frame; although like the casement frame it consists essentially of a head, two jambs, and a sill. In this type of window, the entire side of the frame is called the jamb, but this particular part of it is called a stile. It is nominal 1″ material. The corresponding piece of the frame overhead is called the "yoke." Terminology is not uniform. The terms "side jamb" and "headjamb" are common.

b The stile (and generally the yoke) has a longitudinal groove in the center. In this is placed a rectangular "parting strip," to separate the sash from one another and to form tracks in which the sash move up and down. The parting strip may be of wood or metal, such as aluminum.

c Figure 7.8 illustrates a standard double-hung frame and sash. A blind stop is attached to the outer edges of the frame to form a recess or rabbet into which screen and storm sash can be fitted. Outside of that is the outside casing which extends far enough to overlap the sheathing and be nailed through it (casing nails) to the studs. Sheathing paper is carried past the casing to protect the joint between casing and outside wall finish such as bevel siding or shingles. At the top, the head casing is provided with the same cap or drip molding that is found on casement windows.

d The sill of a double-hung frame has a sloping face to facilitate the shedding of water, and generally this face has one or two breaks in it, one

Windows

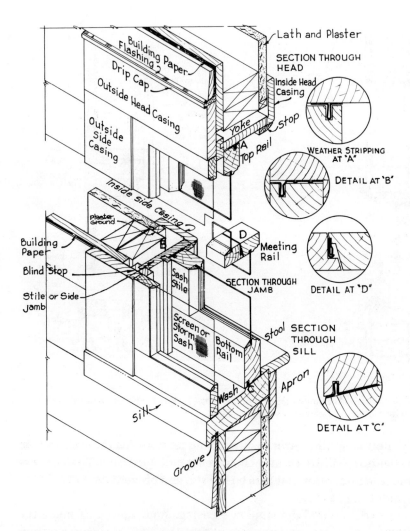

Figure 7.8 Double-Hung Window, Weatherstripped.

Dwelling House Construction   180

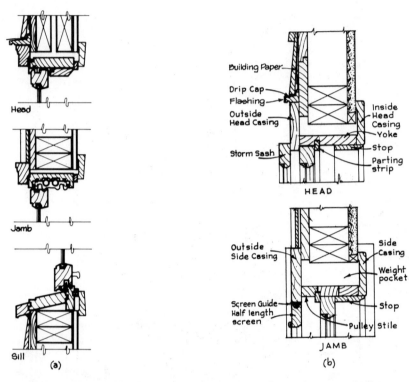

Figure 7.9 (a) Variant of Double-Hung Window with Flexible Jamb Liner and Sill Extension. (b) Simple "Cottage" Frame with Weight Pocket.

immediately under the lower sash, and another between the sash and the outer edge of the sill to form a seat for screens and storm sash. These breaks obstruct the rain water that falls on the sill, and prevent its being driven back under the sash.

e Interior trim around the window—casing, stool, apron—is much the same as on casement windows except that the stool is always present. There is an additional piece called the "stop" which is a thin, narrow molding covering the joint between interior finish casing and frame. It forms the inside edge of the track in which the lower sash travels.

f One variant of the standard wood double-hung window is shown in Figure 7.9a. A spring-loaded flexible vinyl jamb liner is pressed against the sash stiles but is flexible enough to permit the sash to rotate about a pivot for cleaning the outer surface of the double insulating glass. Spring balances are concealed behind the jamb liner. At the sill, the regular sill

Windows

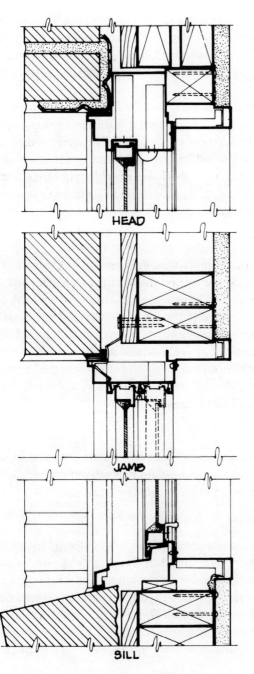

Figure 7.10 Double-Hung Window in Brick Veneer.

does not project, but a secondary sill is attached to provide a lip. Sealing this joint is important, as is the use of preservatively-treated wood. In this particular detail, the stool is omitted at the sill, and inside casing is carried around sides, top, and bottom. Interior wall surface is drywall (Chapter 12).

g Cottage Frame Figure 7.9b illustrates a simpler frame. The blind stop is omitted. This frame has the disadvantage that it is difficult to hang full-length screens or storm sash. Screens are usually half-length, sliding in channels. Storm sash must be placed against the outside casing instead of being fitted into a rabbet. Figure 7.9b shows an old-style frame with weight pocket. The stile is called a pulley stile because it contains a pulley at the top over which the sash cord runs.

7.19 Masonry Veneer (Figure 7.10)

a A metal double-hung window in brick veneer is shown in Figure 7.10. At the head, the brickwork is carried across the opening by a steel lintel. The hollow window frame is shaped to provide a channel for the sash at the head and jamb. The sill is shaped to provide a step for the lower sash and a rabbet to accept the screen, which fits into corresponding rabbets at the jamb and head. Projecting legs at the head and jambs provide nailing to the wood framing. On the inside, metal casing and stool meet the interior wall finish, and nailing is provided below the stool to the wood framing. The metal sill rests on a brick rowlock subsill.

7.20 Mullions (Figure 7.11)

a When several double-hung windows are in one frame, they are separated by mullions. The two stiles can be brought close together, or can be merged into a single, narrow, solid mullion (Figure 7.12).

7.21 Plastic-Covered Windows

a The wood parts of wood windows are good insulators against heat loss because wood has low heat conductivity, but wood shrinks and swells with changes in moisture content, and it must normally be kept painted. Metal windows, unless thermal breaks are introduced, transmit heat readily, are likely to be cold on the inside, and may condense moisture on the metal surfaces. Steel usually requires repainting from time to time and aluminum is subject to attack in some atmospheres. To get around

Windows

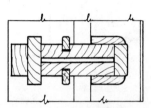 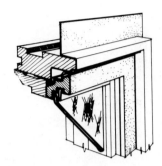

Figure 7.11 Mullion, Double-Hung Window.
Figure 7.12 Plastic-Covered Wood Window. PVC Covers Sash and Exposed Parts of Frame, Provides a Lip for Attachment to Wood Frame and Sheathing, and Flashing at the Head. Double Insulating Glass is Held by Plastic Glazing Gasket.

these limitations, some windows are made with a plastic overlay. A wood window with a vinyl (PVC) overlay is shown in Figure 7.12. Sash rails and stiles are covered with the vinyl, and the exterior exposed parts of the frame are similarly protected. With such complete coverage, changes in moisture content and, therefore, swelling and shrinkage, are minimized. Repainting is unnecessary. The good thermal characteristics, strength, and stiffness of wood are retained. The plastic cover must be carefully formulated and fabricated to remain snugly attached to the wood with changes in temperature because the plastic has a higher coefficient of thermal expansion than the wood, and may tend to wrinkle when hot, or crack when cold, unless properly designed.

b Plastic covers are also employed with metal windows, especially steel. The metal core may, in fact, be a strong spine over which the plastic is extruded to the desired configuration. All-plastic windows are also manufactured, in which there is no wood or metal core. At this writing, these are more common abroad.

7.22 Metal Double-Hung Windows

a A metal plus plastic (aluminum and PVC) double-hung window is shown in Figure 7.13. The essential features are the same as the wood window, with head, jamb, meeting rails, and sill. Because extruded metal and plastic are used, details are in general smaller and finer than in wood, but the functions are similar.

b To avoid rapid conduction of heat outward through the metal, with consequent chilling and condensation, thermal breaks are introduced

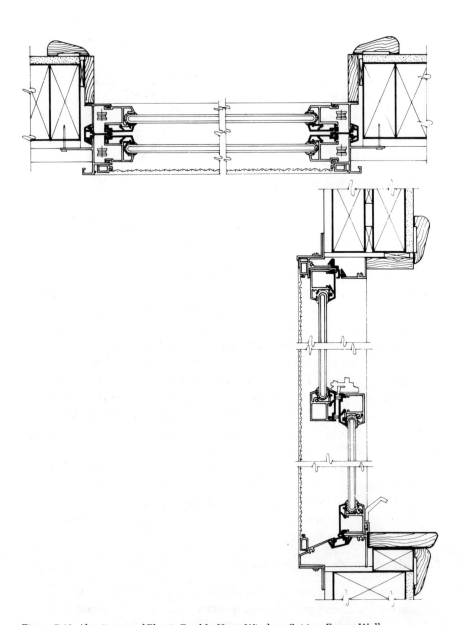

Figure 7.13 Aluminum and Plastic Double-Hung Window, Set in a Frame Wall.

Windows

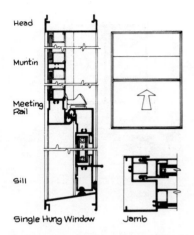

Figure 7.14 Aluminum Single-Hung Window.

by making parts of the sash and frame of rigid vinyl, interlocking with the aluminum. An insect screen fits into rabbets and a step in the sill. At the head, an upstanding lip provides a means of fastening the window to the house frame and acts as flashing over which the exterior finish may be applied. A similar detail exists at the jambs and sill.

c Metal parts may be exposed or have a baked finish. Colors can be matched with the extruded vinyl parts. Interior wood casing conceals the wood framing and edge of the interior wall surface.

d Glass in this instance is double insulating glass (Section 7.28) set in a preformed gasket rather than putty.

7.23 Metal Single-Hung Window

a A metal single-hung window is shown in Figure 7.14. In this particular instance, the upper sash is stationary and the lower one moves. The upper sash is single-glazed; the lower double-glazed. Spring balances are employed. Weatherstripping is a combination of woven pile and compressible vinyl.

7.24 Horizontally Rolling Window

a A horizontally rolling window is shown in Figure 7.15. Many details are similar to those of Figure 7.13, with a combination of aluminum and vinyl extrusions to provide a thermal break. The major difference is that

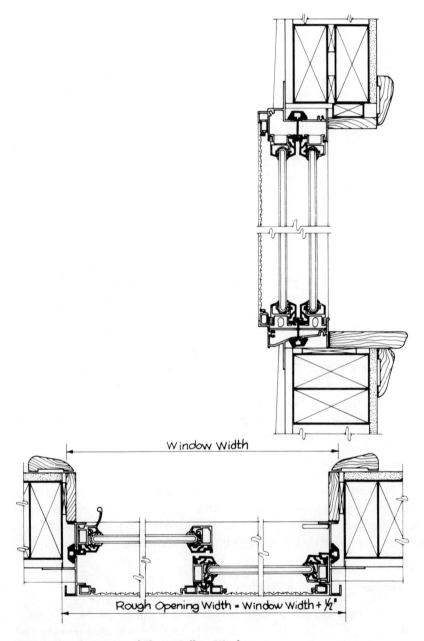

Figure 7.15 Aluminum and Plastic Rolling Window.

Windows

sash roll horizontally instead of moving vertically and sash balances are not needed. Evidently, combinations of moving and fixed sash can be employed.

b A horizontally rolling aluminum window in a masonry wall with brick facing is shown in Figure 7.16. The masonry lintel is carried across the top by steel angle supports. A stepped concrete sill is provided at the bottom. Special anchorages are provided for fastening to the wall. The window jambs and head provide tracks for the rolling sash which contain gasket-mounted double insulating glass (see Section 7.28) The window sill is set in caulking compound where it meets the concrete sill. Metal interior casing and stool meet the interior wall finish.

c By extending the window down to the floor, horizontally rolling doors can be provided which function in much the same manner as the windows shown. Because doors are customarily larger than windows, and traffic damage is more likely, frames, sash, and tracks are generally heavier than for windows.

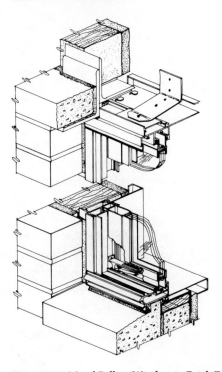

Figure 7.16 Metal Rolling Window in Brick-Faced Masonry Wall.

7.25 Projected Window

a A projected metal window is shown in Figure 7.17. In this instance, the sash and frame parts are solid sections rather than the tubular extrusions shown in Figures 7.13, 14, 15. Single glazing set in glazing compound is shown, but double glazing evidently could be employed. Extending lips of the metal frames can be set in wood or masonry.

7.26 Rubber Gasket

a Fixed lights may be set in heavy rubber gaskets, instead of frame and sash. This is not common in dwelling house construction, but the essentials of such an installation are shown in Figure 7.18. The rubber engages a lip extending from the edge of the window opening. It also engages the glazing. After the glazing has been inserted by prying the gasket open, a "zipper" locking strip is forced into the locking groove to hold the glazing firmly, to lock the attachment to the window opening lip, and to provide watertight seals.

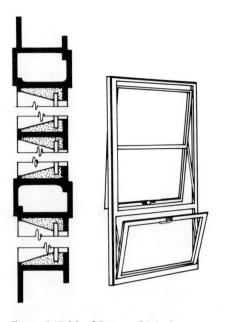
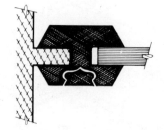

Figure 7.17 Metal Projected Window.
Figure 7.18 Rubber Gasket Fixed-Window Frame.

Windows

Table 7.2 Infiltration through Double-hung Windows, Unlocked on Windward Side

Type of Double-Hung Window	Infiltration (cu ft/min), per Linear Foot of Crack										
	5[a]		10		15		20		25		30	
	No W-Strip	W-Strip	No W-Strip	W-Strip	No W-Strip	W-Strip	No W-Strip	W-Strip	No W-Strip	W-Strip	No W-Strip	W-Strip
Wood Sash												
Average Window	0.12	0.07	0.35	0.22	0.65	0.40	0.98	0.60	1.33	0.82	1.73	1.05
Poorly Fitted Window	0.45	0.10	1.15	0.32	1.85	0.57	2.60	0.85	3.30	1.18	4.20	1.53
Poorly Fitted but with Storm Sash	0.23	0.05	0.57	0.16	0.93	0.29	1.30	0.43	1.60	0.59	2.10	0.76
Metal Sash	0.33	0.10	0.78	0.32	1.23	0.53	1.73	0.77	2.3	1.00	2.8	1.27

[a] Boldface numbers are wind velocities (mph).
Source: ASHAE Tests

Table 7.3 Infiltration through Casement-Type Windows on Windward Side

Type of Casement Window and Typical Crack Size		Infiltration (cu ft/min), per Linear Foot of Crack				
		5[a]	10	15	20	25	30
Rolled Section—Steel Sash						
Architectural Projected	$\frac{1}{32}''$ crack	0.25	0.60	1.03	1.43	1.86	2.3
Architectural Projected	$\frac{3}{64}''$ crack	0.33	0.87	1.47	1.93	2.5	3.0
Residential Casement	$\frac{1}{64}''$ crack	0.10	0.30	0.55	0.78	1.00	1.23
Residential Casement	$\frac{1}{32}''$ crack	0.23	0.53	0.87	1.27	1.67	2.10
Hollow Metal—Vertically Pivoted		0.50	1.46	2.40	3.10	3.70	4.00

[a] Boldface numbers are wind velocities (mph).
Source: ASHAE Tests

7.27 Weatherstripping

a All windows allow some air to filter past them when there is a breeze blowing, but the amount of infiltration depends upon how tightly the windows have been installed. If the sash fit well the leakage is very much less than when they are loose. To reduce infiltration, a number of systems of weatherstripping have been devised to provide a tortuous path and various blocks for the air to traverse, and thereby to reduce the leakage.
b Practically all systems rely upon shaped, thin metal strips which fit tightly against sash and frame or against other metal strips. The simplest consist of thin brass or zinc strips tacked to the frame and bent out to bear against the sash. Because of their inherent springiness they shut off the crack between sash and frame. Double-hung windows have a piece of this strip tacked against one meeting rail so that it bears against the other.

c Somewhat more complex types call for grooves to be cut in the rails and stiles. The stripping consists of U-shaped strips of metal fitted snugly into the grooves and tacked to the frame. Meeting rails are provided with two-member strips, half tacked to each rail, so formed that the two halves engage each other when the window is closed (Figure 7.8).
d Weatherstripping of metal windows is accomplished with strips of stainless steel, soft extruded vinyl plastic, or woven mohair wool pile (generally silicone treated) attached to one or both of the meeting members. Extruded vinyl plastics and treated woven pile are also used with wood windows.
e For comfort and to reduce heat losses, it is advisable to reduce air infiltration, particularly on the exposed sides of the house and in rooms which are used most for living purposes. On the other hand, some infiltration is desirable to provide the fresh air necessary for health and comfort. Furthermore, fireplaces and other heating units require sufficient air for combustion and draft, and it must come from the inside of the house. If doors and windows are too tight, other means of ventilation must be provided or combustion is unsatisfactory.
f Tables 7.2, 3 affords a comparison of infiltration through unweatherstripped windows and weatherstripped.

7.28 Multiple Glazing

a Heat loss through and condensation on glass are serious items, especially condensation in humidified houses (Chapter 11).
b The only practicable way to insulate glass areas against heat loss by conduction is by several panes with air spaces between. Whereas a single thickness has a coefficient of heat transmission (Chapter 11) equal to approximately 1.13, two sheets lower this to approximately 0.60, and three sheets to approximately 0.45.
c There are various ways of installing the extra sheet or sheets of glass. The oldest is the winter window or storm sash, fitting into the screen space, put up in winter and taken down in summer (Figure 7.19a). This installation has the advantage that it not only insulates the glass but breaks the force of the wind, and when snugly fitted largely eliminates the need for weatherstripping. The disadvantages are the necessity for putting on and taking off the sash, and the frequent inconvenience of the installation. Various combination screen and sash installations are available in which

Windows

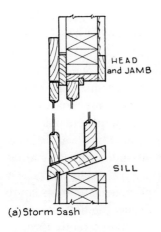
(a) Storm Sash

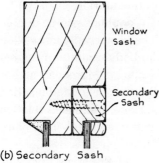
(b) Secondary Sash

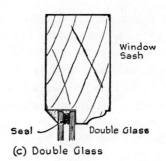
(c) Double Glass

Figure 7.19 Multiple Glazing.

sash and screens interchangeable and small in size, fit into permanent frames installed in the window frame.

d A second method consists of small secondary sash which fit into rabbeted spaces in the regular window sash and are screwed, clipped, or otherwise fastened in place. These can be removed in summer but may be left in place the year around, particularly in houses which have cooling systems for hot weather (Figure 7.19b).

e A third method calls for special double glass made up as a unit. Two sheets are sealed together with strips of sealing material at their edges and an air space approximately $\frac{1}{8}''-\frac{1}{4}''$ thick is left between the sheets. This space is filled with clean dry air just before the edges are sealed. Double-glass panes are set in the sash in the same way as single lights and have the same appearance. They have almost as high insulating value as double sash, and condensation and dust are excluded from the interglass space as long as the seal around the edge remains unbroken (Figure 7.19c). See also Figures 7.9, 10, 12–16.

f Neither double sash nor double glass has any effect upon infiltration around windows, and weatherstripping is required to reduce that source of heat loss. On the other hand, secondary sash are easily removed and cleaned if necessary, and double glass simply remains fixed in place.

g Radiant heat from the sun is not effectively stopped by several sheets of glass. To provide protection against the bright sun, large glass areas must be so oriented and protected by overhangs as to avoid its direct rays in hot weather. The best orientation in the northern hemisphere is south because in summer the sun is nearly overhead during the hottest part of the day, but in winter it is low in the sky and can shine directly through the glass, thereby contributing to the heating of the interior.

h With double glazing, whether by storm sash, double sash, or double glass, the temperature of the inner glass surface is markedly increased. This greatly promotes comfort inasmuch as the body radiation from occupants to cold glass surfaces is considerably reduced. Moreover, higher relative humidities are possible in the building because condensation does not occur readily on the warmer glass surface. Increased relative humidities in otherwise dry interiors promote comfort and probably health in addition to being better for woodwork and furnishings.

i Special types of glass are made to absorb part of the sun's rays and thereby to reduce the amount of radiant energy. These high-iron heat-absorbing glasses absorb most strongly in the invisible infra-red region which constitutes about half of the total radiated solar energy, but they also absorb some of the visible red spectrum, and therefore have a slightly bluish cast. The absorbed solar energy causes the glass to rise in temperature and to re-radiate both inward and outward. The outward portion, constituting about 20 to 25 percent of the incident energy, is excluded from the building. Other glasses are tinted so as to absorb and exclude some of the solar energy in much the same way. Still other glass has an extremely thin mirrored surface which reflects part of the solar energy but still allows a considerable percentage of visible light to pass through. The reflected energy is not absorbed by the glass.

7.29 Skylights

a Dome-shaped acrylic (Chapter 16) skylights provide lighting from above

Windows

Figure 7.20 Bubble-Shaped Skylight.

in place of or supplemental to windows. A typical installation is shown in Figure 7.20. The acrylic dome is vacuum thermoformed to the shape shown. A curb is built up from the roof high enough to prevent rain, snow, or slush washing into the skylight. The dome is clamped down by means of the interior and exterior metal sections shown, with a neoprene-cork gasket to act as a cushion and to permit the dome to expand and contract differentially with respect to the metal. If the dome is shaped properly, condensation forming on the inner surface runs down to the gutter and out through the weep holes.

8 Roofing and Flashing

ROOFING

8.1 General

a House roofs are generally shingled, using the term broadly to include wood, asphalt, slate, asbestos, metal, and tile. Roll roofing, membrane, and metal sheet are infrequently found.

8.2 Wood Shingles

a Wood shingles were formerly split by hand, small units so cut being called shingles and larger pieces, shakes.

b Shingles are now sawn and are usually red or white cedar or redwood. Because the wood is sawn rather than split, the wood cells at the surface are cut open, probably making it easier for water to penetrate into the interior and cause disintegration than in the case of split shingles. To reduce costs and still obtain the advantages of split surfaces, some shakes are made by splitting double-thick pieces and cutting them along the center on a diagonal to form two shakes, each having one split and one sawn side.

8.3 Specification Clauses

a All roof boards shall be 1″ x 4″ square-edge boards, spaced the same distance on centers as the shingles are laid to the weather. Side walls shall be ½″ plywood or sheathing boards laid solid.

SHEATHING PAPER

b Walls sheathed with boards shall be covered with rosin-sized sheathing paper weighing at least 20 pounds per roll of 500 square feet. Joints shall lap 6″.

SHINGLES

c All roof surfaces and all exterior side walls shall be covered with Number 1, 18″ red cedar shingles for all outer courses, and Number 4, 18″ red cedar shingles for under courses where shingles are doubled.

WEATHER, DOUBLING, SPACING

d Shingles shall be laid 5½″ to the weather on roofs and 8″ to the weather on side walls. Side wall courses shall be doubled, and roof shingles shall be doubled at the eaves. Roof shingles shall be spaced apart ¼″ to ⅜″; side wall shingles ⅛″ to ¼″. Joints in the shingles in any one course shall be at least 1½″ away from joints in the two courses next below.

Roofing and Flashing

NAILS

e Each shingle shall be nailed with two full-driven, hot-dipped, heavily zinc-coated 3d shingle nails, except that 5d nails shall be used at hips and ridges.

8.4 Grades and Sizes

a Four standard grades—Nos. 1, 2, 3, and 4—are commonly recognized and three standard lengths—16", 18", and 24"—are commonly manufactured.

b Number 1 shingles are all heartwood vertical or edge grain, and are entirely "clear," that is, free of defects. Number 3 shingles are allowed to contain defects above the standard weather exposure and need not be vertical or edge grain. Number 2 shingles are intermediate between the other two. Number 4 shingles are used for undercoursing or for interiors where their appearance is wanted.

c Number 1 shingles should be employed for roofs in all best quality work. Number 2 may well be used for good but not top quality. Number 3 are useful for economy and secondary buildings.

d Nos. 1, 2, and 3 may be trimmed with exactly parallel edges and butt at exactly right angles, as a special grade.

8.5 Laying (Figure 8.1)

a. Starting Course The starting course is doubled and projects beyond the lowest roof board at least 1½" to form a drip which forces rain water to fall clear of the roof. Along the rake (sloping edge) of a gable roof the end shingles project at least ½" for the same purpose. Joints in the second layer of the starting course and in each successive course must miss ("break") those in the preceding layer or course by at least 1½", otherwise rain water may find its way down through successive courses and eventually through the roof.

b. Succeeding Courses Succeeding courses are set back from the preceding ones a distance which depends upon the length of the shingle and upon the effect desired. Standard 16" and 18" shingles are usually exposed 4" to 5½" to the weather, the exact "weather" depending upon the pitch of the roof and upon the distance from eaves to ridge, which is divided into a full number of courses.

c Courses may be kept straight by laying shingles to a line. This may be done by laying the butts along a chalk line struck for every course, or by

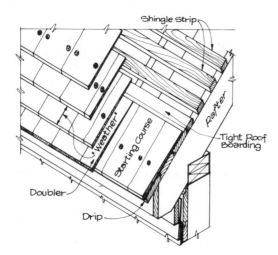

Figure 8.1 Wood Shingle Detail at Eaves.

tacking a straight edge (straight board) to the roof, and laying the shingle butts against this board.

d. Nailing Each shingle should be held with two nails (Chapter 14) driven not more than ¾" from the edge. Nails should be driven just flush with the shingle; driving too far tends to cause splitting, and projecting heads keep the shingles above from lying snugly against those below. The most commonly-used nails are heavily galvanized or aluminum flatheaded nails. Nails must be long enough to penetrate the several thicknesses of shingle and the thickness of the roofers. Three-penny nails are customary for 16" and 18"shingles, 4d for 24". Longer nails may be needed at ridges and hips (see below).

e. Spacing Shingles should be laid with ¼" space between adjacent shingles depending upon their width. These shingles shrink and swell as they dry or become wet and they must have room to move. Shingles on side walls (Chapter 10) are often fitted snugly to each other, but this practice would be positively harmful on a roof because swelling would cause the shingles to buckle and split.

f There are many variations in laying shingles. Sometimes every sixth or seventh course is doubled to give a heavier shadow line and break the plane of the roof, sometimes shingles are laid at random in no set courses at all, sometimes they are staggered so that alternate shingles project beyond their neighbors, sometimes the courses are made wavy to simulate

Roofing and Flashing

thatch and the rakes as well as the eaves of the roof are turned over in heavy curves to emphasize the effect of thatch. These are all merely variations and the basic principles remain the same.

8.6 Ridges and Hips (Figure 8.2)

a Some kind of closure detail is necessary at ridges and hips, and similar details are used at both.

b The simplest type of ridge is the saddle, flashed or unflashed. The top shingles are cut off at the ridge, and a pair of saddle boards is nailed along the two slopes of the ridge. If these are not flashed, the nails should be driven through an elastic cement and the nail heads completely covered with mastic. The better way is to cover the saddle boards with rust-resistant metal and seal the nail heads with solder or to drive the nails through lead washers (Figure 8.2a).

c Analogous to the saddle is the so-called Boston ridge (Figure 8.2b). The shingles are all the same width, and are laid with their long edges parallel to the ridge instead of perpendicular to it. They are lapped the same amount as on the rest of the roof. Alternate shingles lap over the ridge. A piece of flashing is laid under each pair of shingles to make this detail watertight.

d Because of their similarity, hips and ridges can be finished in much the same manner. Saddle boards can be carried down the two slopes of the hip and made tight in much the same manner as the saddle ridge. There is this difference, however. Saddle boards along the hip merely rest on the butts of the shingles and leave wedge-shaped openings which must be filled in with wooden wedges; otherwise the rain can drive in under the saddles. The same holds true of the Boston hip; otherwise it is the same as the Boston ridge (Figure 8.2c). To avoid the use of wedges, the Boston hip may be laid in the same manner as a slate hip (Section 8.14 and Figure 8.6d). This requires close fitting of the shingles and adequate flashing at the juncture of regular shingles and hip shingles.

8.7 Valleys

a Valleys are either "open" or "closed," that is, shingles are either carried right across the valley without a break or they are stopped a few inches short of the valley line, the intervening space being left open and protected by flashing.

Dwelling House Construction 198

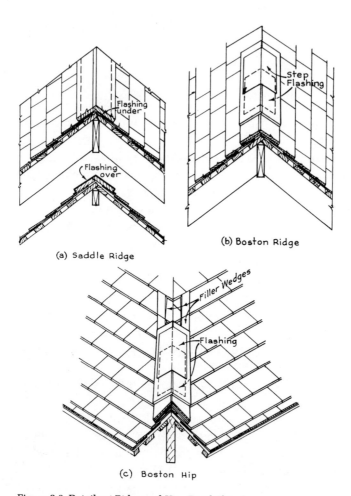

Figure 8.2 Details at Ridge and Hip, Rigid Shingles.

Roofing and Flashing

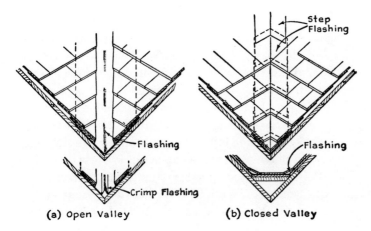

Figure 8.3 Open and Closed Valleys, Rigid Shingles.

b. Open Valleys (Figure 8.3a) Open valleys are much easier to build and to flash than closed valleys. A strip of flashing metal at least 16″ wide for pitches 6/12 or greater, and 20″ wide for lesser slopes, is placed in the valley and turned up on both sides. If the valley lies between two roofs of approximately equal area, the flashing is turned up the same amount on both sides, but if one roof is considerably smaller than the other, the flashing is turned up farther on the smaller roof because the greater volume and force of water coming down the larger roof is likely to force some of the water up the smaller roof. To prevent this, the valley metal is best crimped into an inverted V at the valley center to form a dam.

c With the flashing in place, the shingles are laid so that the exposed portion of the valley is 4″ to 6″ wide at the top. This distance can be increased by 1″ in 8′ toward the foot of the valley to provide a taper which facilitates drainage and minimizes troubles from ice and snow.

d. Closed Valleys (Figure 8.3b) In closed valleys the shingles are carried directly across the valley from one side to the other.

e Each course of shingles must be flashed separately with a sheet of flashing laid with its lower edge just above the butt line of the succeeding course, and carried at least 6″ up onto each valley slope. This dimension is increased on shallow roofs.

f. Shakes Standard lengths of shakes are 18″, 24″ and 32″. Handsplit and resawn shakes are first split and then resawn along a diagonal center line to provide two shakes, each with one split and one sawn face. Butt thicknesses vary from ½″ to ¾″ for the smaller, thinner varieties to ¾″ to 1¼″

or more for the larger ones. Other shakes are split to a taper, or are straight split. Depending upon the pitch of the roof, $\frac{4}{12}$ or steeper, the recommended exposures are $7\frac{1}{2}"$, 10" and 13" for 18", 24" and 32" long shakes respectively. Because shakes do not lie as snugly against each other as sawn shingles, a layer of 30-lb roofing paper is recommended with each course of shakes; the lower edge of the roofing paper held above the butt line at slightly more than double the exposure so that it will not show through the joints between shakes. Roofing felt can be omitted with straight-split and taper-split shakes in snow-free areas if exposures are less than $\frac{1}{3}$ the length of the shake. Other details are similar to those for sawn shingles, including ridges, hips, valleys and doubled or tripled starter courses. The first (under) course for the latter can be special 15" shakes made for the purpose.

8.8 Asphalt Shingles

a Because of their somewhat greater fire resistance and their economy, asphalt shingles have to a large extent taken the place of wood shingles. Many municipalities have legislated against wood shingles because of the fire hazard, but they permit asphalt shingles. The most fire-resistant grades fall in Class A categories; others, mainly in Class C.

b Asphalt shingles are made of asphalt-impregnated felt covered with a layer of colored ceramic granules. The quality and life of the shingle depends upon the quality of the felt and asphalt, and upon the weight of the shingle. Weights range from approximately 350 lb per square (100 sq ft) to approximately 230 lb per square. Although the standard asphalt "shingle" originally was a rectangle 9" x 12", asphalt shingles are applied in strips, usually 36" long and 12" wide from butt to top. These may be embossed to give the appearance of random-width individual shingles, or they may be slotted every 12", with a half-slot at each end. In either case, when the successive courses of shingles are laid so as to cover the ends of the embossed depressions or the slots, the final appearance is of individual shingles.

c Because ordinary asphalt shingles, unlike wood shingles, do not taper to a heavy butt and are fairly thin, they do not give a pronounced shadow line and the roof may consequently be rather uninteresting in appearance. To overcome this drawback, heavy-grade asphalt shingles are sometimes double-coated at the butt end in a rather random pattern to provide horizontal and vertical shadows.

Roofing and Flashing

d To keep shingles from being raised and blown off in high winds, it is customary to provide spots of a thermoplastic mastic-based adhesive along the line which will be just under the butt of the next course above. When the hot sun shines on the roof, these spots soften and adhere to the butts of the succeeding course, bonding it and preventing it from being raised by wind.

e Many of the principles governing the laying of wood shingles apply to asphalt shingles. There are some important differences, however.

8.9 Specification Clauses

ASPHALT SHINGLES

a All shingles upon arrival on the job shall be stored in a cool dry place. The bundles shall be stored with label side up and not over 5′ high when piled up.

b Cover the entire surface to be shingled with water-proof building paper lapped 2″ and securely nailed in place. Use a chalkline to get each course in a true straight line. Secure each strip with a nail at each end and 2 additional nails uniformly spaced 1″ above the butt of the next course. Drive all nails perpendicular to the surface of the roof.

c Lay all shingles 4″ to the weather. At the eaves, use sheet metal or wood shingle drip edge. Directly over this, lay a course of shingles with embossing or slots facing up the roof, then lay a course of shingles directly over the first one, breaking joints and with slots or embossing facing down toward eaves. Break all joints in successive courses one-third the full width of a shingle. No shingles shall be laid on the roof where the pitch is less than 4″ to the foot. Finish hips and ridges with single shingles or strips cut into single shingles with straight edges.

8.10 Laying (Figure 8.4)

a Asphalt shingles are relatively soft and will droop if allowed to project unsupported at the eaves. For this reason a wood shingle starting course or metal drip support is first laid (Figure 8.4a). The roofing felt is brought down over the wood shingles and the starting asphalt course is laid. Over plywood roof boarding, roofing felt is often omitted. The starting course is laid upside down so that the continuous upper edge of the strip lies along the edge of the eave. The first regular course then is laid with the slotted or embossed edge down. Succeeding courses are broken back one-half or one-third the width of shingle, or according to whatever pattern is wanted.

b The amount exposed to the weather can be varied but is approximately 4″ with standard-size shingles. These dimensions can be varied slightly to make the courses come out even at the ridge.

Dwelling House Construction 202

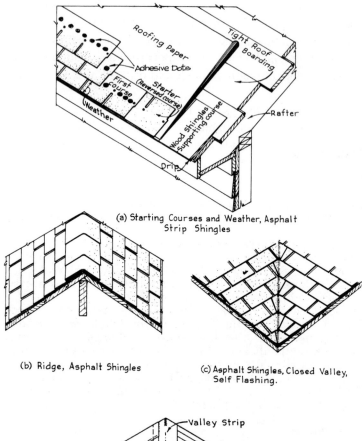

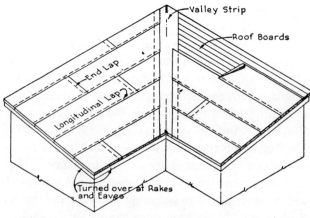

Figure 8.4 Details of Asphalt Shingles.
Figure 8.5 Sheet Roofing, Shallow-Pitched Roof.

Roofing and Flashing

c Shingles should be laid to chalk lines, although with a little care, strip shingles can be made self-aligning. One nail is driven at the end of each strip and two nails equally spaced between.

d In addition to the method of laying just described, there are numerous others, such as interlocking, which are variations of the standard method.

8.11 Ridges and Hips (Figure 8.4b)

a Ridge and hip details are almost identical with those employed for wood shingles except that the Boston hip and ridge are made by using single shingles instead of pairs, the one shingle simply being bent over the ridge or hip and nailed down on both sides. No flashing is necessary.

8.12 Valleys (Figure 8.3, 4c)

a Open valleys are the same as for wood shingles (Section 8.7, Figure 8.3), except that a strip of roll roofing (Section 8.13) may replace the metal flashing. Closed valleys on roofs covered with individual asphalt shingles are handled in much the same manner as wood shingles, but strip asphalt shingles frequently are run into the valley from one side and a short distance up the roof on the other, the procedure sometimes alternating from course to course as shown in Figure 8.4c. The shingles are thus self-flashing.

8.13 Sheet Roofing (Figure 8.5)

a Old-style roll roofing is the same material as asphalt shingles, but made up in rolls. It is simply applied by rolling it out flat, lapping the edges of adjacent bands $2''$ to $3''$ and the ends $6''$ to $8''$, sealing all laps with liquid roofers' cement, and nailing with flat headed shingle nails (Chapter 14) placed $2''$ to $3''$ on centers. At eaves and rakes, the edges are turned over the edges of the roof boards and nailed. Strips $9''$ to $12''$ wide are nailed over hips and in valleys. The roofing is merely turned over at ridges.

b Newer sheet roofing materials are based upon elastomeric polymeric materials such as butyl rubber and chlorinated polyethylene. White surfaces are available in addition to the usual black. Thicknesses range approximately from $\frac{1}{32}''$ to $\frac{3}{32}''$. The materials are proprietary and must be applied with manufacturers' adhesives, sealing tapes and other accessories. They may be applied on surfaces having any slope from horizontal to vertical. They may have a thin foamed plastic backing for resilience and some thermal insulating value, as well as to cover small irregularities in the substrate. With any thin adhered sheet, such irregularities are likely to show ("telegraph") through and the substrate should be smoothed by filling in joints and depressions and levelling projections. Large substrate sheets with minimum joints such as plywood or other boards are preferred to narrow strips such as roof boards for this reason.

8.14 Slate

a Slate is one of the best and most durable of all roofing materials. Whereas both wood and asphalt shingles have to be replaced from time to time, a good slate roof should outlast the building. Because of its laminar structure, slate can be split readily into thin sheets. For roofing purposes it ought to be at least $3/16''$ thick and often is much heavier.

b Slate used for roofing and other purposes comes mainly from Pennsylvania and nearby areas, and from Vermont, with some of the Vermont beds extending into New York. Pennsylvania slates are dark gray and blue gray permanent colors, but often show streaks of darker color, called ribbons, apparently the same composition as the rest of the slate.

c Vermont slates are green, purple, red, and mottled. Ribbons are not found in Vermont slate.

d Because of the weight, framing for a slate roof must be more substantial than for wood or asphalt, or for any other kind of roofing except tile. Roof boards are laid tight, instead of in strips as is customary for wood shingles. Roof should pitch not less than 6" per foot.

e Before slate are laid, the roof is covered with 30-lb asphalt-impregnated slater's felt, lapped at the edges and ends 2" to 3" and nailed with large-headed shingle nails (Chapter 14) or metal discs.

f Slate must be provided with nail holes. This should be done at the factory, where holes can be drilled and countersunk. Attempting to punch holes on the job, especially if a special punching jig is not available, is apt to cause fine cracks to radiate from the holes. After the slate is in place, cracks are likely to work entirely across it, causing the lower portion to come loose, fall, and leave an unsightly leaky spot, very difficult to repair.

8.15 Laying (Figure 8.6)

a Slate are laid in much the same manner as wood shingles, with some important differences. At the eaves (Figure 8.6a) there is a cant strip thick

Roofing and Flashing

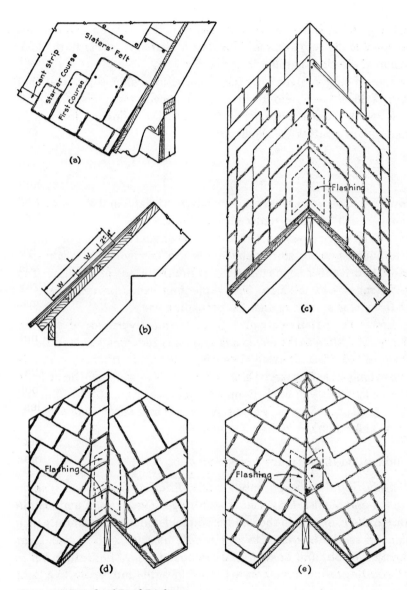

Figure 8.6 Details of Rigid Roofing.

enough to give the starter course the same slant as subsequent courses with respect to the roof surface. The starter course is laid with the long dimension of the slate parallel to the eaves.

b The second course above any particular course should lap over that course by at least 2″–3″, and the first course above should lap half the remaining length (Figure 8.6b). From this rule, the following formula to determine the "weather" is deduced:

$$W = \frac{L - x}{2}$$

where W = exposure to weather, L = length of slate, and x = length of head lap—2″ to 3″ on steep roofs, 4″ on shallow slopes.

c To match the permanence of the slate, nails should be copper or copper alloy, although heavily zinc-coated nails are common. Care must be exercised to drive nail-heads just flush with the surface of the slate. Too heavy driving cracks the slate, and projecting heads prevent the next course from lying snugly against the preceding one.

d On large roofs, various sizes of slate are frequently employed, starting with large thick slate at the eaves and gradually reducing size and thickness toward the ridge. Starting slate are, in this case, often as much as 2′ or more long and 1″ or more thick, whereas ridge slate may be as little as 10″ long and only $3/16$″ thick. Standard sizes vary from 26″ x 14″ to 10″ x 6″. This kind of roof not only gives a variation in texture, it also causes the roof to look much larger than it actually is.

8.16 Ridges and Hips (Figure 8.6c, d, e)

a Ridge and hip details are only slightly different from wood shingle (Figure 8.2 and Section 8.6). The Boston ridge (Figure 8.6c) and saddle ridge are quite common. In the saddle, slat laid end to end generally take the place of saddle boards, although saddle boards can be used. Tight (see Section 8.16b) and Boston hips are both found. The Boston hip is made by cutting the course slates to fit the hip slates and results in a tight joint, provided that plenty of cement is used under the slates at that point and that it is well flashed (Figure 8.6d).

b A hip detail sometimes found on steep roofs is the "close" or "tight" hip (Figure 8.6e). Slate courses are laid to the hip and the end slates are merely mitered to the slope of the hip. Flashing and liberal quantities of elastic cement are necessary to make this detail watertight.

Roofing and Flashing

8.17 Valleys

a Both open and closed valleys are handled in the same manner as wood shingles (Section 8.7, Figure 8.3a, b). The transition from one side of a closed valley to the other is sometimes made more gradual by laying a secondary roof board in the valley as shown in the lower detail of Figure 8.3b. Slate are cut to fit into the trough of the valley and make the change from one side to the other without a sharp break.

8.18 Asbestos Shingles

a Asbestos shingles are a mixture of portland cement and asbestos fibers formed under high pressure into the desired shapes. They are factory punched or drilled for nails and are similar to slate in application.

8.19 Laying

a The ordinary "American Style" asbestos shingle is an individual rectangular shingle much the same size as slate and is applied in almost exactly the same way. It has the same starting details, the same coursing rule, and the same precautions regarding plenty of cement at joints such as at ridges, hips, and adjoining walls. Most manufacturers supply specially formed half-cylindrical pieces which are used for ridge and hip rolls, although the usual Boston, tight, or saddle details can be employed equally well.

b In order to reduce the amount of material required to cover a given roof area, makers of asbestos shingles have devised other shapes and methods of application which do away with the large amount of concealed shingle which is inherent in the "American" method of laying shingles. Such variations are known as Diamond, Hexagonal, Dutch Lap, and so forth. All are characterized by the fact that only a small part of each shingle is covered by the lapping over of other shingles, so that economy is achieved in the amount of material required to cover a given area. Special clips are required to hold the exposed edges of shingles down so that no water can drive up under the small lap and penetrate to the roof boards. Care must be exercised in laying to align these shingles properly and to take fullest advantage of the small amount of lap provided.

8.20 Tile

a Tile are the aristocrats of roofing materials. They are heavier, need stouter framing, and require more care in their laying than do other roof coverings. All kinds require a tightly sheathed roof covered by the best quality 30 to 40 lb roofing felt. At important points such as hips, valleys, ridges, and ōther breaks, and on low-pitched roofs, the felt must be doubled. On top of this it may be necessary to install horizontal wood roofing strips or strips running from eave to ridge, depending upon the kinds of tile employed. Some combination of cant strip and roofing strip is required with almost every kind of tile. Nails are copper or other durable metal to match the longevity of the tile.

b Although most tile are made of burnt clay and therefore have its familiar red color, some are of concrete, and colors depend upon the pigments added. Clay tile are also frequently colored, green being a favorite hue. When glazed, clay tile may be almost any color and usually are quite brilliant.

c. Shingle Tile (Figure 8.7) The simplest kind of tile is a flat rectangular block which is applied to the roof in a manner similar to slate. Tile are nailed near their top edges, and nail holes must be formed during manufacture. Copper nails are used, their length depending upon the thickness of the tile, which varies from $3/8''$ to $1''$ or more.

d. Interlocking (Figure 8.7) For the same reason that asbestos shingles are made in various shapes to provide the greatest coverage with minimum material, shingle tile are also made in a variety of interlocking shapes designed to cover the roof with a small amount of head and side lap. The interlocking designs are usually so made that a small gutter is formed at the side to carry off water which works down between tile, and an upward projecting lip at the head of the tile provides both a bulkhead against water driven upward by high winds and a key to hold two adjacent courses together. A special under-eaves piece or wood cant strip is needed at the eaves. A similar closure tile is needed at the rake. Additional stripping is needed at the ridge to support the special ridge roll which covers that point. Another special piece is needed to cover the hip. At hips and valleys special triangular pieces, made to fit the pitch of the roof, are required to make the joint.

e. French Tile (Figure 8.7) A variation of the flat shingle tile is the French tile, which is heavily grooved on top and has corresponding projecting lugs on the lower side which fit into the grooves of the next lower course. These tile interlock and thus provide maximum coverage with minimum material. Special return pieces are required at the rakes, special hip and

Roofing and Flashing

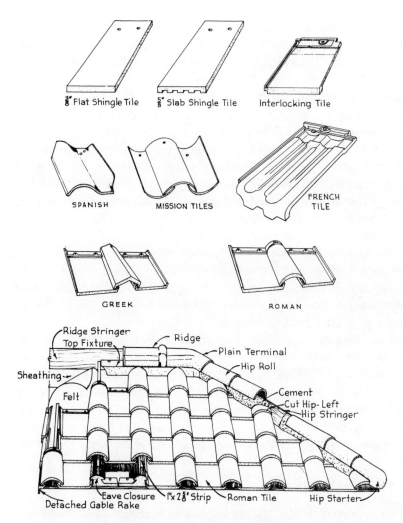

Figure 8.7 Tile Roofs: Shingle, Interlocking, Spanish, Mission, French, Greek, and Roman. Details of Roman Tile Roof Are Shown; Other Tile Roofs Are Similar.

ridge rolls are needed, hip and valley tile must be made to a triangle depending upon the pitch of the roof, and special wood strips or stringers are called for under hip and ridge rolls. Cement is used plentifully at these points.

h. Mission and Spanish (Figure 8.7) These are the familiar curved tiles frequently seen on houses showing tropical architectural influence. Of the two the mission is the simpler inasmuch as it is merely a curved piece. The roof is covered by turning adjacent tile alternately concave ("pans") and convex ("covers"), the covers fitting down over the upturned edges of the pans. Both are nailed with a single nail at the upper end. The convex tile rest on 1″ x 4″ wood strips set on edge, running from eave to ridge. At the eaves, special closure pieces cover the opening which would otherwise occur under the convex tile. The rake of the roof is usually covered with a tile turned down over the edge, but special rake pieces are also employed. Ridge and hip require special rolls, and special tile are needed at hips and valleys. Spanish tile differ from the mission type in that the convex and concave parts are formed in one S-shaped tile. Details of laying are otherwise much the same.

j. Greek and Roman (Figure 8.7) These are both combinations of flat tile with upstanding outer edges and curved or angular covers. Covers are tapered so that succeeding courses fit snugly, and edges of flat tile are tapered to fit. Special eave closures, top or ridge fixtures, ridge and hip roll, gable rake closures, and triangular hip and valley tile are required, in addition to the various wood stringers which support the covers and rolls. Details are shown in Figure 8.7. Details for other types of tiles are similar.

8.21 Liquid-Applied Roofing

a Liquid-applied roofing systems based upon synthetic elastomeric rubbers are available, e.g., roofing composed of neoprene-hypalon rubber, and silicone rubber. Both make light colors possible.

b As is true of any thin surfacing materials, the substrates must be carefully smoothed or imperfections will show through. Plywood or similar boards should have the joints filled with cement and smoothed; taped joints show through. An overall layer of glass mat helps to conceal joints, although not entirely. Surfaces must be clean, dust-free and dry and, with most materials, if light colors are to be used, there must be no trace of tar or asphalt.

c For neoprene, a primer is required on surfaces such as plywood and insulating board. This is followed by a coat of neoprene in turn covered by hypalon for light colors. Total dry thickness is usually 16–28 mils, depending upon whether mat is or is not used.

d Silicone rubber application is similar. A primer is first applied, followed by the catalyzed silicone liquid, which must be used immediately once it is mixed. After it has cured (about 12–18 hours) a second coat is applied.

e For surface texture, small ceramic granules may be embedded (e.g., by using low-pressure sand-blasting equipment) in the surface coat before it has hardened.

8.22 Built-Up Roofing (Figure 8.8)

a Built-up roofing is commonly found on "flat" roofed commercial and similar buildings, but may also be employed on shallow-pitched or flat roofs of dwellings.

b In essence, it consists of layers of an asphaltic or bituminous roofing cement in which are embedded sheets of roofing felt as shown in Figure 8.8. Depending upon the number of plies wanted, successive layers of felt are offset from the one below by $\frac{1}{2}$, $\frac{1}{3}$, $\frac{1}{4}$ the width of the roll (2-ply, 3-ply, 4-ply, etc.). If the substrate is wood, such as roof boards, a first layer of rosin-sized sheathing paper is commonly employed, and the felt is nailed to the deck.

c For appearance and for reflection of the sun's rays, a layer of light-colored granules of stone or ceramic chips is embedded in the top or flood coat of asphalt while it is still liquid.

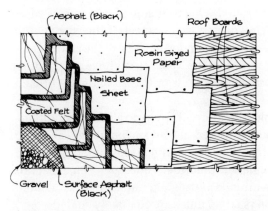

Figure 8.8 Built-Up Roofing. As Shown, Roof Boards Are Covered With Rosin-Sized Paper. Nailed Base Sheet, Sheets of Roofing Felt Are Embedded in Roofing Cement and Topped With Granules Embedded in Top Coat.

FLASHING

8.23 Specification Clauses

FLASHING

a Flash and counter-flash all usual and required places with clean roofing-temper 16-ounce copper, using sufficient metal for perfect tightness.
b Chimney flashing shall be through flashing, turned up 1″ between flue lining and brickwork. Where roofs strike brick walls flashing shall be carried up on walls 6″, and thoroughly counter-flashed.
c Flash all rakes over edges of roof boarding, starting copper 2″ in from upper edges of roof boarding, bending around and under edges and back to brick, and down onto brickwork 1″ with hemmed edge, properly secured to under side of roof boarding.

8.24 General

a Wherever joints occur in which dissimilar materials come together or where the possibility of leakage arises because of the type of construction, it is usually necessary to insert flashing—sheets or membranes of waterproof materials—to turn back the water.
b Flashing is usually metal, but may be heavy impregnated fabric or felt, or sheet plastic. Metals most commonly used are copper, aluminum, stainless steel, monel, terne, zinc, lead, leaded copper, and galvanized iron.
c Care must be taken to keep dissimilar metals from being in contact because water acts an an electrolyte to set up electrolytic action similar to that occurring in a battery. A small electrical current is set up and the more active metal is corroded. If copper and aluminum, for example, are in contact, the aluminum is corroded. This principle of galvanic or sacrificial protection operates in zinc-coated iron or steel. If the zinc coating is broken, it continues to protect the iron underneath until the zinc is eaten away so far that it can no longer protect the iron. Tin coatings, on the other hand, may accelerate the corrosion of the iron if the coating is broken, because the iron now acts sacrificially.
d Some metals are susceptible to acidic and others to alkaline conditions. Iron and steel, for example, are corroded in acid surroundings, and even stainless steel should be protected with a coating if in contact with concrete or mortar containing calcium chloride. Aluminum, on the other hand, corrodes in alkaline conditions and should not be in direct contact with concrete, mortar, or lime plasters unless protectively coated.
e No matter what the material may be, the details of flashing are prac-

Roofing and Flashing 213

tically identical. This discussion is based upon the use of copper, keeping in mind that other materials are handled in much the same way.

f When using metal exposed to the changes in temperature found on building exteriors, the first point to be remembered is that metals have fairly high coefficients of thermal expansion, and that they must be free to move with changes in temperature. If they are confined, the stresses set up in trying to change dimensions eventually cause fatigue cracks. Short narrow strips, less than 10″ to 12″ wide, can be fastened along both edges without much danger; wider sheets must not be confined along two opposite edges, but must be free to move. Coefficients of expension of some common materials are given in Table 8.1.

g The second point to remember is that flashing must be installed in such a way that water is shed over any unsealed joints in it, i.e., joints must be so made that water could work through them only against the force of gravity. A further qualification is that it must be impossible for driving winds to force water through. This usually requires that the joint either provide a tortuous path in which the driving force of the wind is dissipated, or that the laps in the joint be so long that water cannot possibly be driven through.

h Copper flashing is generally 16-oz "roofing temper" (R.T.), i.e., it weighs 16 oz per sq ft and is soft and pliable. Heavier copper is required for heavy-duty flashing such as around certain kinds of tile roofs, whereas lighter (not less than 14-oz) can be used in relatively protected points. Table

Table 8.1 Coefficients of Expansion (Approximate)

Material	Coefficient of Expansion 10^{-6} in./in./ °F
Wood (Fir), parallel grain	2.1
Brick masonry	3.3
Sheet glass	4.7
Concrete, standard structural	5.5
Galvanized steel (Type 1008)	6.5
Terne (Type 1008)	6.5
Monel (Type 400)	7.5
Copper (Type 100)	9.3
Stainless steel (Type 304)	9.6
Aluminum (Type 3003)	12.8
Hard lead	15.1

8.2 gives thicknesses of some other metals for comparable applications, together with a few other flashing materials.

8.25 Chimney Flashing (Figure 8.9)

a Where chimneys penetrate roofs, the juncture must be flashed carefully or leaks are almost sure to occur. Flashing consists of cap and base flashing, a detail common where roofs and vertical masonry surface intersect. The flashing is in two parts (Figure 8.9a), the lower part or base flashing fitted into the shingles, and the upper or cap flashing built into the brickwork. The downstanding portion of the cap flashing folds down to lap 4″ or more over the upstanding portion of the base flashing. One piece of base flashing is required for each course of shingles, and is laid with its lower edge just above the butt line of the succeeding course of shingles. It should extend onto the roof at least 4″, and 5″ or 6″ is better. Cap flashing is usually one brick (8″) or more wide, and its lower edge is cut on a diagonal to fit the pitch of the roof. Cap flashing must be built into the chimney as the brickwork is laid and should extend into the mortar joint a minimum of 3″, in fact, many specifications require it to be built in the full 4″ and to be turned up against the flue lining 1″ or 2″.

b At the lower edge of the chimney the base flashing becomes a single sheet or apron which rests on top of the shingles, and the cap flashing is a continuous piece which laps down over the base flashing.

c At the rear of the chimney, unless it straddles the ridge, a small gable

Table 8.2 Flashing Materials

Material	Thickness	Gauge or Weight
Copper	0.022	16 oz
Stainless steel (303)	0.018	26 ga.
Monel (400)	0.021	24 ga.
Aluminum (3003)	0.032	20 ga.
Galvanized steel (1¼-oz coating)	0.028	24 ga.
Terne (40-lb coating)	0.024	24 ga.
Plastic sheeting	0.010	11 oz/sq yd
	0.020	22 oz/sq yd
PVC, or other thermoplastic elastomeric formulation	0.030	33 oz/sq yd
	0.056	60 oz/sq yd

Note: Combination flashings also occur, e.g., copper plus fabric flashing. This is made of 2, 3, 5, or 7 oz copper plus asphalt-saturated fabric both sides. Other formulations employ combinations of copper, copper-lead, and facings of impregnated fabric, paper, glass fiber, and felt.

Roofing and Flashing

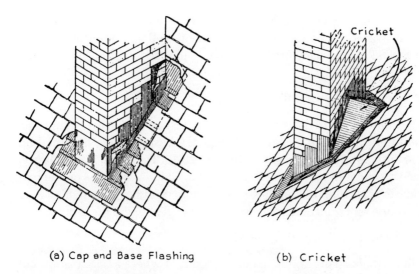

(a) Cap and Base Flashing (b) Cricket

Figure 8.9 Chimney Flashing.

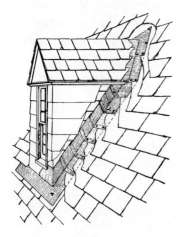

Figure 8.10 Dormer Flashing.

called a cricket should be built to keep snow and dirt from collecting. This is covered with a flashing metal carried well up under the shingles and turned up the back of the chimney under the cap flashing (Figure 8.9b).

8.26 Dormers (Figure 8.10)

a The intersection of dormers and roofs is flashed similarly to chimneys, except that cap flashing is omitted since the base flashing is turned up under the wall covering, as shown in Figure 8.10. The apron in front is also turned up under the wall covering, and if the window sill rests on the roof the apron is carried back under the sill where it is turned up behind it an inch to stop rain from being driven under the sill and into the interior.

8.27 Changes in Pitch (Figure 8.11)

a Where roofs change pitch (as in gambrels, or where a porch or ell roof adjoins a steeper roof), it is necessary to flash the break if the shingles cannot be bent to conform. Generally, the flashing is carried under the upper course of shingles and out on top of the lower course, but it may be completely concealed by doubling the lower course. In this instance, nails driven through the lower portion of the flashing should be embedded in lead washers to prevent leakage.

8.28 Vents (Figure 8.12)

a Special flashing is required where the upper ends of plumbing stacks extend through the roof, or where any other pipe protrudes. Flashing consists of a sheet of metal approximately 16" square in which is cut a circular opening. To the opening is soldered a collar the proper size to fit snugly around the pipe. This is slipped over the pipe, and the collar is brought over the top and down the inside. The sheet forms an apron resting on top of the shingles below the pipe and fitted under the shingles above.

8.29 Rakes (Figure 8.13)

a Thick inflexible roofing units such as heavy slate and tile leave large wedge-shaped openings between courses at the rakes of gable roofs, and these must be filled in either with liberal quantities of roofing cement, with special closure pieces, or with individual pieces of flashing turned down over the edge of the rake at each course. Flashing should extend in under the shingles at least 4" and be bedded in cement.

Roofing and Flashing

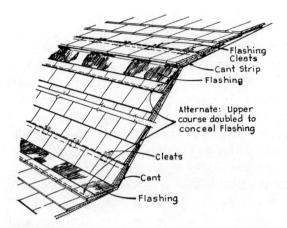

Figure 8.11 Flashing at Changes in Pitch of Gambrel Roof.

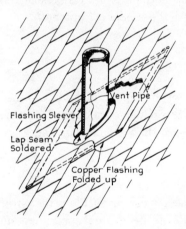

Figure 8.12 Flashing around Vent Pipe.

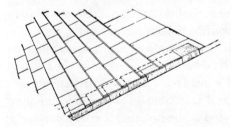

Figure 8.13 Flashing at Rake.

9 Cornices, Gutters, and Leaders

9.1 General

a Although cornices and gutters are here considered together, cornices in actuality are almost always built before roofing is applied, and gutters are usually not hung until painting is finished unless the gutter is an integral part of the cornice detail. As a matter of fact, the exact sequence in which windows, roofing, and cornices are installed depends very much upon their details. Roofing may not be applied until the house is practically finished, with windows, exterior finish, cornices, and gutters all in, and the interior well along. In such an event heavy roofing felt must be applied to keep the roof tight until the final roofing is in place, and be securely fastened and sealed against leakage at all joints and seams (Chapter 8).

b The usual sequence is: cornice, roofing, windows, exterior doors, and exterior finish. An exception occurs when exterior finish is brick veneer, which may have to be applied before the cornice is built, because the lower part of the cornice laps over the top of the veneer (Chapter 10).

CORNICES

9.2 General

a Cornices may be classified as open or closed, depending upon whether or not the ends of rafters are exposed to view. Open cornices are usually, but not necessarily, simpler in their details than are closed. There are many variations of both kinds.

9.3 Open

a. Eaves (Figure 9.1) Rafter ends are exposed at the eaves. Since the rafter ends are open to view, they should look well, be straight, and show no defects. Because the grade of material used for rafters is apt to have knots and other imperfections, rafter ends may be false and are cut off a short distance back of the wall line. The actual rafters terminate at the plate and are concealed by the closure boards. In addition to looking well, false rafters can be uniformly spaced along the length of the building, allowing the spacing of the actual rafters to be whatever is required by the framing.

b The outer ends of rafters usually are faced with a board known as a facia or fascia. To prevent wind and rain from sweeping into the attic, the spaces between rafters are sealed at the building line by a vertical or

Cornices, Gutters, and Leaders

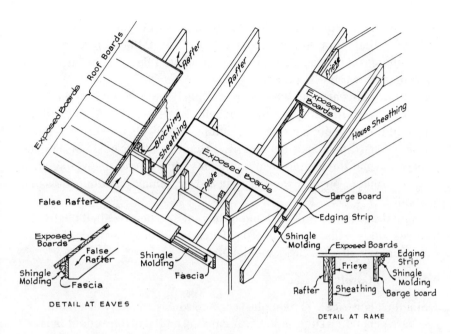

Figure 9.1 Open Cornice, Construction Details.

inclined closure board (frieze), or by a combination of closure board and molding to form an ornamental detail. To obtain tight and substantial construction, short blocks are nailed to the sides of the rafters to provide nailing and firm support for the closure boards.

c Since the roof boards are exposed underneath, they must be of good quality clear material. Better grade stock is therefore substituted for the ordinary roofers, and frequently this material is patterned. These boards are run up beyond the line of the house wall and the ordinary roofers begin there. Other exterior-type boards such as plywood may be employed.

d. **Rake** (Figure 9.1) If the roof is a gable, gambrel, or shed, substantially the same detail as at the eaves is repeated along the rake, except that no rafter ends protrude from the wall. An outer rafter, variously known as a "flying rafter," "verge board," and "barge board," is employed along the edge of the roof.

e The projecting ends of the roof boards are exposed to view and are therefore better stock than are ordinary roof boards. The latter are mostly cut off at the edge of the building but at frequent intervals are cut back to the first rafter behind the edge so that the patterned or other boards can continue back that far to support the barge board in cantilever fashion. Wide overhanging rakes may be carried on brackets fastened to the wall or framed as shown in Chapter 5.

f For better appearance, the ends of roofing boards are cut back an inch and a narrow edging strip is nailed to their ends the full length of the rake. The joint between edging strip and ceiling boards occurs directly above the barge board, or shingle molding, if a shingle mold is used. Roof shingles project beyond the edging strip approximately $\frac{1}{2}''$, and the edging strip in turn projects beyond the barge board or shingle molding the same amount.

g Along the building wall, just under the roof, is a board called the frieze. It corresponds to the closure board or frieze between rafter ends along the eaves and usually has the same details.

9.4 Closed Cornice

a. Simple (Figure 9.2) The simplest cornice consists merely of a plain fascia board or molding such as "crown" molding nailed to the ends of the rafters and carried down against the wall (Figure 9.2). The rafters are cut off at the outer edge of the plate or project beyond it only $1''$ or $2''$. The fascia or molding is wide enough to cover the ends of the rafters and may run past the lowest roof board to the under side of the shingles, or it may butt against the lower side of the lowermost roof board, which projects beyond it to form a drip. Where the molding bears against the wall there may or may not be a frieze board.

b. Boxed (Figure 9.3) This kind of cornice is also called "box" and "boxed rafter." Although there may be additional ornament, it consists essentially of three members. They are (a) fascia, a vertical piece fastened to the outer ends of the rafters, (b) plancia, a piece which fastens either directly to the lower edges of the rafters or to secondary horizontal pieces called "lookouts," and (c) frieze, a vertical piece which fastens to the building wall directly under the plancia. The fascia projects down beyond the outer end of the plancia, thus forming a drip, the plancia runs from fascia to building wall, and the frieze is brought up against the plancia to make a tight joint against the wall.

c When the gutter is fastened to the eaves it covers a large part of the fascia and the fascia is just a simple board carried up to the underside of

Cornices, Gutters, and Leaders 221

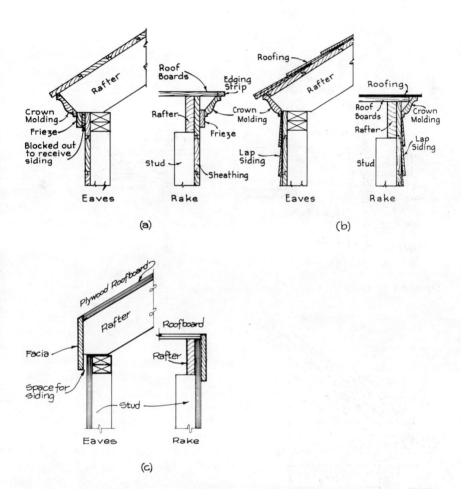

Figure 9.2 Simple Closed Cornice, Construction Details. (a) Molding Plus Frieze. (b) Molding Alone. (c) Flat.

Dwelling House Construction

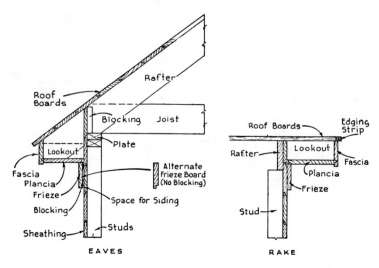

Figure 9.3 Simple Box Cornice, Construction Details.

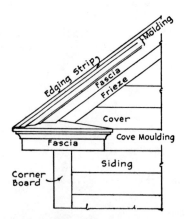

Figure 9.4 Return, Box Cornice.

the lowest roofing board. When the gutter is free-standing, a shingle molding may be run along the upper edge of the fascia and may either bear against the lower side of the lowest roofing board or run up beyond it to bear against the underside of the shingles.

d The plancia may be a single board if the cornice is not very wide or it may be several boards wide, or may be plywood or other exterior wallboard.

e Friezes may be simple, consisting of a piece of molding or narrow board, or they may be wide and ornate, made up of several boards of varying thickness with moldings of several kinds added to provide the proper details and shadow lines.

f Eaves and rakes are usually the same in detail but somewhat different in construction. Along the eaves, short horizontal lookouts run from rafter ends to house wall to support the plancia, unless the plancia follows the slope of the rafters. Along the rake, lookouts are necessary to support the fascia which is nailed directly to their outer ends. Since the lookouts act as small cantilevers, they must be fastened especially well to the wall. See Chapter 5 for additional framing for projecting eaves.

g At building corners, the eaves cornice may be carried a short distance (approximately equal to the width of the cornice) along the end walls. The rake cornice is brought down on top of the eaves cornice at this point. This detail is known as the "return" (Figure 9.4).

GUTTERS AND LEADERS

9.5 Gutters

a Gutters or eavetroughs catch rainwater from roofs and carry it to leaders or downspouts extending to the ground where water is delivered to cast-iron or terra-cotta drain pipes discharging into dry wells or storm sewers. Without this system of drains the rainwater would fall off the roof at the eaves, splash on the ground, and cause serious erosion as well as inconvenience to anyone at that point. On small buildings such as sheds gutters are seldom used, and on small houses they are sometimes omitted in favor of a concrete, masonry, or gravel splash strip set in the ground under the eaves to break the fall of the water.

b Gutters are of many kinds and shapes. They are commonly wood, metal, or plastic and all give satisfactory service if properly installed. Certain basic requirements must be met by all:

1. They must be large enough to handle the discharge.
2. They must have sufficient pitch to carry the water off quickly and not leave any pockets of standing water.
3. They must not leak.
4. There must be no obstructions to free flow.
5. They must be so installed as to avoid danger of backing up snow and ice under the roofing material with consequent leakage into the building.

9.6 Wood Gutters (Figure 9.5)

a. Specification Clause Install fir or redwood gutters of the sizes indicated on the plans. Joints of gutters shall be leaded flush with the insides of the gutters. The sub-contractor for carpentry shall furnish lead goosenecks.

b The simplest wood gutter is a vee formed by fastening two boards together. To be satisfactory it must be lined with metal or other waterproof material such as heavy asphalt-impregnated roofing felt, because the bottom of the vee otherwise soon opens and lets the water through. This is probably the oldest type of wood gutter, and is still found on old farmhouses and their replicas. Otherwise it is seldom used.

c The commonest wood gutter is molded from a solid piece of Douglas fir, cypress, redwood, cedar, or other durable species. It has an ornamental molding (ogee) on the outside and a semicircular channel on the inside. This type of gutter usually forms a part of the cornice detail, and thus performs a double function.

d Figures 9.5a and b show cross-sections and splicing details of such gutters. Wherever a splice is necessary, the two ends are carefully and snugly butted, after which the channel is cut away deep enough to take a small sheet of lead or soft copper which is fitted over the joint, bedded in elastic cement, and tacked down securely. All joints are finally filled with elastic cement. The same procedure is followed wherever a corner occurs; the ends are carefully mitered and the joint is lined with metal bedded in cement. Similar lining, cut to fit around the head of the leader, forms the joint between it and the gutter.

e Although such a gutter can be completely lined with metal, it is neither necessary nor desirable. The danger of rotting is small, provided the joints are sealed against the penetration of water, and provided the gutter pitches sufficiently and uniformly enough to prevent water pockets from

Cornices, Gutters, and Leaders

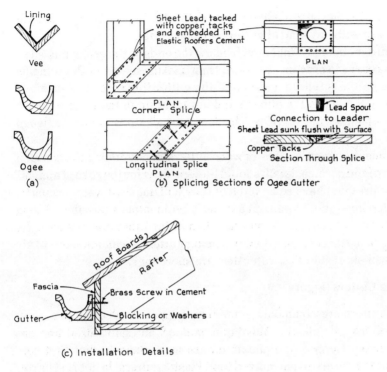

Figure 9.5 Wood Gutter.

forming after a rain. The periodic wetting in rains is soon dried, and the sun beating down on the gutter during the rest of the time keeps the wood too dry to rot. Standing water and water seeping into joints, however, may quickly cause decay.

f Wood gutters are usually fastened to the fascia. Certain precautions must be taken:

1. Fastenings should be brass or other noncorroding metal, bedded in elastic cement to prevent leakage around the fastenings, preferably screws.

2. The gutter should not be fastened directly to the fascia, but should be held away from it $\frac{1}{4}''$ by small wood strips or by metal washers (Figure 9.5c). This allows ventilation of the joint between fascia and back of gutter and reduces the decay hazard. Moreover, if the gutter becomes stopped up by ice and snow in the winter, and melting water overflows or is backed against the back of the gutter, the water can seep over the back and drip

to the ground. Otherwise it is likely to work into the house. This precaution is necessary with any kind of gutter, wood or metal.

3. The gutter should be far enough below the ends of the projecting shingles at the eaves to prevent any water from backing up under the shingles when the gutter is full. In addition, if the gutter fills with ice and snow, it should spill over on the outside and not build up to the point where it can back up under the shingles. This is true of all kinds of gutters, wood or metal.

4. Distances between leaders for wood gutters, or any other gutters which form part of the cornice detail, should be kept short for the sake of appearance. Gutters must have some pitch or they will not shed water properly, but cornice lines should be kept as nearly horizontal as possible. A long gutter is noticeably higher at one end than at the other and destroys the appearance of the cornice. Moreover, long gutters are much more likely to form pockets of standing water than are short ones.

9.7 Hung Gutters (Figure 9.6)

a Hung gutters are commonly aluminum, copper, leaded copper, zinc, galvanized iron, or plastic. Aluminum, copper, and galvanized iron are most common. Copper and aluminum are more durable, but first cost often dictates the use of galvanized iron. Plastic gutters do not rust or rot, but may degrade otherwise unless of proven weather-resistant materials.

9.8 Specification Clauses

MATERIAL

a All leaders, eavetroughs, and molded hanging gutters shall be 16-ounce hard (cornice temper) copper.

EAVETROUGHS AND HANGERS

b Eavetroughs or half-round hanging gutters, the size and type shown, shall be installed where shown on the drawings. They shall be in 10-foot lengths and shall be joined by 1" lapped and soldered or slip joints. They shall be supported by wrought copper strap hangers of approved design, or by bronze or copper shank-and-cinch type hangers adjustable for slope. Hangers shall be not more than 36" apart and shall be secured by brass screws.

LEADERS, STRAINERS

c Leaders shall be installed where shown on the drawings, of the sizes

Cornices, Gutters, and Leaders

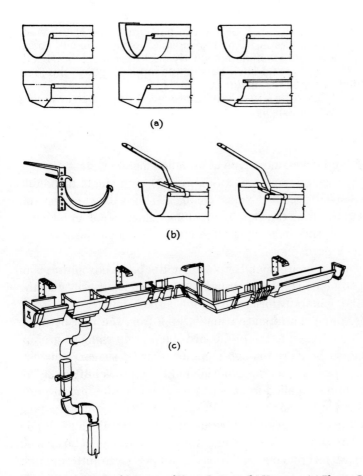

Figure 9.6 (a) Typical Sections of Hung Gutters. (b) Hangers. (c) Plastic Gutters and Leaders.

and shapes indicated. They shall be held in position, clear of the wall, by heavy brass or copper straps, $\frac{1}{8}''$ by $1\frac{1}{2}''$, spaced not more than 6' apart, soldered to the leaders, and fastened to woodwork by brass screws. Leaders shall be in 10' lengths and shall have joints lapped. A $1\frac{1}{2}''$ slip joint shall be provided every 20' of leader.

d All gutter outlets shall be fitted with approved copper wire strainers of the basket type set in loosely.

9.9 Attached Gutters

a The simplest and commonest type of attached metal or plastic gutter is the simple half-round eavetrough (Figure 9.7) with single or double bead. The single bead type is the most common. This is the cheapest and most efficient gutter from the standpoint of maximum use of metal since it is easiest to form and the semi-circular cross-section carries the most water for a given perimeter.

b The other common type of metal or plastic gutter is the box gutter, usually with an ogee molded face and straight bottom and back. This, like the ogee wood gutter, generally forms part of the cornice detail.

c Attached or "hung" gutters are usually hung from the roof by strap hangers which are nailed to the roof boards before the shingles are applied. They should be close enough together to avoid sagging between them. Spacing depends upon the material, but is in the vicinity of 30" to 36". They should be installed to form a straight line with uniform pitch not less than $\frac{1}{16}''$ per foot so that when the gutters are subsequently hung to them proper run-off is provided. Wire hangers, cheaper and lighter than strap hangers, are common with small gutters, but these must be spaced closer together. Metal brackets may be used to support large gutters which carry fairly heavy quantities of water.

d Hung gutters should ordinarily be held away from the fascia for the reasons already given under wood gutters. If the gutter must be set directly against the fascia, the back of the gutter should be carried up in one piece well under the lowest courses of shingles so that any water, ice, or snow which is backed up cannot work through the roof (Figure 9.7). This becomes essentially a molded gutter.

e Attached gutters must be free to expand and contract with changes in temperature. This is accomplished most simply by allowing one length of gutter to lap the next in the direction in which water is to flow. A somewhat more watertight joint is secured with a separate short section in

Cornices, Gutters, and Leaders

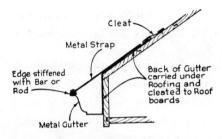

Figure 9.7 Molded Metal Gutter.

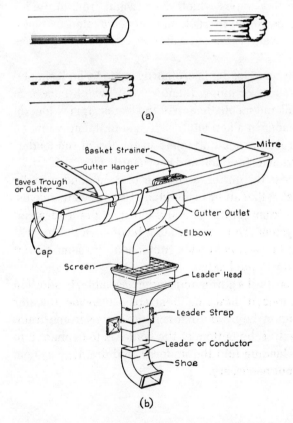

Figure 9.8 (a) Conductors. (b) Gutter and Leader Details.

which both lengths of gutter fit and slide (Figure 9.6). Slip joints should be provided every 25 to 50 feet; intermediate joints can be soldered together.

f Materials that have high coefficients of thermal expansion, such as aluminum and, especially, plastics must be permitted to move or buckling and splitting may occur. A variety of motion-permitting joints has been devised for these materials, including heavy flexible rubber which can stretch and retract as the adjacent gutter sections move.

9.10 Leaders (Figure 9.8)

a Leaders are simply vertical pipes which carry water from gutter to ground drain. They may be round or rectangular, corrugated or plain, and must be large enough to carry the roof water away as fast as it comes (Figure 9.8a).

b The joint between gutter and leader is ordinarily made by an S-shaped piece called the "goose-neck" or elbow (Figure 9.8b), which curves in from the gutter to the wall and meets the top of the leader. It may join directly to the top of the leader in a slip joint, or it may first empty into an ornamental box called a leader head attached to the top of the leader. The box performs no particular function and is primarily an ornament.

c Where the leader pierces the gutter, a strainer of some kind, usually a small wire inverted basket-like affair is inserted. This prevents leaves, twigs, and other rubbish from getting into and clogging the leader and underground drainage system. The strainer must be of such design that it will not readily stop up but will hold back rubbish; at the same time it must not slip down and jam in the leader.

d Figure 9.8b shows the parts of a gutter and leader assembly. In addition to the parts already described, it shows leader straps and a shoe. If water is to be allowed to splash upon the ground or to run into some surface drain, a shoe should be provided. If the bottom of the leader is to connect into a vertical cast-iron pipe leading into the underground drainage system (Chapter 4), the shoe is not necessary.

10 Exterior Finishes. Water Tables

10.1 General

a The usual exterior finishes for frame construction are:
Siding
  Drop (Novelty)
  Lap, Clapboard
  Vertical Boarding
  Horizontal Flush Boarding
Shingles
  Wood
  Asbestos
  Asphalt
Sheet Materials (see Chapter 12)
Masonry veneer
  Brick
  Stone
   Rubble
   Ashlar
Stucco (see Chapter 12)
Each of these calls for its own materials and techniques.

b At the base of the frame, where foundation and superstructure meet, a special detail called the water table may be found. It may be simple or ornate, depending upon the detailing.

SIDING

10.2 Drop (Novelty) Siding (Figure 10.1)

a This is probably the simplest of exterior finishes because it combines sheathing and siding in one piece. Since it must afford a weathertight wall in one thickness of wood, the individual pieces are tongued and grooved and must be driven up tightly against each other when they are nailed in place.

b After studs are up, drop siding is nailed into place to form the finished wall (Figure 10.1). Extra weathertightness may be afforded by stretching building paper across the studs before siding is nailed on, but the paper is usually omitted altogether. Siding is run from sill to plate, and is cut at window and door openings, the same as sheathing. Window and door frames are installed after the siding is on, which makes it difficult to flash

Dwelling House Construction

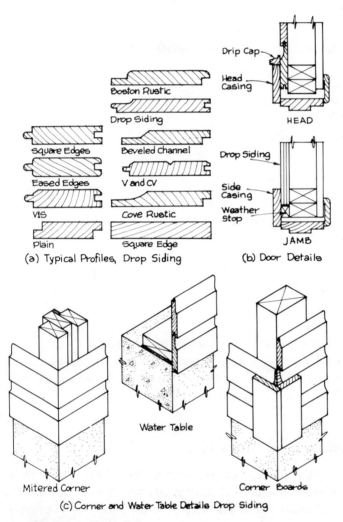

Figure 10.1 Drop Siding, Details.

Exterior Finishes. Water Tables

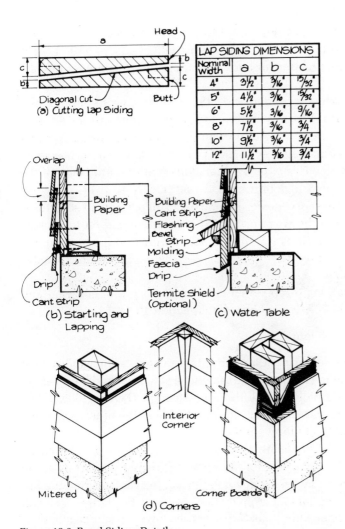

Figure 10.2 Bevel Siding, Details.

the heads of these frames (Figure 10.1b). At the corners the ends are roughly cut off and the corner is finished by nailing on a pair of vertical boards called corner boards, or ends are mitered at the corners (Figure 10.1c).

c Drop siding is usually beaded and otherwise shaped to various patterns (Figure 10.1a). Where window and door casings are nailed over the siding, the beads and cutaway portions leave openings through which wind and rain can find their way unless they are sealed. Sealing is accomplished by nailing wood strips or weather stops over the cutoff ends of the siding before window and door frames are set (Figure 10.1b).

d Drop siding alone is not considered good finish for permanent structures or those which are to be weathertight. It is mostly found on temporary structures, or on structures which must merely provide the minimum of shelter, such as storage sheds for non-perishable goods, inexpensive garages and the like.

e Used over sheathing, drop siding is a satisfactory wall covering and falls into the class of horizontal boarding (Section 10.5a, b).

10.3 Lap Siding   (Figure 10.2)

a This is also known as bevel siding and, when narrow, is commonly called clapboards. Lap siding is made by rip-sawing a square-edged board along its length and diagonally across the cross-section (Figure 10.2a). The result is two boards narrow at one edge and wide at the other. The thick edge may be rabbeted to form shiplap.

b Siding is applied after sheathing and building paper are on. Ordinarily it is put on in bands as high as men can conveniently reach, starting at the top of the building and working down to the sill, to avoid marring the finished surface by scaffolding. Application starts at the bottom of each band since each succeeding board laps over the lower one much like shingles. The amount of overlap depends both upon the width of the boards and upon the amount which is to be exposed to the weather, i.e., upon the spacing. Narrow boards lap each other approximately $\frac{1}{2}''$ to $1''$ whereas wide ones lap $1''$ to more than $2''$ although the lap is sometimes less than $1''$.

c Spacing is varied to suit conditions. On the side of a house, for example, there may be one spacing from foundation line to first story window sills, another between the first story window sills and the window heads, another one for the band between first story window heads and second story window sills, another from those sills to the window heads, and still another from window heads to eaves. On a gable end there may be still further changes in spacing to the peak of the gable. Usually the variation in spacing is quite small and is not noticeable, particularly with narrow siding. This is one of the advantages of lap siding over drop siding, since the latter cannot be varied in its spacing.

d Individual lap siding boards are nailed along the lower edge, so that the nails drive through the butt of the upper board and through the top of the lower one. An exception is narrow clapboards, which frequently are nailed along the top instead of at the butt. Nails are usually 6, 7, or 8 penny box or siding nails, either plain or coated (Chapter 14).

e At the eaves, the top siding board slips under the lower edge of the frieze board to make a weathertight joint (Chapter 9). The frieze is made to allow space for the top edge of the siding, either by rabbeting the lower back edge of the frieze board or by furring out the frieze from the wall. Along the rake the siding boards are carefully cut to fit against the lower edge of the frieze, which is not furred.

f Corner boards may or may not be employed with lap siding. If they are, they are nailed on first and the siding is carefully cut to fit snugly against them (Figure 10.2d). It must similarly be fitted snugly against door and window casings. At all these points building paper (Chapter 8) must be carried continuously across the joint so that any water which may drive between siding and casing or corner boards is prevented from working through the wall. When corner boards are omitted, the siding boards from the two adjoining walls are brought together in a miter joint (Figure 10.2d). Interior corners may employ corner strips (Figure 10.2) or siding boards may be butted alternately left and right.

g Figure 10.2c illustrates a typical water table for lap siding. The sheathing is brought down flush with the edge of the foundation wall. To it is nailed the water-table fascia. The lower edge of the fascia extends down beyond the top of the foundation wall and should be grooved or bevelled to form a drip. If a termite shield is employed, the bottom of the fascia is bevelled to fit the top of the termite shield, which projects beyond it approximately an inch.

h The top of the fascia is bevelled and is topped with another member which projects beyond it and in turn forms a drip. If this member projects very far, or if the water table is to be accentuated, a piece of molding can be nailed under it. The top strip is covered with flashing carried up the wall 2″ or 3″ under the lower edge of the building paper. If thick-butt siding is used, the butt of the lowest siding board should be beveled to fit snugly to the top of the water table.

i Usually, there is no water table. The lower edge of the lowest siding board is brought down approximately an inch below the top of the foundation wall. Before it is nailed on, narrow wooden "cant" strips (wood lath are excellent) are tacked to the lower edge of the sheathing to give the lowest siding board the same flare as the upper ones (Figure 10.2b).

j Aluminum, plastic (PVC), plywood, and hardboard are also employed for siding.

k Aluminum siding is formed to have the appearance of lap siding (Figure 10.3) and is finished with anodized or baked-on finish in plain or embossed patterns, white or colored. The upper and lower edges are crimped to interlock as the successive courses are applied, and slots are provided at the upper edges for nailing. Because metal expands and contracts with changes in temperature, the slots allow for such motion. Nails, therefore, should be driven snugly but not so tightly as to prevent motion.

l Because metal is impervious to the passage of water vapor, and because a wall should be allowed to "breathe," provision should be made, usually by opening along the bottom of the lower or butt edge, to permit ventilation.

m The insulating value of thin metal is negligible. Consequently, the aluminum siding may be backed with insulating board such as plastic foam or fiber board. The denser fiber boards also provide backing for the metal to help to prevent denting.

n Accessories include corner boards for exterior and interior corners, and window and door trim which permit the ends of siding strips to slip into a slot, and to move with changes in temperature. Starter strips are employed to anchor the first course.

o Extruded PVC (Chapter 16) is shaped to have the appearance of lap siding. White or other colors are integral rather than applied. Surfaces may be plain or embossed. Upper and lower edges are formed to interlock as successive courses are applied (Figure 10.4), and slots are provided at the upper edge for nailing. Plastics generally, including PVC, have higher coefficients of expansion than metals, so it is essential to allow movement to occur with changes in temperature. Nails should, therefore, be driven snugly but not so tightly as to prevent motion, or buckling in hot weather and possible cracking in cold weather may occur.

Exterior Finishes. Water Tables

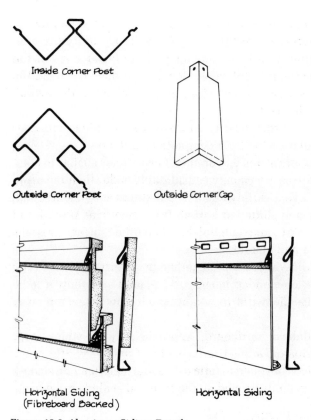

Figure 10.3 Aluminum Siding, Details.

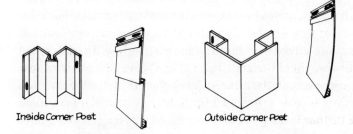

Figure 10.4 Plastic Siding, Details.

p Plastics siding, like metal, is essentially impervious to the passage of water vapor. It is, therefore, essential to permit the siding to breathe, usually by incorporating ventilation openings in the lower edge of the siding strip. Unlike metal, it is not easily dented and does not need the backer for that purpose, but an insulating backer enhances the overall thermal insulation of the wall.

q Accessories include corner boards and closures, window and door trim, and similar details needed for a complete installation.

r Plywood may be cut into strips and applied in a manner similar to standard lap siding. The strips are usually considerably wider than the usual lap siding, and may have a variety of surface textures such as brushed, striated, rough sawn, and slotted or kerfed. It is sometimes prefinished with factory-applied paint, baked-on finishes, and other surface coatings including colored film and metal foil.

s Where adjacent siding strips butt together in a vertical joint, wood shingle wedges are recommended to support the joint and allow it to be firmly nailed; otherwise the two strips might tend to bow in and out away from each other.

t Tempered lignocellulose hardboard, approximately $\frac{1}{2}''$ thick, may likewise be cut into strips and applied as lap siding. Common widths range from 6″ to 12″ with edges bevelled to form a drip. Surfaces may be smooth or textured, and are available primed on the face and sealed on the back.

10.4 Vertical Boarding (Figure 10.5)

a The vertical boarding frequently found on walls of barns, sheds, and the like is also used quite commonly as exterior covering for houses. In this instance extra precautions are taken to make the joints weathertight.

b As usual, the sheathing is covered with building paper and then tongued and grooved or square-edged boards are nailed vertically, with each board running the entire height of the wall if possible. Figure 10.1 shows a few patterns. Joints are best coated with paint just before the boards are nailed on, and succeeding boards are driven up tightly against the preceding ones. If nothing further is done, a plane unbroken surface results, unless the edges of the boards are bevelled to accentuate the joint (Figure 10.5).

c In order to break the surface and further to protect the joints, battens are frequently nailed over the joints. Before they are applied their backs are best coated with paint and the boards themselves are coated similarly,

Exterior Finishes. Water Tables 239

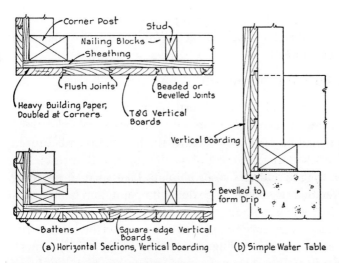

(a) Horizontal Sections, Vertical Boarding   (b) Simple Water Table

Figure 10.5 Vertical Boarding, Details.

unless they are not to be painted later, so that when the battens are applied, the joint is filled and sealed (Figure 10.5a). If joints are not to be covered with battens, the vertical boarding should be tongued and grooved; if battens are employed, square-edged material is satisfactory.

d Generally no special water table is employed with vertical boarding. The bottom ends of the boards are merely carried down an inch or so past the top of the foundation wall and are cut off on a bevel to form a drip (Figure 10.3b). If a water table is desired, the bottom ends of the boards are brought down and fitted to the sloping top of the top member of the water table (Figure 10.2c). Flashing of the usual type is required at water table and at door and window heads (Chapter 7).

e Aluminum strips, similar to those used for lap siding, may be employed as vertical siding. Instead of the bevelled appearance of horizontal siding, profiles resembling board and batten or V-groove or both may be employed. Round or trapezoidal corrugations provide other profiles. The same precautions respecting thermal motion and ventilation must be observed as for horizontal siding. Usually, these strips are open at the bottom and can be vented at the top with appropriate closure strips or eaves details.

f Plywood, hardboard, particle board, and cement-asbestos board sheets are frequently grooved, striated, formed to simulate board and batten or other surface textures, painted, provided with baked-on surfaces, or faced with film, foil, or both. These are applied vertically to provide appearances similar to vertical wood boarding. Wallboards are discussed at greater length in Chapter 12.

10.5 Horizontal Boarding (Figure 10.6)

a Boards may be horizontal instead of vertical, in which case they are in practically the same category as drop siding applied over sheathing. Various patterns may be employed (Figure 10.1). If the boards are not patterned but have square corners, a plane unbroken wall results. This material may be tongue and groove or shiplap. Like vertical boarding, the joints between boards should be coated with paint, and boards must be driven up tightly when nailed.

b A water table may or may not be employed. If it is called for, its construction is much the same as for lap siding; it is flashed on top, and the lowest board is beveled to fit snugly against the top member (Figure 10.2c). Flashing at door and window heads is the same as for lap siding (Chapter 7).

SHINGLES

10.6 General (Figure 10.7) (See also Chapter 8)

a There is little difference between the application of shingles to roofs and to side walls, whether the shingles be wood, asphalt, asbestos, or any other kind.

b Unlike lap siding, shingles are started at the house sill and are carried up the wall instead of being applied in bands starting at the top and working down.

c Because water is less likely to work through shingles on side walls than on roofs, the exposure to weather can be increased. Moreover, instead of laying all courses with the same weather, successive courses may be made wide and narrow, or some other pattern worked out to break the regular coursing. Asbestos and asphalt shingles are made in a variety of forms and shapes and are applied in different ways to correspond to these shapes. Some asbestos shingles, in order to reduce the amount of material required, are made up in strips and are mounted on waterproof building paper. Only the exposed part of the strip is asbestos board whereas the upper half of the strip is paper. Wood shingles for side walls are frequently uniform in width so that regular patterns can be worked out readily.

Exterior Finishes. Water Tables

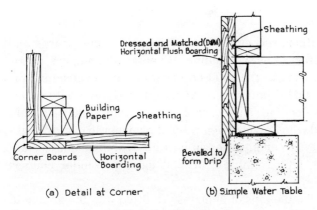

Figure 10.6 Horizontal Flush Boarding, Details.

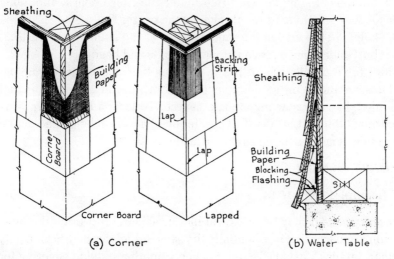

Figure 10.7 Shingled Wall, Details.

10.7 Wood Shingles

a The same precautions respecting snug joints at corner boards and casings must be observed as with siding. When wood shingles are returned at the corners, unlike the miter corner return in bevel siding, the shingles are butted and the butt joint is carried into the corner alternately from each side in succeeding courses. Each pair of corner shingles should be flashed or backed with a strip of heavy building paper to insure a tight corner (Figure 10.7a). To emphasize the butt lines, shingle courses are sometimes doubled, with Number 3 or 4 shingles for under courses (Chapter 8).

b Under window sills, the tops of shingles are fitted into the groove in the lower face of the sill, in much the same manner as siding. Drip caps above the head casing are flashed in the usual manner (Chapter 7). At the eaves, the tops of shingles are fitted under the frieze board, and along the rakes they are cut to fit snugly against the frieze. The spacing is varied up the height of the wall in the same manner as lap siding (Section 10.3) so that butt lines come out even at window sills and heads.

c Water table details are ordinarily simpler than those for lap siding. The lowest shingle course simply extends down beyond the sill line and is doubled, like any starting shingle course. To form a better drip, it is furred out at the butt $\frac{1}{4}''$ to $\frac{1}{2}''$. If greater emphasis is to be placed on the water table, it may be flared as shown in Figure 10.5b.

MASONRY VENEER

10.8 General (Figure 10.8)

a Masonry veneer may be brick, rubble, or ashlar, with brick the most common. Construction is essentially the same for all three, differing only in minor details. Masonry veneer is simply another outside covering for a frame building, and does not alter its essential character.

10.9 Details

a. Foundations Because standard brick are $3\frac{3}{4}''$ wide and because space must be left between brick and sheathing to take up any irregularities in construction, the sheathing line is held back $4\frac{1}{4}''$ to $4\frac{1}{2}''$ from the outer face of the foundation wall. Brick veneer, therefore, requires a foundation wall thick enough to carry the wood sill in addition to the veneer. The thickness must be made greater for stone veneer because ashlar is at least $5''$ thick and rubble runs $6''$ to $8''$ or more.

Exterior Finishes. Water Tables

b. Paper Differences of opinion exist as to whether the space between veneer and sheathing should be filled in solidly with mortar or should be left open. However, this space really is never quite open at the bottom inasmuch as mortar falls in during the course of construction; hence it should be built as if it were solidly filled. This requires that heaviest grade waterproof building paper be applied over the sheathing before brick are laid. The paper is needed not only to keep the frame dry as the wall is built, but even more to keep rainwater from penetrating, because thin masonry veneer cannot be expected to be entirely weathertight.

c. Brick and Mortar Brick should be well-burned, hard, and absorb 5 to 14 percent of water by weight. If the absorption is lower, good bond is not obtained between brick and mortar; if higher, the water is drawn out of the mortar into the brick and a dry non-adherent layer is left between brick and mortar. Very light-colored (salmon) brick are underburned and likely to be overly absorbent. Over-burned brick are dark, look glazed, and usually have low absorption. Mortar may be a prepared brick mortar, or may be mixed on the job. Standard practice is to use approximately half and half lime and portland cement. A common mix is one part lime, one part cement, and six parts sand, with sufficient water to make the mortar workable. The exact mix depends largely upon the sand, particularly the range and proportions of the particle sizes. No matter what kind of mortar is used, a wall cannot be expected to be reasonably weathertight or structurally sound unless all mortar joints, both horizontal and vertical, are completely filled.

d. Ties Since a 4″ to 8″ wall cannot stand unsupported for wall heights, the veneer must be tied to the sheathing at frequent intervals by wall ties or brick anchors, which are small strips of corrugated metal about 1″ wide and 6″ long. Best metal is copper, but zinc-coated iron is most common. One end of the tie is laid in a mortar joint, the other is turned up against the wall and nailed to the sheathing. Nails should be the same metal as the anchor to avoid electrolytic action. Ties are needed every 4 or 5 courses in height and every 2′ along the mortar joint (Figure 10.6).

e. Construction If there is no special water table, the first course of brick or stone is laid directly on the foundation wall and, unless it is random rubble, is carefully levelled to form a straight, horizontal base for succeeding courses. Before the first course is laid, flashing is placed on top of the foundation wall with its outer edge turned down a short distance over

Dwelling House Construction   244

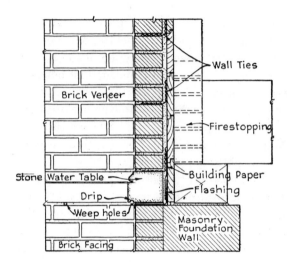

Figure 10.8 Brick Veneer.

the edge of the wall and its inner edge turned up against the sheathing several inches. Building paper is brought down over, not under, the flashing so as to shed any water which may drive through the veneer and trickle down the building paper to the sill. It is good practice, moreover, to expedite drainage of this water by providing "weep holes"—small metal tubes—several feet apart on top of the flashing, their outer ends flush with the face of the wall and their inner ends in the space between brick and sheathing.

f Brick courses, like spacing of siding and shingles, are laid out to come out even at window sills and heads or any other horizontal details in the exterior finish. This is done by slightly varying the thickness of mortar joints. Brick or stone sills are set under the wood sills of window frames (Chapter 7). Best practice is to lay up the veneer to the window line, set the windows in the usual way, with wood sills on top of the masonry sills, and continue with the veneer.

g Veneer is carried over the heads of windows and doors on steel angles, usually 3" x 4" or 4" x 4", which rest in the veneer at each end (Chapter 7). Heads of this kind must be flashed in the same way as the first course at the foundation. Here the flashing is laid on top of the outstanding leg of the angle and is carried up the upstanding leg to the sheathing. Building paper laps over the back edge of the flashing.

10.10 Water Table

a Figure 10.8 shows an ornamental stone water table sometimes found, not only in brick veneer, but in solid masonry. It may or may not project beyond the face of the foundation. Frequently the water table is a soldier course, i.e., brick laid with the long dimension vertical instead of horizontal. The water table may be a course of molded brick or other specially formed burnt clay unit. An ogee (cyma reversa, Chapter 13) is a favorite detail. Like the soldier course, it does not project beyond the foundation wall but curves back and up from the face of the foundation wall. The face of the veneer is, consequently, set back several inches from the face of the foundation, which must be made even thicker than is ordinarily necessary for veneer.

10.11 Wallboards

a These are discussed briefly in Section 10.4 and at greater length in Chapter 12.

11 Insulation

11.1 General

a The practice of insulating houses against excessive heat loss or gain is now practically standard procedure. It therefore becomes necessary to consider what insulation may be expected to accomplish, where it does the most good, the various materials used for insulation and how they accomplish their task, and the ways in which they are installed.

11.2 Heat Losses

a Heat losses from a building bear a direct relationship to the fuel bill. However, the total heat inherent in every pound of fuel is by no means lost through the structure of the building. Imperfect combustion in the burner usually accounts for a very large part of the fuel consumption. Unless the draft is regulated to supply the correct amount of air to the fire, and unless burners are kept in condition, large quantities of soot and only partially burned gasses are liberated. These constitute a distinct loss. Unless the heat exchanging surfaces in the heater are properly designed for the particular types of fuel burned, only partial transfer of heat from burning zone to house heating medium (air, hot water) takes place and a large part of the heat evolved escapes up the chimney. Proper care of the heater can frequently reduce heat losses more, at smaller cost, than can additional insulation.

b Heat losses through the structure occur chiefly at the following zones:
1. Foundation Walls
2. Basement Floors and Floors on Grade
3. Outside Walls (above grade)
4. Windows and Doors
   Infiltration of cold air
   Loss through the glass
5. Roof

c Some approximate typical rates of heat loss or conductance U in Btu transmitted through the construction in 1 hr per sq ft of surface per °F difference, are given in Table 11.1. Resistance is the reciprocal of conductance $(1/U)$. These values include the effect of the films of air absorbed on the interior and exterior surfaces in addition to the values of the materials themselves.

d Glass and concrete are relatively poor insulators, whereas the ordinary frame wall is, relatively, much better than any other part of the structure.

Insulation

Table 11.1 Thermal Conductance and Resistance

	U	R
10-inch solid concrete wall, unplastered	0.60	1.68
4-inch solid concrete floor	1.00	1.00
Frame wall, 5/16″ plywood sheathing, bevel siding, plastered on 3/8″ gypsum board	0.29	3.45
Same with 2″ fibrous insulation	0.09	11.11
Wood shingles on wood strips, rafters bare below	0.48	2.08
Single thickness of glass	1.13	0.88

From the standpoint of insulation, the ordinary roof is poorer than the ordinary frame wall, but it is better than the usual single thickness of glass in the windows, or the ordinary concrete wall or concrete floor. On the other hand, the differential between basement temperature and outside soil temperature is much less than the temperature differential between upper rooms or attic and the outside air. Soil temperatures rarely go much below 40° to 50°F, whereas outside air temperatures may drop well below zero.

e In house construction, basement losses are commonly ignored and insulation is concentrated in the superstructure. On the other hand, in slab-on-grade construction, losses through the slab are appreciable, particularly at the periphery, where the high conductivity of concrete results in chilled zones in the floor and at the floor-wall intersection. This can cause not only discomfort but condensation on the floor and lower portions of walls. (See Section 11.11c)

f At first glance, the figures given in Table 11.1 seem to indicate that glass is the most important area to insulate, but the total area of glass is usually much smaller than the roof or the net wall area. Moreover, because warm air tends to rise, the warm zones likely to be found near the top of the house build up the temperature differentials at these zones and, consequently, increase the heat loss. Furthermore, insulation works both in summer and winter, and in summer the hot sun heats the attic unless insulation is provided. For these reasons, one of the most important areas to insulate is almost always the roof or ceiling under the attic space.

g Total heat losses through outside walls and through windows are usually about the same in most houses. From the comfort standpoint, however, the cold glass areas need more attention than do the relatively warm interior surfaces of walls. Therefore, the glass areas, taken by and large, are at least as important from the insulation standpoint as are the walls. In houses having very large window areas, insulation of the glass may easily overshadow that of the walls in importance. Orientation of windows with respect to the sun becomes extremely important.

11.3 Mechanics of Heat Transfer. Reduction of Heat Loss

a Heat is transferred by convection, conduction, and radiation. All three of these play a part in the loss of heat from buildings, and various methods of insulating combat one or more of these three sources of loss. Convection currents in the air transfer the heat from warm zones to cold. Solid parts of the structure transmit heat by conduction. Warm surfaces emit radiant heat energy which passes through intervening air to colder surfaces.

b Comfort for the occupants of a building depends chiefly upon three factors: the temperature of the air, the difference between the surface temperatures of the bodies of the occupants and the surface temperatures of the exterior walls or windows, and the relative humidity of the air. The temperature of the air is less important than is often realized. It is possible to be comfortable even if the air is at a temperature much lower than usually considered comfortable, provided the surrounding walls are sufficiently warm to reduce the rate of radiation from body surface to walls. As a matter of fact, radiant heating systems keep the walls or floors warmer than usual, and comfortable conditions are obtained at air temperatures much less than are ordinarily considered to be necessary. Any method of raising the surface temperature of surrounding walls, floors, ceilings, and window areas by insulation reduces the air temperature required for comfort.

c Dry air must be warmer than moist air to provide the same degree of comfort because the rapid evaporation of moisture from the surface of the body in dry air has a chilling effect. Hence, the relative humidity must be kept fairly high for comfort, but sufficiently high relative humidity in turn allows the air temperatures to be lowered for the same degree of comfort. Humidification is limited by the surface temperatures of walls and of glass areas. If these are cold, the moisture in the air becomes chilled and condenses even at low relative humidities. Since excessive condensation on walls and windows leads to deterioration of finishes, corrosion,

Insulation

and possible decay of woodwork, the amount of humidification permissible is limited by the temperatures of interior wall and glass surfaces. Proper insulation is therefore needed to keep these surface temperatures at higher levels.

d Heat loss through a wall is a combination of radiation, convection, and conduction. Heat is transferred to the inner surface of the wall largely by convection currents and by radiation from warmer bodies, with conduction playing only a small part at this stage. The heat passes through the inner wall surface (plaster or other covering) by conduction. Across the stud space from inner wall covering to sheathing the transfer takes place chiefly by convection and radiation, with conduction playing only a small part (mainly at the studs) because wood is a poor conductor. Heat is transferred through the sheathing and outside wall covering by conduction, and from exterior surfaces of the wall by radiation and convection.

11.4 Insulation against Convection (Figure 11.1)

a Without using any insulating material as such, convection losses can be reduced considerably in the stud spaces by proper construction of the house, particularly the frame.

b. Flues Long vertical flues in the stud spaces should be blocked off. Such construction considerably reduces the convection currents in the stud spaces, since the higher the flue, the more rapidly the convection current travels. In Figure 11.1a the inside wall is not only warmer than the outside, but the temperature T_2 near the ceiling is higher than T_1 at the floor. Cold air drops along the sheathing and warmed air rises along the inner surface. The effect is noticeably stronger in the high stud space than in the blocked-off space (Figure 11.1b).

c. Continuity There should be no continuity between inter-stud spaces and inter-joist spaces but the joist spaces should be blocked off at the exterior walls. Otherwise, particularly at the second story ceilings, convection currents can sweep up from between studs and across under the attic floor, carrying heat to the cold attic and thence out through the roof. Figure 11.1c illustrates an extreme case of continuity which not only promotes heat loss by convection but is a bad fire hazard. In Figure 11.1d the condition has been corrected by blocking and firestops (see also Chapter 5).

Dwelling House Construction 250

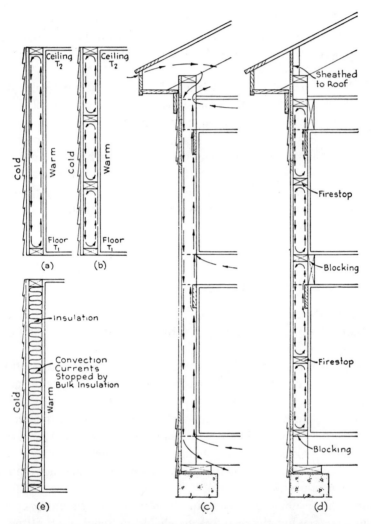

Figure 11.1 Air Currents in Wall. (a, b) Open Stud Height vs. Headers. (c, d) Open Sweep from Basement to Attic vs. Blocking. (e) Bulk Insulation.

Insulation

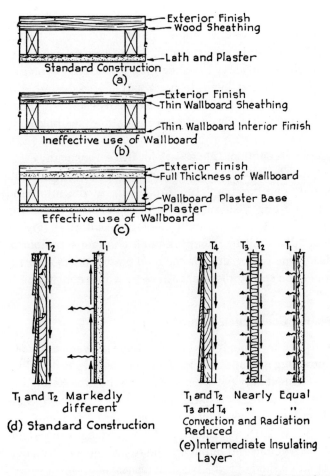

Figure 11.2 (a–c) Standard Construction vs. Wallboard. (d, e) Radiation and Convection, Standard Construction vs. Blanket.

d. Cornices Cornices, which are likely to develop cracks through which cold air from the outside can seep, should be blocked to prevent the air sweeping in between joists and under the attic floor, or up into the attic itself. One simple method is to carry the sheathing up to the roof boards (Figures 11.1c, d). On the other hand, in hot weather it is desirable to allow air to sweep through ventilation openings in the eaves soffits and into the attic space. This makes it desirable to put insulation (see below) in the ceiling.

e These rather simple precautions, necessary for fire safety in any case, can substantially reduce heat losses without any extra insulating material.

f Convection currents can be completely broken up in the inter-stud space by packing it with bulk insulating material. Quiet air is one of the poorest conductors of heat, and the bulk insulating materials (Figure 11.1e) provide quiet "dead" air by substituting an enormous number of very small air cells in which convection is reduced practically to zero, for the large spaces between studs, joists, or rafters, in which convection is appreciable.

11.5 Insulation against Conduction (Figure 11.2)

a Conduction losses occur through the interior finish of walls and ceilings, the exterior covering of walls, and through roof boards and roofing on roofs, as well as through any covering which may be attached to the undersides of rafters. By using materials which are poor conductors for these coverings, heat losses caused by conduction can be reduced. Because the heat must be carried by conduction to those surfaces from which it is transferred by convection and radiation, the retarding of conduction becomes an important method of reducing heat losses.

b Wood is in itself an excellent insulator, but numerous wall boards exist which, for the same thickness, have lower transmission coefficients and hence are better insulators. Whereas the coefficients of commonly used species of wood range approximately from 0.80 (many soft woods) to 1.00 Btu per hr per in. of thickness per sq ft per °F temperature difference, the coefficients of most wallboards are about 0.30 to 0.65. Insulating boards, like wood itself, provide a large number of dead air cells. When applied in the same thickness as wood they form more retardant paths than does wood, and much more retardant paths than do masonry materials or metals. It should be noted that a good insulator must not only provide

Insulation

the dead air cells which are poor conductors, but that the solid portion of the insulating material must itself be a poor conductor. Thus cellular metal would not be a good insulator even if it provided an enormous number of minute cells, because heat would be conducted readily through the metallic portion of the mass.

c Practically all insulating wallboards are designed to take the place of wood sheathing and many are made to be used in place of lath and plaster as well (Chapter 12). Their efficacy must, therefore, be judged upon their replacements, not as additional insulation. For instance, if ½" wall board with a conductance of 0.90 is substituted for ¾" wood with a conductance of 1.06, there is a small net gain in insulating value. Using ¾" insulating board, of course, shows greater net gain as far as conduction is concerned. For example, if in the uninsulated frame wall of Table 11.1, ¾" insulating board is substituted for ⁵⁄₁₆" plywood, the U (conductance) value drops from 0.29 to 0.19. The same observations hold true of, for instance ½" unplastered wall board used as interior finish as compared with ¾" or ⅞" lath and plaster. Not only the coefficients per inch of thickness, but the actual thicknesses must be compared to find the net gain (Figure 11.2a, b, c).

d A combination of insulation against convection and radiation is provided by placing an intermediate layer of insulating material halfway through the stud space (Figure 11.2d, e). This provides a pair of intermediate surfaces, the inner one at nearly the same temperature as the back of the interior wall surface and the outer at nearly the same temperature as the back of the sheathing. Since the surfaces facing each other across each of the two air spaces are at nearly the same temperature, both convection in and radiation across the air spaces are reduced. Coupled with this is the low rate of conduction through the insulating layer. The effectiveness of this combination is much enhanced if the surface of the insulating layer is bright metal.

11.6 Insulation against Radiation (Figure 11.3)

a The nature of a radiant surface has a great deal to do with the rate of heat loss. A bright surface, speaking very generally, is a poorer radiator than is a dull one; therefore, if surfaces of the materials composing wall coverings are made bright, they radiate less heat than do dull or dark ones.

Dwelling House Construction

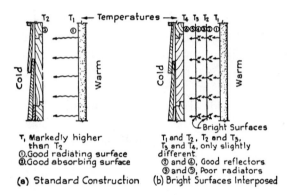

Figure 11.3 Bright Metal Foil Insulation.

This is merely another way of saying that their emissive coefficients are low. Moreover, again speaking very generally, bright surfaces are good reflectors and poor absorbers of radiant heat, so if the surfaces in a wall to which heat is being radiated are bright they absorb relatively little of the heat but reflect much of it.

b Since radiation plays an important part in losses through walls, it follows that bright surfaces interposed in the path of heat flow through a structure such as a wall or roof reduce the heat loss through that structure. The bright surfaces must, however, be free-standing to be effective, because wherever they are in contact with any part of the structure they may act merely as conductors. Bright metal surfaces, for instance, are apt to be good reflectors and poor emitters, but the metals are excellent conductors.

11.7 Insulating Materials

a Insulating materials used in house construction fall into five chief categories:
Wall Boards (Insulating)
Blankets
Sprayed-on Insulation
Fill
Bright Metal

11.8 Wall Boards

a Synthetic wall boards are made of a variety of materials but many use pulped or mechanically separated wood fibers or vegetable fibers.

Insulation

Others are based on mineral or glass fibers. The fibers are recombined by matting and felting to form the many small air cells which give the boards their insulating properties.

b The rigidity and strength of a fiber board are functions of the pressure under which it was formed; the greater the pressure the denser and stronger the board, but the smaller its insulating value. Most insulating boards are therefore made under only moderate pressure and are much softer than wood.

c Although insulating wall boards are softer than wood, they impart rigidity to a wall, when substituted for sheathing, because of their size (Chapter 5). Commonly they are 4' wide and 8' or more long, and act as large webs when nailed to the wall. They can be used satisfactorily under siding, brick veneer, and the like, but if the walls are to be shingled it may be necessary to nail wood shingle straps to the walls first because shingle nails may not hold well in the soft matrix of insulating boards.

11.9 Blankets (Figure 11.4a, b, c)

a Frequently, blankets differ from insulating wall boards only in the degree of stiffness they possess and not in any inherent difference in material or manufacture. For instance, blankets may be formed of pulped or mechanically shredded wood fibers, or they may consist of vegetable fibers, in both cases the same material as used for wall boards. In one case, the mass is compacted enough to form a structural board, in the other it is loosely felted, will not stand unsupported, and must be attached to the frame. Blankets differ from wall boards, moreover, in that the loose material must be inserted between coverings of paper or cloth to hold the mass in blanket form, and it may be quilted to hold it together better. Materials much used for blankets are mineral wools made of rock, slag, or glass; vegetable fibers; and wood flour (shredded wood).

b Blankets are usually inserted in the stud space between inside and outside and therefore may have an air space on each side. Each air space has practically the same insulating value in itself as the full air space originally in the inter-stud space, which means that in addition to the insulating value of the blanket, another air-space insulator has been added. This is a decided advantage. Between joists and rafters insulating blankets are also placed so as to have an air space on each side (See also Section 11.5d).

c Blankets are often attached by bending the edges against the studs and tacking them in place with wood strips. Many blankets are provided with special flaps at the edges which bend over the edges of studs and are tacked or stapled. Still others, which are stiff enough to stand with little support, are merely attached with small clips along the stud.

11.10 Fill-Type Insulating Materials (Figure 11.4e, f)

a This class includes all insulating materials which completely fill the inter-stud space, and which fill a large part of the inter-joist or inter-rafter space. The materials may either be loose and poured in place, or they may be loosely matted blocks, called bats, which are placed as units.

b A great many materials fall into this category and historically it is the oldest kind of insulation. The oldtime icehouses with their double outer walls filled with sawdust are an example. Any loose, porous material which can be poured or packed into a space will do provided it does not gradually settle with time and vibration, leaving the upper portion of the wall open and producing a dense mass in the lower portion whose insulating value has decreased. Flake or cellular gypsum, crumbled cork, and vermiculite are examples of this kind of insulation. Loose mineral wool is not usually poured into place in new construction but is used as auxiliary packing in odd spaces. In existing uninsulated houses mineral wool can be placed by removing some of the exterior wall covering, cutting holes in the sheathing, and blowing in the wool through a large hose.

11.11 Foam Insulation

a Plastics, glass, and concrete may be formed into efficient foams for thermal insulation. Glass and concrete foams are employed mainly for industrial and commercial buildings, but may also be used in houses. Glass foam is made into blocks, typically employed in the cavities of cavity masonry and as roof insulation over concrete or steel decks. Foamed concrete is frequently foamed and poured into place on decks or into irregular spaces.

b Foamed plastics are either prefoamed or foamed in place. (Chapter 16.11). The most common are polystyrene and polyurethane.

c In house construction, one common use of prefoamed polystyrene planks and boards is as perimeter insulation in slab-on-ground construction (Figure 11.5), where otherwise heat loss through the concrete would cause chilled floors and lower walls. The essential feature is to make sure

Insulation

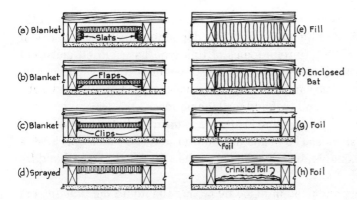

Figure 11.4 Methods of Applying Insulation.

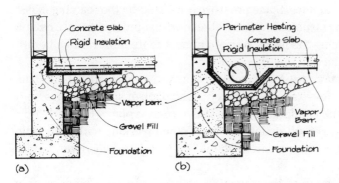

Figure 11.5 Slab-On-Grade Insulation. (a) Boards Should Be Placed to Provide a Thermal Barrier Under Periphery of the Slab and Toward the Outside. (b) Insulation Carried Under Perimeter Heating. Vapor Barrier Interposed.

that concrete does not extend through to the outside, and that the outer one to several feet of floor are insulated, depending upon the severity of the cold outside. Polystyrene foams of this type have closed cells and are reportedly impervious to the passage of water or water vapor; consequently, they need no vapor barrier. Other uses for such prefoamed slabs are as plaster bases, and roof and wall insulation.

d Expanding-bead polystyrene insulation is used for filling irregular and hard-to-reach spaces. The beads are poured into the space and heat is supplied, for example, by a live steam probe. The beads expand and fill the space.

e Polyurethane is foamed in place by mixing the liquid ingredients and immediately pouring the mixture into the space to be insulated where it can rise, fill it, and solidify. This is particularly convenient for irregular and hard-to-reach spaces. Some pressure is exerted by the foaming mass, so the surrounding structure must be able to withstand it. Pressure can be lessened by allowing the liquid to froth almost to its final volume before pouring it into place.

f Polyurethane may also be mixed and sprayed in one operation. This makes it possible, for example, to spray to the back of sheathing in a stud space and allow the insulation to foam and harden in place (Figure 11.4d). Cracks and joints are well sealed this way. When working with the spray, appropriate masks and inhalators should be worn, and good ventilation provided.

g Other plastics, such as the phenolics, can be sprayed. The sprayable insulating materials can also be cast and cut into slabs and boards.

h As is true of vegetable and wood-based fibrous insulation, the plastic foams can be destroyed by fire. Depending upon formulation, they may burn readily, slowly, or be self-extinguishing when flame is removed. Variable amounts of smoke and noxious or possibly toxic gases are given off, smoke ranging from moderate to dense, depending upon composition of the foam and depending upon the ready availability of air. This is true of combustible materials generally.

11.12 Bright Metal (Figure 11.4g, h)

a Radiation losses are combatted by bright metallic surfaces, generally aluminum foil several thousandths inch thick. The bright surface of aluminum foil possesses high reflecting power and low emissivity. This combination is made effective by interposing several sheets of the foil in

Insulation

the inter-stud space with air spaces between sheets. Heat radiated from the back of the plaster is largely reflected, and that portion which is absorbed is not readily emitted from the other side of the sheet. Each sheet thus forms a barrier against the transmission of heat across the stud space by radiation. At the same time convection currents are reduced because of the constricted air spaces and because the temperature differential across any one space is fairly small. A free-standing sheet provides one reflecting surface and one low-emission surface, no matter whether heat is traveling outward or inward. When foil is pasted to some backing, the value of one surface is lost.

b Foil is placed in various ways. The sheets can simply be turned over at the edges and fastened to the studs with staples and cardboard strips. This method requires considerable care because foil is fragile. When the foil is pasted to a stout paper backing, it can be placed easily and quickly but to possess the same effectiveness as a sheet of unmounted foil the paper must be coated on both sides. To obtain the full effectiveness of each sheet of foil and at the same time to decrease the labor of placing, therefore, several sheets are often mounted on special paper frames or backs in such a way that the foil is not in contact with the paper except at the edges or at a few intermediate points. One such backing is a plain paper sheet to which are attached several sheets of crinkled foil. The sheets of foil come into contact with each other and the paper backing only where the crinkles happen to meet. Another form is an accordian-fold. Both are stapled to adjacent studs.

11.13 Humidification and Condensation (Figure 11.6)

a. Problem The simultaneous use of insulation and humidification raises the potentially serious problem of condensation in exterior walls.

b In any exterior wall a temperature difference exists from inside to outside in cold weather, beginning at the indoor temperature and descending in a broken line through the various parts of the wall until it reaches the exterior temperature at a point just outside the wall (Figure 11.6a). The slope of actual temperature is steep in good insulating materials and is shallow in good conductors. In uninsulated frame walls the gradient falls off most sharply through the sheathing and exterior finish. In insulated walls the steep portions of the gradient are found in the insulating material (Figure 11.6b), which means that the temperature of the inside of the sheathing is lower in insulated walls than it is in uninsulated.

Dwelling House Construction

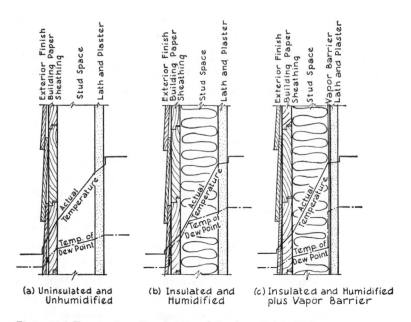

Figure 11.6 Temperature, Dew Point, and Condensation in Walls.

Because of a thin layer of quiet air held at surfaces of materials, there is a temperature drop at those surfaces.

c In addition to the temperature gradient there is an absolute humidity difference which also starts at a high value in the interior and decreases through the wall until it reaches the absolute humidity of the exterior. The absolute humidity of the interior is higher because the warm air is capable of holding much more moisture than the cold outside air, and evaporation from objects and persons within the building, even if there is no humidification system, raises the humidity of the inside air. The higher vapor pressure inside the building generates a tendency for vapor to diffuse outward through the wall. Ordinary building materials such as insulation, plaster, wood, and brick offer relatively slight resistance to the diffusion of vapor but there is enough resistance to cause the absolute humidity to decrease through the wall.

d Cold air cannot hold as much vapor as can warm, and when air at any temperature is saturated with all the vapor it can hold it has reached the dew point and any excess vapor must condense. In a building wall two tendencies are at work. As the temperature of the air in the wall decreases, its capacity for holding vapor drops off in proportion to the temperature gradient (Figure 11.6a, b). The actual amount of vapor in the air also decreases and if it decreases so rapidly that at no point is the air in the wall saturated, condensation does not occur because the dew-point temperature is not reached (Figure 11.6a). On the other hand, if the moisture content does not drop off rapidly enough, dew point is reached somewhere and condensation can occur (Figure 11.6b).

e In ordinary uninsulated walls and in unhumidified buildings, the dew point is apt to be well out in the sheathing and condensation is hardly likely to occur within the stud space (Figure 11.6a). Experience shows that such walls have not caused trouble. In insulated walls, however, the dew point is brought in much farther, and if the building is humidified in addition, so that the dew-point temperature starts at a much higher value, the likelihood of condensation is much enhanced (Figure 11.6b). If condensation does occur, it generally results in blistered paint, discolored plaster, and warped interior trim. It can lead to decay in the frame.

f. Remedy Any system of construction which depresses the humidity sufficiently to prevent the dew-point temperature from coinciding with the actual temperature automatically prevents condensation from occurring. It follows that when moisture is prevented from escaping from the interior of the building into the wall, the moisture gradient falls off very steeply, the dew point is at a much lower temperature, and condensation is much less likely to occur. The logical conclusion is that some sort of barrier must be erected within the wall, close to the interior, to stop the passage of vapor.

g Metal is probably the only practicable material which to all intents and purposes completely stops the flow of vapor. However, complete stoppage is not necessary; greatly retarded flow answers just as well. The ordinary tar paper and light felt commonly used for building paper do not offer much resistance to the passage of vapor, but it has been found that heavy, glossy surfaced, asphalt-saturated paper, 50-lb or more, is a satisfactory barrier.

h The best position for the barrier is immediately behind the plaster or other interior wall finish (Figure 11.6c). Vapor works its way through the interior surface and is stopped or retarded at the barrier, whose temperature is well above the dew point so that no condensation occurs. The small amount of vapor which does pass through is not sufficient to raise the relative humidity at any point in the wall to the dew point.

i The barrier must be installed carefully. There must be no breaks in it, because vapor will flow toward and escape through any breaks which might occur. Any seams in the barrier sheets must therefore be sealed or the seams must be made at studs, joists, or other framing members so that they can be securely fastened and closed down tight when the wall covering is applied. Aluminum foil insulation forms its own vapor barrier but care must be taken to see that the foil is snugly attached to the framing members. The easiest way to install building paper as a seal is to nail it to the studs in vertical strips long enough to run the full height of the room and turn out on floor and ceiling several inches. Seams are made at studs. If the walls are to be plastered on wood or metal lath, the paper must be bellied back into the inter-stud space slightly so that the plaster will have room to push through and key itself to the lath. Foil-coated paper is applied in this manner also, but must be placed with the foil facing the stud space to keep plaster from coming in contact with it because alkalies attack aluminum readily.

j Some insulating materials, especially the bats and blankets, are provided with vapor barriers. For example, vaporseal paper or foil is attached to one side and porous paper on the other, and when the insulation is installed, the vaporseal side is turned toward the interior. The efficacy of these attached barriers must be judged, first, on the effectiveness of the barrier as a vapor retardant and second, on whether the barrier is so arranged as to provide no breaks through which vapor can escape.

k When building paper is applied to the inside of the wall as a barrier it is not necessary to apply more paper on the outside of the sheathing. In fact, it is better that the outer layer be omitted in insulated humidified houses because it is desirable that any vapor which gets into the inter-stud space be allowed to escape through the exterior of the wall as rapidly as possible. If outside paper is desired as an additional wind seal, it should be completely permeable to vapor.

l Certain wall boards have the faculty of retarding vapor sufficiently to form their own barriers. Ordinary plywood, for instance, is not a good barrier but the plywoods made with the synthetic resin adhesives have been found to be good barriers because of the resin layers. Ordinary gypsum board is a poor barrier but when it is backed with a sheet of foil it is a good barrier provided the joints between sheets are sealed.

12 Wallboard. Lath and Plaster

12.1 General

a Interior surfaces of walls and surfaces of partitions and ceilings in dwelling houses are mainly finished with a variety of wallboards, sometimes called "dry wall" construction. Lath and plaster forms the other most important interior finish material.

WALLBOARD, OR "DRY WALL"

12.2 General

a Wallboards used for interior finish include gypsum board, fiber boards, plywood, hardboard, and cement-asbestos board. Of these, gypsum board is most widely employed.

12.3 Specification Clauses

GYPSUM WALLBOARD

a On partitions, wallboard shall be regular gypsum wallboard with square or rounded edges as called for on plans; on outside walls, wallboard shall be insulating board.
b For base for wall tiles in bathrooms, water-resistant grade gypsum wallboard shall be used.

SUPPORTING FRAMING

c Framing shall be at least $2'' \times 4''$ studs set $16''$ o.c., except that framing for closets may be $2'' \times 3''$ studs set $16''$ o.c. or $24''$ o.c.
d As an alternate, metal studs may be employed in place of wood. Metal studs may be cold-rolled channels, dry-wall furring channels, or resilient furring channels.

APPLICATION

e Wherever practicable, gypsum wallboards shall be applied horizontally, i.e., with long edges perpendicular to supporting members. Nails shall be $1\frac{3}{8}''$ long, $\frac{1}{4}''$ flat head annular ringed nails, placed $\frac{3}{8}''$ from edges and ends, spaced $7''$ o.c. on ceilings and $8''$ o.c. on walls. Floating angles shall be employed at ceiling-wall and wall-wall corners.

12.4 Gypsum Board

a Gypsum is a naturally-occurring material, calcium sulfate, that contains two molecules of water per molecule of calcium sulfate. This is heated (calcined) to drive off most of the water, cooled, and ground to a fine powder. When water is added, the calcium sulfate recombines with the water to form a hard material that has good fire resistance, partly because the water bound up in it must be driven off.

b The various forms of gypsum board are flat panels consisting of a gypsum core between surfaces of paper or other sheet materials. Because of the gypsum core, they have good fire resistance, and because of their weight, they can provide sound isolation, depending upon thickness and method of installation. Depending upon the surfacing material employed, they can provide various degrees of abrasion resistance. Their dimensional stability is good. They should, however, be employed in dry locations because persistent moisture can cause disintegration of most gypsum boards.

c Gypsum board is made in the following most important types:

1. *Gypsum wallboard* (Figure 12.1), used for the surface layer on interior walls and ceilings. It has a gray paper back, but a special paper on the facing and edges. The paper is usually cream colored and provides a smooth, even surface ready for finishing and decorating. When a layer of aluminum foil is applied to the back, it becomes insulating wallboard. Pre-decorated wallboard has a decorative paper or vinyl sheet bonded to the front, or exposed face, and does not require further decorating. These wallboards are $\frac{1}{4}''$, $\frac{3}{8}''$, $\frac{1}{2}''$, and $\frac{5}{8}''$ thick. The $\frac{1}{4}''$ thickness is used mainly for lining old surfaces or for direct application to substrates such as masonry. The $\frac{3}{8}''$ is used mainly as the outer layer of a two-ply system. The $\frac{1}{2}''$ is used as a single layer for walls and ceilings in new construction, and the $\frac{5}{8}''$ thickness is used for best quality and enhanced resistance to fire and the passage of sound.

2. *Backing board*, used as the base layer where two or more layers of gypsum board are applied. Backing board has gray paper on all sides and edges. When supplied with a layer of bright aluminum foil on one side, it becomes insulating backing board.

3. *Core board* is 1″ thick, either as a single panel or two $\frac{1}{2}''$ panels laminated together, employed for solid gypsum wallboard partitions. It forms the core and to it are applied additional layers of gypsum board to build out the required thickness of partition.

4. *Type X*, similar in every respect to the preceding boards, except that the core has been made more fire resistant by the addition of glass fiber reinforcements and other materials. It is used where the greatest degree of fire resistance is required.

5. *Water-resistant backing board*, more resistant to the absorption of

Wallboard, Lath and Plaster

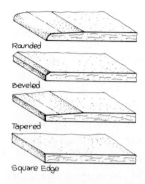

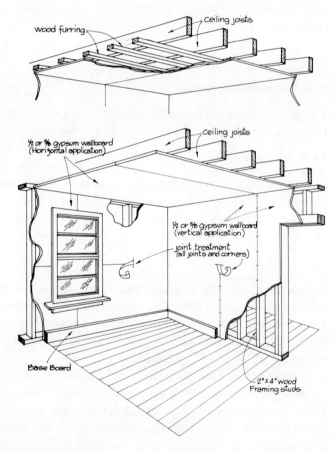

Figure 12.1 Types of Gypsum Wallboard.
Figure 12.2 Application of Gypsum Wallboard on Wood Framing. Horizontal Application on Left Wall and Ceiling, Vertical on Right Wall. Alternate Application on Furred Ceiling Shown Above. Joints Are Taped.

water than ordinary, untreated gypsum board, has water-repellant face papers and water-resistant gypsum core. It is used as a base for wall tile in baths, showers, and other wet areas.

12.5 Application

a Gypsum board may be applied in single or multiple layers (Figure 12.2 and Section 12.13) but in dwelling house construction the single layer, employing regular gypsum wallboard, is most common. The gypsum board can be applied over any firm, flat base, such as wood or metal wall studs, ceiling joists or roof trusses, or furring, such as wood strips or metal channels, supported by underlying construction. Gypsum boards can also be applied directly to masonry or concrete, but furring is generally preferred, especially on outside walls, to separate the wallboard from possible dampness in the masonry.

12.6 Supporting Structure

a Gypsum boards are applied to supporting structures. In dwelling houses this is mainly wood framing and furring, but metal supports are also found.

12.7 Wood Framing

a Framing must be accurately placed so that facings are all in the same plane to allow the gypsum board to fit flat against it. Irregularities leading to high spots and low spots, such as bowed or twisted studs, misaligned plates, and other imperfections, should be corrected before the gypsum board is applied, to avoid loose spots, bulges, cracks, and other blemishes (Figure 12.3).
b Spacing of framing members should not exceed the maximum recommendations given in Table 12.1. Here the term "horizontal" means that the long edges of the wallboard are at right angles to the directions of the supporting members, whereas the term "vertical" means that the long edges are parallel to the supporting members.
c Supports should be sufficiently rigid to prevent buckling or cracking of wallboard. Headers over openings, for example, should be strong and rigid enough to avoid excessive deflection under superimposed loads. Special construction should be provided to support wall-hung equipment and fixtures.
d If spacing between framing members is greater than the maximum recommended for a particular board thickness, or if the surface provided

Wallboard, Lath and Plaster

Table 12.1 Maximum Framing Spacing for Application of Gypsum Board

Gypsum Board Thickness				Maximum Spacing	Two Layers	
Base Layer	Face Layer	Location	Application	One Layer Only	Fasteners Only	Adhesive Between Layers
⅜″	—	Ceilings	Horizontal	16″ o.c.	16″ o.c.	16″ o.c.
⅜″	⅜″	Ceilings	Horizontal	NA	16″	16″ o.c.
⅜″	⅜″	Ceilings	Vertical	NA	NR	16″ o.c.
½″	—	Ceilings	Horizontal	24″	24″ o.c.	24″ o.c.
½″	—	Ceilings	Vertical	16″ o.c.	16″ o.c.	16″ o.c.
½″	⅜″	Ceilings	Horizontal	NA	16″ o.c.	24″ o.c.
½″	⅜″	Ceilings	Vertical	NA	NR	24″ o.c.
½″	½″	Ceilings	Horizontal	NA	24″ o.c.	24″ o.c.
½″	½″	Ceilings	Vertical	NA	16″ o.c.	24″ o.c.
⅝″	—	Ceilings	Horizontal	24″ o.c.	24″ o.c.	24″ o.c.
⅝″	—	Ceilings	Vertical	16″ o.c.	16″ o.c.	24″ o.c.
⅝″	⅜″	Ceilings	Horizontal	NA	16″ o.c.	24″ o.c.
⅝″	⅜″	Ceilings	Vertical	NA	NR	24″ o.c.
⅝″	½″ or ⅝″	Ceilings	Horizontal	NA	24″ o.c.	24″ o.c.
⅝″	½″ or ⅝″	Ceilings	Vertical	NA	16″ o.c.	24″ o.c.
¼″	—	Walls	Vertical	NR	16″ o.c.	16″ o.c.
¼″	⅜″	Walls	NR	NA	NR	NR
¼″	½″ or ⅝″	Walls	Horizontal or vertical	NA	16″ o.c.	16″ o.c.
⅜″	—	Walls	Horizontal or vertical	16ᵃ o.c.	16″ o.c.	24ᵃ o.c.
⅜″	⅜″ or ½″ or ⅝″	Walls	Horizontal or vertical	NA	16″ o.c.	24″ o.c.
½″ or ⅝″	—	Walls	Horizontal or vertical	24″ o.c.	24″ o.c.	24″ o.c.
½″ or ⅝″	⅜″ or ½″ or ⅝″	Walls	Horizontal or vertical	NA	24″ o.c.	24″ o.c.

NA—Not Applicable
NR—Not Recommended

Spacing for Fasteners, Single Ply

Location	Nails	Screws
Ceiling	7″ o.c.	12″ o.c.
Walls and Partitions	8″ o.c.	12″ o.c.

ᵃFor single nailing. For double nailing, end nailing same; field nails in pairs 2″ apart, spaced approximately 12″ between centers of pairs.

Spacing For Fasteners Used With Stud Adhesives

Location	Support Spacing	Fastener Spacing	
		Nails	Screws
Ceiling	16 in. o.c.	16 in. o.c.	16 in. o.c.
	24 in. o.c.	12 in. o.c.	
Load-bearing Walls and partitions	16 in. o.c.	16 in. o.c.	24 in. o.c.
	24 in. o.c.	12 in. o.c.	16 in. o.c.
Non-load-bearing Walls and partitions	16 in. o.c.	24 in. o.c.	24 in. o.c.
	24 in. o.c.	16 in. o.c.	

Source: *Using Gypsum Board for Walls and Ceilings*, Gypsum Association, Chicago, Illinois.

by the framing members is not sufficiently flat, furring should be used (Figure 12.2). The furring surface supporting the gypsum board should be at least 1½″ wide if wood and 1¼″ wide if metal.

e To provide sufficient stiffness, wood furring is best at least nominal 2″ x 2″ material although nominal 1″ x 3″ is frequently used when wallboard is attached to the furring with screws rather than nails. If the furring is backed by solid construction such as masonry, wood furring can be as little as ⅝″ thick by 1½″ wide.

f Wood studs in loadbearing partitions are best nominal 2″ x 4″, and 2″ x 3″ in non-loadbearing partitions. In staggered partitions, 2″ x 3″ studs are employed at 16″ on centers, staggered 8″ apart in opposite rows, spaced 1″ apart. If two-hour fire resistance is required, studs must be 2″ x 4″.

g Ceiling joists should be evenly spaced with bottom edges aligned in a level plane. Excessively bent and crooked joists should be avoided. Those that have a crown should be applied with the crown upward. Joists somewhat out of line can be brought into line by stringers or bracing members (Figure 12.4). Where wide variations in joist spacing occur or where joists are not in a level plane, furring should be applied at right angles to joists at proper distances on centers.

h Where roof trusses are employed and particularly where some roof trusses extend their full length without intermediate partitions, whereas others have intermediate partitions under them, it is highly desirable to have the entire roof and other loads applied before the partitions are put into place. In this way, the roof trusses will have deflected their full amount; otherwise, the subsequent greater deflection of full-span trusses as compared with partially-supported trusses causes unevenness and cracking in the wallboard ceiling. Intermittent snow loads still can cause difficulties.

12.8 Metal Framing

a Although wood framing is most common in dwelling houses, metal framing is found, particularly in metal stud partitions (Figure 12.5).

b Various configurations of metal studs are found. These are inserted between channel-shaped runners top and bottom that take the place of the plates in wood partitions. The spacing for metal framing is given in Table 12.1. Metal furring members are commonly such shapes as rolled channels, special dry wall channels, hat section channels, or dry-wall steel studs.

Wallboard, Lath and Plaster

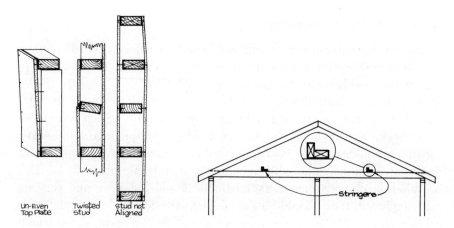

Figure 12.3 Faulty Framing, to Be Avoided.
Figure 12.4 Method of Aligning Ceiling Joists with Horizontal and Vertical Stringers.

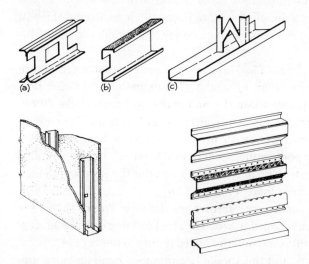

Figure 12.5 (Top) Types of Metal Studs and Runners. (Lower Left) Metal Studs with Applied Board. (Lower Right) Types of Furring: Top, Plain; Center, Two Types of Resilient; Bottom, Cold-Rolled.

12.9 Fasteners, Attachments

a Nails and screws are commonly used to attach gypsum board in either single or multiple layer installations. Clips and staples are used only to attach the base layer in multiple-layer applications.

b Special dry-wall adhesives can be used to bond single-ply gypsum board onto framing, furring, masonry, and concrete. Such adhesives can also be used to laminate the face layer of multiple layers to backer board, sound-deadening board, or rigid foam insulation. Adhesives are generally supplemented with some mechanical fasteners.

c When mechanical fasteners, particularly nails, staples, and screws, are employed, they should be the special varieties made especially for gypsum board. Ordinary wood screws, sheet metal screws, and common nails are not designed to hold the board tightly and to countersink neatly.

d Any fasteners should be placed at least $3/8''$ from the edges and ends of boards. It is best to begin fastening at the center of the board and to work outwards toward the ends and edges. Nails should be driven with a crown-headed hammer to form a dimple not more than $1/32''$ deep around the nail head. It is important not to break the face paper or to crush the core with too heavy a blow.

e. Nails Several types of nails used for gypsum board are shown in Figure 12.6. Heads should be between $1/4''$ and $5/16''$ in diameter and tapered to avoid cutting the face paper when the nail is driven home. Nails should be long enough to go through the wallboard layer or layers and far enough into the supporting construction to provide adequate holding power. For smooth-shank nails penetration should be at least $7/8''$, but for annular-ring nails, $3/4''$ is adequate. The less the penetration, the less the chance that nail popping will occur (see Section 12.15).

f. Screws Several types of dry-wall screws are shown in Figure 12.6. They are used to attach gypsum board to wood or steel framing or to other gypsum boards. Screws of this type pull the board tightly to the supports without damaging the board, and the specially contoured head makes a uniform depression without breaking the paper or providing ragged edges.

g Type W screws are designed for fastening gypsum board to wood framing or furring. The recommended penetration into these supports is $5/8''$.

h Type S screws are designed for fastening gypsum board to metal studs or furring. They are self-drilling and self-tapping; the hardened drill point penetrates sheet metal with little pressure.

Wallboard, Lath and Plaster 271

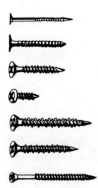

Figure 12.6 Fasteners. (Top) Flat Head Annular Ring Nail; (Second from Top) Flat Head Ratchet Nail. Screws: Type S Bugle, Type S Pan, Type G Bugle, Type W Bugle, Type S Trim.

i Type G screws are used for fastening gypsum board panels to gypsum backing boards. Regular dry-wall screws should not be used for this purpose because they lack sufficient holding power. Nails or longer screws should be driven through both layers into the supporting wood or metal construction.

j Staples are recommended only for attaching the base ply to wood members in multi-layered structures. These should be 16-gauge flattened galvanized wire with minimum $7/16''$ wide crown and spreading points. They should penetrate at least $5/8''$ into the wood.

k. Adhesives Adhesives are employed to bond single layers of gypsum board directly to wood and metal framing, or to other substrates, or to laminate wallboard to such base layers as backing board. Adhesives are generally used in combination with nails or screws to provide supplemental support.

l Three principal types of adhesives are:
1. stud adhesives,
2. laminating adhesives,
3. contact or modified adhesives.

m Stud adhesives are similar to caulking compounds or sealants in consistency. They must possess sufficient strength, bridging ability, aging, and other qualities for long-term performance. Many of these are solvent-based; they harden and stiffen by evaporation of the solvent.

n Laminating adhesives are frequently casein-based and are mixed with water as needed, only in sufficient quantity to be usable in the recommended working time.

o Contact adhesives have a synthetic rubber base, with synthetic resins sometimes added. They are used primarily to laminate gypsum boards to each other or to bond wallboards to metal studs. They have the advantage of providing an immediate bond, with good long-term strength and resistance to fatigue. However, they have little gap-filling ability and, consequently, cannot tolerate significant irregularities in the surfaces of the supporting members. Furthermore, once contact has been made, it is difficult or impossible to shift the wallboard to a new position. For this purpose, modified contact adhesives have been developed to provide a longer placement time, but they may require temporary supports for the wallboard, as is true of laminating and stud adhesives, until their strength has developed.

12.10 Preparation

a Job conditions are important, especially temperature and humidity, because these can affect the performance of joint treatments, the quality of the joint, and the adhesive bond when adhesives are employed. During the winter season, buildings should have controlled temperatures of at least 55°F maintained for at least 24 hours before installation, and afterwards until permanent heat is installed. During warm months, the building need not be completely glazed, but the material should be protected from the weather. Ventilation may have to be provided if humidity is excessive. Windows should be kept open to provide air circulation, and, in enclosed areas, fans should be used. On the other hand, in hot, dry weather, drafts should be avoided to prevent excessive and rapid drying of joint compounds.

b Because lumber shrinks across the grain as it dries, the moisture content should be low enough to avoid excessive shrinkage and nail popping (Section 12.15). Lumber should be no higher than 19 percent moisture content and preferably should be at the final moisture content it will have when the building is in use.

c Gypsum board should be delivered at the time of installation so that it need not be stored for excessive periods of time. When it is stored, it should be stored flat and in its original packing in reasonably dry conditions to avoid moisture pickup.

12.11 Cutting and Fitting

a Careful attention and planning are necessary for successful gypsum

Wallboard. Lath and Plaster

board installation. Boards, if at all possible, should span the entire length or width of ceilings and walls to avoid butt-end joints which are difficult to finish. If butt-end joints must be employed, they should be well staggered, preferably away from conspicuous spots such as centers of walls and ceilings. Ends and edges of boards should fall on supporting members such as studs and joists and not between them. If they must fall between, additional blocking should be supplied. If possible, the long edges of wallboards should be installed at right angles to supporting members because greater stiffness results. On long walls, boards of maximum practical length should be used to minimize the number of end joints. Proper measuring, cutting, and fitting are important, and measurements should be made for each edge or end of a board. Inaccuracies in framing and furring should be corrected before boards are applied.

b Straight-line cuts across the full length and width are made by scoring the face paper, snapping the core, and then cutting the back paper with a sharp knife. Gypsum board can also be sawn by cutting from the face side to preserve a smooth paper edge.

12.12 Single-Ply Application

a. Nail Attachment Wall board may be single-nailed or double-nailed (Figure 12.7). Nail spacing is determined more by requirements for fire resistance and rigidity than for sufficient strength to support the wallboard.

b Ceiling panels should be installed first and then the walls. Joints should not be forced, but should only be loosely butted. Tapered edges should adjoin and square-cut ends should adjoin. Tapered and square-cut ends and edges should not adjoin because a ridge results, difficult to conceal (Figure 12.2).

c On ceilings, nails should be spaced 7" on centers, and on walls, 8" on centers along the supports (Figure 12.7).

d If double nailing is employed, the first set of nails is driven 12" on centers in the middle of the panel and a second set, 2" from each of the nails of the first set, which should be driven again to reset them firmly. Edges and ends falling over supports should be single nailed 7" on centers on ceilings and 8" on centers on walls (Figure 12.7).

e Loose attachment, that is, nailing in which the board is not driven firmly in contact with studs or other supports, should be avoided because it can

Dwelling House Construction 274

lead to nail pops, or nail heads projecting beyond the surface of the wallboard. Loose attachment is avoided by having the supports in the same plane, avoiding too-tight joints, and making certain that the boards are driven tightly against the supporting surfaces.

f. Screws or Screw Attachments When screws are employed, fewer fasteners are required (Table 12.1). They should be spaced 12" on centers on ceilings and 16" on centers on walls where framing members are 16" on centers: or a maximum of 12" on centers on walls and ceilings where framing members are 24" on centers.

g. Adhesives and Nails Adhesives and nails may both be employed to apply gypsum board, and result in a stronger, more rigid system than is true of nails or screws alone. Fewer nails are employed with less chance of defects and blemishes than is true of all-nail attachments (Table 12.1).

h Stud adhesives are applied with a caulking gun, either as a straight bead $\frac{1}{4}$" in diameter or, especially where two adjacent panels join over a supporting member, a serpentine or zigzag bead is applied for ordinary wallboard in which the joint will be subsequently treated and covered (Figure 12.8). If pre-decorated wallboard is used, two parallel beads of adhesives should be applied, one near each edge of the supporting member so that when adjoining pieces of wallboard are pressed against the surface, the adhesive will not squeeze out into the joint.

i Where wallboards are applied to studs with adhesives, only perimeter fasteners are required. These should be spaced 16" on centers along edges and ends that fall along supporting members, and at each crossing of a supporting member. For ceiling applications, the same perimeter fastenings are employed but additional fasteners are spaced 24" on centers in the middle of the field.

j If pre-decorated wallboard is applied with adhesives, it is desirable not to use mechanical fastenings because these may show. In this case, it is desirable to prebow the wallboard so that as it is flattened against the supporting surfaces, it comes into close contact with them. Temporary bracing is required to hold the wallboard in place while the adhesive hardens. Bracing should remain in place at least 24 hours.

12.13 Double-Ply Application (Figure 12.9)

a Double-ply construction is seldom used in dwellings because of the increased cost. It does, however, result in sturdier construction, greater fire resistance, and enhanced acoustical isolation because of the greater

Wallboard, Lath and Plaster

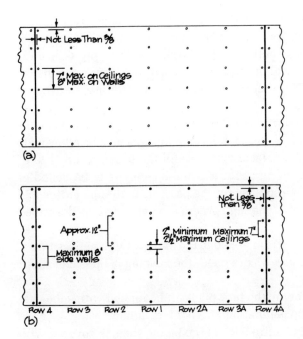

Figure 12.7 Nailing. (a) Single. (b) Double.

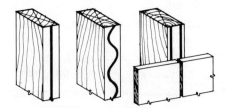

Figure 12.8 Adhesive Beads.

mass, provided that other sources of acoustical leakage are stopped. Nail-popping (see below) and other blemishes are reduced.

b When double-ply construction is employed, the first ply is usually backer board; although wallboard may be used, it is more expensive. Special gypsum board and other sound-deadening boards may be used for enhanced sound isolation.

c The base ply is attached with nails, screws or staples; nails and staples for wood framing, and screws for metal. If the face ply is to be adhesively laminated to the base ply, nails and screws for the base ply are 7″ to 8″ on centers, and screws 16″ on centers. If the face ply is to be nailed or screwed, the nail or staple spacing for the base ply is 16″ on centers and screws are 24″ on centers. Floating corners are employed, as described below.

d The face ply of wallboard is fastened to the base ply with nails, screws, or adhesives. If nails or screws are used, the spacing is the same as for single-ply attachment.

e In double-ply construction, the face ply is often applied with adhesive, by sheet lamination, strip lamination, or spot lamination.

f In sheet lamination, the entire back of the facing sheet is covered with laminating adhesive applied in a ridged configuration with a notched spreader, and the sheet is pressed firmly against the base. Any squeeze-out at the edges and ends is promptly removed.

g Strip and spot lamination are usually preferred in sound-rated construction because sound deadening is enhanced. In strip lamination, adhesive is applied in notched ribbons approximately 6″ wide, 16″ to 24″ on centers. In spot lamination, spots of adhesive approximately 2″–3″ in diameter are applied 8″ to 10″ apart. The board is pressed firmly against the base.

h Temporary supports are needed to hold the facing board in place until the adhesive has hardened. Such supports may, for example, be double-headed nails, or temporary bracing. For fire-rated walls, permanent mechanical fasteners such as nails or screws are required around the perimeters of the boards and in the field.

12.14 Treatment of Joints and Fasteners

a Joints between edges and ends of wallboard and joints at interior corners are usually reinforced with a paper tape embedded in a compound

Wallboard. Lath and Plaster

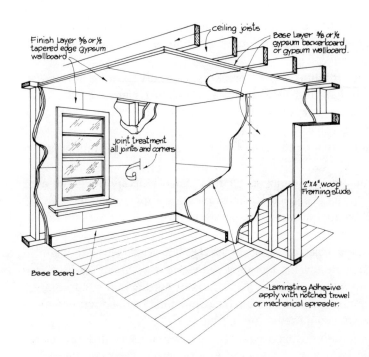

Figure 12.9 Double-Layer Application. Surface Layer of Gypsum Wallboard on Base Layer of Backing Board.

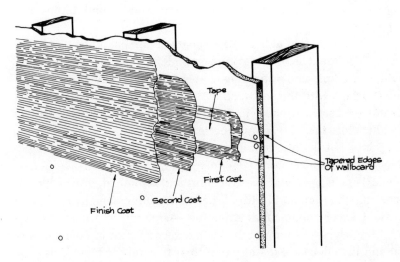

Figure 12.10 Taping Joints. Tape Is Embedded in First Coat; Then a Smooth Surface Is Built Up with Second and Finish Coats.

to provide a smooth, inconspicuous monolithic appearance. Exterior corners and exposed edges are protected with corner and edge trim for appearance and protection.

b The tape used for joint reinforcement is generally a strong fiber paper with feathered edges. It resists tensile stresses across and along the joints (Figure 12.10; also Figures 12.2, 9). The paper is embedded in a compound or compounds, as follows:

1. taping or joint compound, used to embed and bond the taping at the joint,
2. finishing compound, employed for final smoothing and levelling over the joints to provide a smooth, inconspicuous appearance, or
3. All-purpose compounds, combining features of both taping and finishing compounds.

c Dry compounds based on casein or vinyl resins are mixed with water to the proper consistency for handling with a heavy, broad knife. Some harden by evaporation of water, and others by setting of the ingredients; the latter frequently have a gypsum base and must be used within specified time limits to avoid premature setting.

d Taping compounds are used to bond tapes over joints, to fill depressions, and to conceal fasteners and edge and corner trim. They are, therefore, sometimes called "embedding" compounds. Topping, or finishing, compounds are primarily used to conceal and smooth over embedded tapes, fasteners and trim. All-purpose compounds are used for both taping and topping purposes.

e It is recommended that at least three coats of compound be employed with all taped joints. The first coat is an embedding coat to bond the tape, after which two finishing or topping coats are applied and smoothed off to a level surface with the surrounding board as shown in Figure 12.10. To prevent excessive shrinkage, each coat should be allowed to dry thoroughly before the next coat is applied.

f. Flush Joints For flush joints, taping compound is spread into the depression formed by the tapered edges of adjacent boards, and over butted end joints. Tape is centered over the joint and smoothed, after which it is pressed into the compound with a broad knife under enough pressure to squeeze out excess compound. After the compound is dry, it may be sanded if the surface is too rough. Then a second coat of compound is applied and the edges are feathered 2″ to 4″ beyond the tape. This coat is also sanded after drying, if necessary, and a third coat of compound is

Wallboard. Lath and Plaster

spread with its edges feathered 2″ to 4″ beyond the second. It is blended smoothly with the wallboard surface. Taping and spreading of compound can be done by hand or by machine.

g. Interior Corner Joints At interior corners, taping compound is applied to both sides of the joint, after which the tape is folded along a center crease and embedded snugly in the corner to form a right angle. The surfaces are then finished as described for flush joints.

h. Edge and Corner Trim Exposed corners of wallboard need to be protected against damage. This is often accomplished by wood trim, especially at window and door frames and in the form of baseboards at floors.

i Metal trim, such as corner beads and casing beads, is commonly employed to protect exterior corners and exposed edges of wallboard. A number of standard shapes are shown in Figure 12.11. Metal corner beads protect wallboard corners from damage, and other trim protects edges. Trim is generally nailed or screwed approximately 6″ on centers to the supporting construction. These various metal trim shapes require, at least, three-coat finishing with joint compound if a smooth surface is to be obtained and the adjacent fasteners, such as nails or screws, are to be concealed.

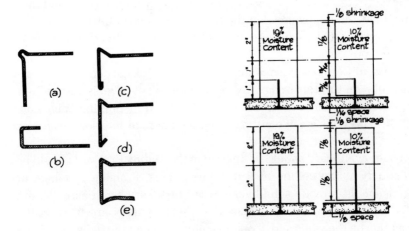

Figure 12.11 Types of Corner Beads and Protective Edging Strips. (a) Corner. (b) U. (c) L. (d) LK. (e) LC.
Figure 12.12 Nail Popping. If Wood Shrinks After Wallboard Is Nailed, Long Nails (Bottom) Cause Separation and Allow Nail Heads to Push Out (Pop). Short Nails Minimize Separation.

j. Fasteners The heads of fasteners, such as nails or screws, must be concealed, and this is generally done with joint compound at the same time that the joints are taped or otherwise protected. This is commonly done with a broad knife and the compound not only fills the depression at the fastener, but is feathered out to the surrounding board. Second and third coats are generally applied at the same time as the second and third coats to joints.

12.15 Damage and Blemishes

a. Nail Popping This may occur any time wallboards become loose. Looseness may be caused by shrinkage of the wood studs, joists, or other supporting members, if the moisture content of the wood is too high to begin with and the wood dries in use (Figure 12.12). For this reason, the shorter the nail used, the better, and the special nails for gypsum board described above are, therefore, recommended. Twisted, crooked boards out of line also lead to looseness and nail popping. The usual repair is to drive a new nail of the proper size about $1\frac{1}{2}''$ from the popped nail by holding the board tightly against the framing. Then the popped nail can be re-seated or withdrawn. The damaged part and the new fastening should be filled smoothly with compound.

b. Cracks Cracks may be caused by a variety of circumstances, and they may be filled, but if movement of the frame is a cause of the cracking, the cracks probably will open again and will need to be refilled from time to time.

c. Cracking at Corners It is not uncommon for cracks to occur in corners because of structural stresses where walls meet walls or walls and ceilings meet. Where this may occur, it may be minimized by floating-angle construction. This consists of omitting some fasteners at interior corners as shown in Figure 12.13.

d As usual, wallboard should be applied to the ceiling first and should be fitted snugly at the ceiling to wall intersections. Framing members or blocking should provide solid backing, but fasteners, such as nails, are omitted at the corners. This means that the first nails are $8''$ from the intersection on the walls and $7''$ on the ceilings. When the gypsum boards are applied to the walls, they butt against the ends of the ceiling panels and the wall panels are nailed along the edges.

e When two walls intersect, the panels applied to the first wall are snugly fitted into the corner, but the nails are omitted along the edge. When

Wallboard. Lath and Plaster

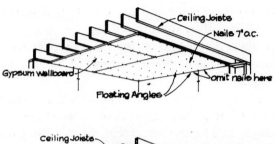

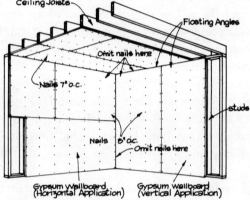

Figure 12.13 Floating Angle Construction. Lower Figure Shows Vertical Application of Wallboard on Right Wall and Ceiling. Horizontal Application Is Shown on Left Wall and Upper Figure. Nails Are Omitted at Edge of Ceiling Wallboard at Corner; Also, First Wallboard on Walls to be Fitted into Corner.

panels are subsequently applied to the other wall, these are butted against the panels of the first wall and are nailed along the edges.

f By this procedure, at least one of the boards abutting the corner has some ability to move as the corner moves, and, therefore, to avoid building up stresses that can lead to cracking.

g Ridges may occur, particularly where ends of boards meet. The reasons may be expansion and contraction of the framing or other movement. It is just as well to allow enough time for the ridges to develop completely before attempting to correct them. Correction consists of sanding the ridges down to the reinforcing tape without actually sanding through the tape, and filling the sanded area with topping compound. After it has dried, it is blended into the repaired area with a light film of additional topping compound as necessary to bring it out to a smooth surface.

h The most effective preventive measure is to use multi-layer gypsum board systems, but this is not commonly done in dwelling house construction. For single-ply construction, back blocking or strip reinforcing is employed to reinforce joints and to prevent the stresses that cause ridging.

i Back blocking is employed when gypsum board ends do not fall on framing members or furring. Gypsum board back blocks extending the full distance between supports are laminated to the back of the wallboard to reinforce the joint and to prevent flexing or movement.

j In strip reinforcing, strips of $3/8''$ backing board or scrap wallboard $8''$ wide are installed $24''$ on centers perpendicular to the framing members as cross stripping to support the edges and provide intermediate support in the field. Where butt ends occur additional short strips are secured directly over the supporting members between the cross stripping.

k Laminating adhesive is applied to the strips in parallel beads, preferably with a split laminating head on the applicator. Wallboard is applied perpendicular to the framing members so that the long edges fall on the cross stripping, and ends are supported on the short intermediate strips.

l. Shadowing Fasteners such as nails and screws are better thermal conductors than the surrounding wallboard and wood studs. Consequently, dust and moisture tend to collect on the surface of the compound over nail and screw heads. This causes a situation known as "shadowing" and is simply a surface discoloration, but often looks like popping. Shadowing is most likely to occur on exterior walls and under roofs or ceilings

Wallboard. Lath and Plaster

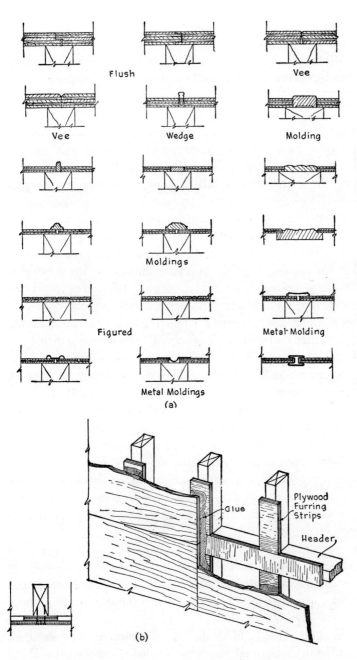

Figure 12.14 (a) Joint Details, Plywood. (b) Reinforced Plywood Joints.

where marked indoor-outdoor temperature variations occur. Shadowing is best avoided by the installation of adequate insulation to prevent large temperature differences, by good ventilation, or by using multi-layer installation. It can be corrected by periodic washing or decorating.

12.16 Plywood (Figure 12.14)

a Plywood consists of an odd number of layers or "plies" of wood veneers glued together with the grain of adjacent plies oriented at right angles. This construction gives the resulting board practically the same properties in all directions, allows the manufacture of large thin boards, and reduces the shrinkage in all directions to a small amount because the longitudinal grain in both directions resists the tendency to shrink or swell in the transverse direction. Splitting is markedly reduced so that nails or screws can be driven close to the edge with little danger. Structural plywood commonly is made in thicknesses from $3/16''$ to several inches, but the most commonly used are all under one inch, with $1/4''$, $3/8''$, $1/2''$, $5/8''$, and $3/4''$ predominating. Three-ply, five-ply, and seven-ply are the most widely employed.

b For ordinary work no special pains are taken to select the wood for the face plies, except to see that no blemishes occur if the material is to be used for a finished face. The most common species used for ordinary plywood is Douglas fir (see Chapter 5). If surfaces are to be exposed they are sanded, if not, sanding is unnecessary. For walls which are to be finished "natural," i.e., which are to show the grain of the wood, specially-selected face veneers may be used in place of the ordinary species because the latter may have an uninteresting figure. Numerous fine veneers exist which may be selected for this purpose, ranging from the most expensive exotic woods to the more common native species such as birch and white pine. These in turn may be cut in any one of a variety of ways to produce the best figures and may be arranged or "matched" in various patterns to produce the effects desired.

c Various glues are employed to bond the plywood, depending upon the degree of exposure to be encountered and upon the possibility of staining thin face veneers. Intermediate-resistant glues are commonly employed if no moisture conditions are anticipated, and phenolic or melamine-formaldehyde glues for severe conditions.

d Plywood is applied directly to studs, joists, strapping, or furring. It should be nailed with small finish nails, such as 4d (Chapter 14) and the heads "set" (driven below the surface) if the joints are not to be covered.

Wallboard. Lath and Plaster

If wood battens are to be nailed over the joints, ordinary flat-headed nails such as 5d or 6d box may be used. Nails should be fairly close together—not over 6" apart—and all edges must be nailed, even if intermediate blocking is required. In addition, the plywood sheets should be nailed to intermediate framing members and to the fire-stopping or similar blocks let in between the studs. For additional rigidity, the edges of studs may be covered with glue before the plywood sheets are applied. Glue for the purpose should be water-resistant.

e Like all wall boards which do not cover the entire wall in one sheet, the treatment of joints between sheets is an important problem. Numerous methods have been tried, of which the following are the most common (Figure 12.14):

1. *Wood battens.* This is the oldest and perhaps the commonest method. Molded or plain wood battens are nailed over the joint, and therefore form raised strips wherever joints occur. By proper arrangement of the sheets on walls and ceilings and by using additional battens even if no joints are to be covered, the effect of panelled areas can be secured. To obtain good panelled effects usually requires strapping, furring, or a solid base, so that the sheets can be applied without regard to the studs.

2. *Accentuated joint.* Instead of attempting to conceal the joint, it may be emphasized and made a part of the decorative treatment of the room by cutting it in a vee, a bead, or otherwise making it prominent. When this is done, it is essential that edges be cut straight so that a good fit between sheets is obtained. By scoring or beading the sheets at other points than at the edges, and by arranging the sheets properly, decorative effects may be obtained.

3. *Moldings and Beads.* Instead of wood battens, small metal moldings may be applied over joints, or small metal or wood beads may be inserted between sheets of plywood and allowed to project from the wall, to provide the reverse of the sunken joint typical of the vee joint.

f Figure 12.14 illustrates a method of resisting small movements at the joints which may be caused by changes in moisture content of the studs or other framing members. It is to fasten furring strips of plywood to the edges of the framing members before the plywood panels are applied. Dimensional changes in the cross-grain direction in the framing members are resisted by the face plies of the plywood furring strips, and joints in the plywood panels are therefore prevented from opening.

Dwelling House Construction 286

12.17 Fiber and Pulp Boards

a Several different vegetable fibers are employed in the manufacture of fiber boards; a common one is bagasse, the pressed-out stalks of sugar cane. Wood pulp may be the product of mechanically-shredded wood, or may be chemically-separated wood fibers, In any event, the base fibers are usually mixed with other materials which help act as binders and to some extent as preservatives, and the resulting wet pulpy mass is rolled or flowed out into sheet form, pressed between rollers to dry and consolidate it, and finally dried in large sheets, usually from $3/8''$ to $1''$ thick. The hardness and weight of the board are functions of the pressures employed in the consolidating operation.Usually they are fairly soft and porous, which gives them better heat-insulating qualities than the denser and stiffer boards such as cement-asbestos and plaster boards. On the other hand, soft boards do not withstand rough usage so well as the harder varieties.

b Application is much the same as plywood, with the same provisions for abundant nailing and blocking behind all edges. If nail heads must be concealed, finish nails are employed; if they are to be covered, flat-headed nails can be used. The latter are to be preferred because these boards do not have the nail-holding power of plywood in which the wood fibers are spread apart by and press against the nails.

c Treatment of joints is an important problem, and is met in largely the same way as for plywood joints, except that no attempt is ordinarily made to provide a concealed joint. Joints are commonly covered with wood battens, battens of the same material as the board, or with fine metal moldings. Frequently the edges are bevelled and the rest of the sheet may be scored with vees of the same shape as the bevelled edges to form geometrical patterns on walls and ceilings. Panelled effects are obtained with moldings of the same shape as the battens.

12.18 Hardboard

a Exploded wood fiber boards are somewhat different in composition from the wood pulp boards. Wood chips are subjected to steam under high pressure which is suddenly released and causes explosive separation of the fibers. These are recombined under heat and pressure into large sheets the density and weight of which depend upon the pressure employed. Soft insulating board is seldom used as a finished surface where there is any likelihood of contact with persons or objects, but the hard board

Wallboard. Lath and Plaster

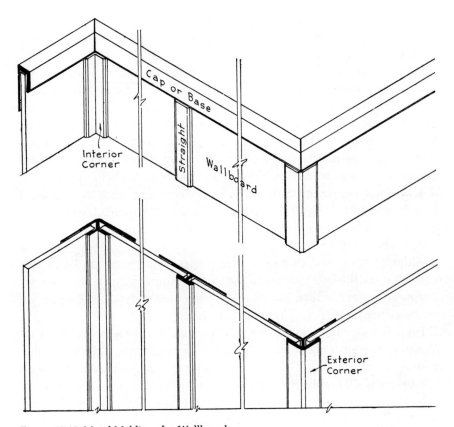

Figure 12.15 Metal Moldings for Wallboards.

provides a wear-resistant finished surface. Application and joint treatments are much the same as for plywood or other boards.

12.19 Cement-Asbestos Boards

a Boards of this type have much the same composition as asbestos shingles. They may be made by flowing a mixture of cement and asbestos fibers into forms and curing under steam, or they may be formed in presses. The former method gives a monolithic sheet which is homogeneous in all directions, the latter a laminated sheet, but one which has a certain amount of flexibility lacking in the monolithic type.

b These boards may be used in their natural gray color, they may have pigments added in the original composition to provide different colors, or they may have a surface coating of enamel, lacquer, or other finish to give a glossy, hard, and possibly figured surface which may simulate marble, tile, or other materials. The natural boards can be painted or otherwise decorated if necessary.

c The thinner ($1/8''$) flexible boards can be nailed without drilling although a certain amount of breaking away of material on the far side is to be expected. Heavier boards ($3/16''$) and the monolithic boards must be drilled. Boards are applied directly to the frame or to strapping, the same as other wall boards. Although the $1/8''$ board can be used without backing on ceilings, it is recommended that it be backed with plywood or similar board on side walls where a severe blow might otherwise crack it.

d Joint details are much the same as for plywood and other hardsurfaced boards, except that battens are not so commonly employed, while metal moldings and beads are much more common. A few typical metal moldings are shown in Figure 12.15.

e Although these boards can be nailed with finish nails (in the thicker varieties) and the heads set below the surface, round-headed drive screws are much better because, like all wallboards except plywood, the cement-asbestos board does not grip nails firmly.

f Typical installation details for wallboards are shown in Figure 12.16. Vee-jointed boards are shown attached to blocked studs. Battens or moldings are shown covering the joints of wallboards applied over furring on a masonry wall. Lightweight fibrous sound-absorbing acoustical tile are attached to furring strips on the ceiling. Baseboards at the floor and corner molding at the ceiling cover the intersections with the walls.

LATH AND PLASTER

12.20 General

a Lath and plaster has been the traditional method of finishing interior walls, partitions, and ceilings. The use of plaster goes back to ancient times. There are two principal constituents, the base (lath, masonry surfaces, and so forth) and the plaster, applied and adhering to the base.

12.21 Specifications Clauses

GYPSUM LATH

a Gypsum lath and veneer plaster lath shall conform to the "Standard Specifications for Gypsum Lath" ASTM C37, except that dimensions do not apply to veneer lath. They shall be plain on partitions, and insulating on walls, as shown on plans.

Wallboard. Lath and Plaster

METAL LATH

b Metal bases for gypsum and portland cement plaster shall comply with "Federal Specifications for Lath, Metal, QQ-L-101a."
c All metal lath shall be fabricated from copper-bearing steel sheets and covered with rust-inhibitive paint after fabrication. Flat rib lath shall weigh 2.5 lb per sq yd, and self-furring diamond mesh 3.4 lb per sq yd.

CORNERITE, STRIP REINFORCEMENT

d Cornerite* and strip reinforcement shall be minimum 1.75 lb per sq yd metal lath coated with rust-inhibitive paint.

PLASTER

e Gypsum plasters shall be mill-mixed bond, gauging, or gypsum plasters as shown on plans. They shall conform to "Standard Specifications for Gypsum Plasters," ASTM C28.
f Hydrated lime shall conform to "Standard Specifications for Special Hydrated Lime," ASTM C206.
g Portland cement for plaster shall conform to "Standard Specifications for Portland Cement," ASTM C150, Type I, II, or III.

SAND

h Shall be clean and well graded from coarse to fine, meeting requirements of ASTM C144.

WATER

i Shall be clean, fresh, suitable for human consumption.

12.22 Framing

a Framing requirements for lath and plaster are similar to those for wallboard. In dwellings, this is mainly wood, but metal framing, especially studs and furring, is becoming more common (Figures 12.2, 5, 9). Framing must be sturdy, rigid enough to avoid excessive deflection, and reasonably plane and level to avoid the requirement for excessive straightening by the lath and plaster, although lath and plaster can accommodate greater deviations from line and plane than can wallboard. A rule of thumb respecting rigidity is that a ceiling or wall should not deflect under imposed (live) load more than $\frac{1}{360}$th part of the span, e.g., in a 10'-0" (120") span, the deflection should not be more than $\frac{1}{3}$".

*See Section 12.24d for a description of Cornerite.

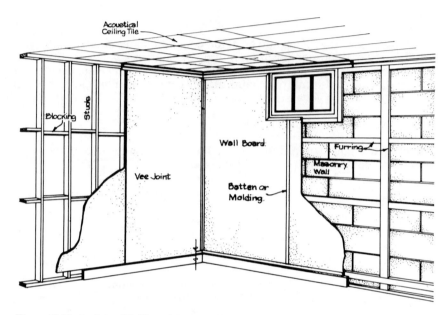

Figure 12.16 Applying Wallboard Over Studs and Furring.

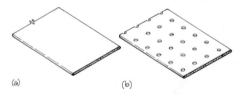

Figure 12.17 Gypsum Lath. (Left) Plain. (Right) Perforated.

Wallboard. Lath and Plaster

12.23 Lath

a At one time, most plaster was applied to four-foot wood lath. Although it has now virtually disappeared, its influence still lingers on in the four-foot dimensions of most wallboards and the 16″ or 24″ spacing of studs and other framing members. Today, when plaster is used in dwelling houses, it is applied mainly to gypsum lath and, occasionally, to metal lath.

12.24 Gypsum Lath

a Gypsum lath is similar to the gypsum wallboards already described, except that the lath is made of gray porous paper on both sides and the edges are rounded. The most commonly used thickness is $3/8″$ and the common sizes are 16″ by 48″, although 96″ is also common, and lath up to 12′-0″ in length can be obtained. Veneer plaster (Section 12.32), on the other hand, is commonly applied to boards 4′-0″ wide by 8′-0″ long. Gypsum lath may be plain, or perforated with holes to provide a mechanical key in addition to the adhesion of the plaster to the paper surface (Figure 12.17). Insulating lath, like insulating wallboard, has a sheet of bright aluminum foil attached to the back, that is, the side applied to the studs or other framing members.

b Lath should be applied with its long dimension perpendicular to the framing members. For $3/8″$ lath, framing members should be not more than 16″ on centers, and for $1/2″$ lath, 24″ on centers. The ends of lath must have bearing on a framing member unless special clip attachments are used. The folded or lapped edges of the paper should be placed toward the framing member. Edges and ends should be in moderate contact. If spaces greater than $1/2″$ occur between edges and ends, these should be reinforced with metal-lath strips.

c Each piece of lath should be fastened with four nails, staples, or screws to each framing member. Fasteners should be driven home so that the head is just below the paper surface without breaking that surface. If the lath is not tightly driven against the framing member, the resulting looseness may result in nailhead popping (Section 12.13). Holes for electrical outlets, plumbing, and so forth must be neatly cut.

d Interior corners should be reinforced with "Cornerite" (Figure 12.18) which consists of a strip of metal lath bent into the corner and secured to the lath with staples or wire. Cornerite and other metal lath strips (Figure 12.18) used as reinforcing should be secured to the lath and not by nailing through the lath into the framing, particularly at corners where warping or twisting of the framing caused by motion can be transmitted into the plaster coats and induce plaster cracks.

e Exterior angles are finished with corner beads (Figure 12.19) set accurately to line. Casing beads are used at windows and doors, and metal or wood grounds are used at openings, trim, and other places where it is essential that the finished plaster be brought to true thickness and a straight level surface. Picture molds are employed where pictures are to be hung. Corner beads, grounds, and so forth should be fastened to the framing or furring.

12.25 Metal Lath (Figure 12.20)

a Metal lath consists of expanded metal or of wire fabric. For expanded metal, sheets of steel, preferably copper-bearing, are pierced with lines of offset splits which provide a diamond mesh pattern when the sheets are pulled laterally. The principal types are:
1. Flat and self-furring diamond mesh weighing 2.5 and 3.5 lb per sq yd. Self-furring mesh is dimpled at intervals so as to hold the mesh away from the flat substrate.
2. Flat rib lath, weighing 2.75 and 3.4 lb per sq yd, has shallow stiffening ribs at intervals.
3. Three-eighths inch rib lath, weighing 3.4 and 4.0 lb per sq yd, is similar to flat rib lath, except for the ⅜" ribs.
4. Three-quarter-inch rib lath is still heavier and weighs 0.60 and 0.75 lb per sq ft.

b All of the foregoing laths should be coated with rust-inhibitive paint, at least, after fabrication.

c Wire fabric lath is fabricated of copper-bearing cold drawn steel wire and is galvanized or coated with a rust-inhibitive paint after weaving or welding, or is fabricated from galvanized wire.
1. Standard welded wire fabric lath, fabricated of not lighter than 16-gauge wire, with a mesh not less than 2" square, is intended for interior surfaces or protected horizontal exterior surfaces.
2. Welded wire fabric may have a paper backing to help contain the plaster.
3. Stucco netting is intended to be used for exterior surfaces with or without backing. It is woven of steel wire with openings ranging from 1"

Wallboard. Lath and Plaster

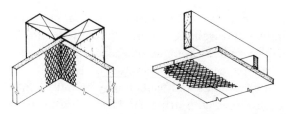

Figure 12.18 (Left) Cornerite. (Right) Strip Reinforcement.

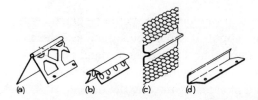

Figure 12.19 (a, b) Corner Beads. (c) Base or Parting Screed. (d) Square Casing Bead.

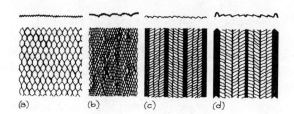

Figure 12.20 Metal Lath. (a) Flat Diamond Mesh. (b) Self-Furring. (c) Flat Rib. (d) ⅜″ Rib.

to $2\frac{1}{4}$″ in size, but not more than 4 sq in. per opening. The minimum size of wire is 18-ga. for 1″ openings, 17-ga. for $1\frac{1}{2}$″ openings, and 16-ga. for 2″ openings.

4. Wire cloth is fabricated of not lighter than 19-ga. wire, and not more than $2\frac{1}{2}$ meshes to the inch. It is galvanized or coated with rust-inhibitive paint.

d Table 12.2 gives recommended weights and spacing of supports for metal lath.

e Metal lath should be erected with the long dimension of the sheet across the supports; rib lath, either wire or expanded metal, with the ribs across the supports.

f Lath should first be applied to the ceilings and the sheets carried down 6″ on the walls and partitions. If metal lath is not used on the walls, the ceiling lath may be bent and carried down 6″ on the wall, so that no joints occur at the juncture of ceiling and walls. On walls, the lath should be started one stud away from the corner, be bent into the corner, and carried over the abutting wall so as to avoid a joint at the juncture of the walls.

g Where $\frac{3}{8}″$ rib lath is used on ceilings and walls it may be butted into all corners, and the corners reinforced with strips of flat lath—Cornerite—not less than 8″ wide, bent into the shape of an "L," and securely wired along each edge. Cornerite should not be fastened in the corner but only along each edge.

h Lath must be securely attached to supports and should be placed with the lower sheet lapping over the upper. It should be lapped at sides not less than $\frac{1}{2}″$ and at the ends not less than 1″, and there should be a wire tie on side laps half way between supports. End laps should occur only over supports. Rib lath is lapped at sides by nesting the outside ribs. At ends, the lap should be 1″.

12.26 Masonry Bases

a A number of masonry materials provide adequate bases for plaster.

b Gypsum partition tile bonds well to plaster and is also provided with a rough face for a mechanical key. Only gypsum plaster should be applied to gypsum block because lime and portland cement do not adhere well to gypsum.

c Brick and clay tile form good bases for plaster provided they are sufficiently porous to allow for absorption and the development of a good bond. Hard-surface glazed tile and brick should not be used. When clay tiles are employed as plaster base, they preferably should be scored and otherwise roughened to provide a mechanical key.

d Most concrete blocks provide a satisfactory base for plaster when the blocks are properly aged. On walls, if the block's surface is reasonably rough and porous, gypsum plaster can be applied directly, but on ceilings, if concrete filler blocks are used, only gypsum bond plaster should be employed. Portland-cement plaster bonds well to concrete block. Plaster may be applied successfully to monolithic concrete depending upon the surface of that concrete. High-strength, dense, vibrated concrete providing smooth, non-porous surfaces may not be satisfactory for the application of plaster. Rough-surfaced concrete, made with rough-sawn form boards and devoid of oil or other parting materials may provide a satisfactory base. However, if there is any chance of moisture getting into any

Wallboard. Lath and Plaster

Table 12.2 Metal Lath

Type of Lath[a]	Minimum weight of lath (lb/sq yd)	Maximum allowable spacing of supports (in.)			
		Vertical Supports		Horizontal Supports	
		Wood	Metal	Wood or Concrete	Metal
Flat expanded[b] metal lath	2.5	16	12	0	0
	3.4	16	16	16	13½
Flat rib metal lath	2.75	16	16	16	12
	3.4	19	19	19	19
⅜-in. rib metal lath[b]	3.4	24	24	24	19
	4.0	24	24	24	24
Sheet metal lath	4.5	24	24	24	24
Wire lath	2.48	16	16	13½	13½
V-stiffened wire lath	3.3	24	24	19	19
Wire fabric	c	16	16	16	16

[a] Lath may be used on any spacings, center to center, up to the maximum shown for each type and weight.
[b] Rod-stiffened or V-stiffened flat expanded metal lath of equal rigidity and weight is permissible on the same spacings as ⅜-in. rib metal lath.
[c] Paper-backed wire fabric, 16-ga. wire, 2 x 2-in. mesh, with stiffener.

Table 12.3 Gradation of Aggregates (ASTM C-35)

	Percentage Retained on Each Sieve					
	Sand by Weight		Perlite by Volume		Vermiculite by Volume	
Sieve Size	Max.	Min.	Max.	Min.	Max.	Min.
No. 4 (4,760-micron)	0	—	0	—	0	—
No. 8 (2,380-micron)	5	0	5	0	10	0
No. 16 (1,190-micron)	30	5	60	10	75	40
No. 30 (590-micron)	65	30	95	45	95	65
No. 50 (297-micron)	95	65	98	75	98	75
No. 100 (149-micron)	100	90	100	88	100	90

Source: *Manual of Gypsum Lathing and Plastering*, Gypsum Association, Chicago, Illinois

Table 12.4 Proportions for Plaster Mixes

Plaster Base	Sand damp, loose		Water	Perlite or Vermiculite	Water
	Volume (cu ft)	Weight (lb)	Weight (lb)	Volume (cu. ft.)	Weight (lb)
Base Coats, per 100 lb plaster					
Gypsum Lath					
Two-coat work	2½	250	62	2	64
Three-coat work					
scratch	2	200	57	2	64
brown	3	300	68	2 [a]	64
Metal Lath					
Three-coat work					
scratch	2	200	57	2	64
brown	3	300	68	2 [a]	64
Masonry					
Two- or three-coat	3	300	68	3	76

[a] Where the plaster is 1 in. or more in total thickness, the proportions for the brown coat may be increased to 3 cu ft.

Note: For three-coat work over gypsum or metal lath the proportions of sanded plaster only may be 1:2½ for both scratch and brown coats in lieu of above.

Finish Coats

Manufacturer's specifications to be followed, some approximate proportions are:

Trowel

Gypsum-lime putty

100 lb dry gauging gypsum plaster, 200 lb dry hydrated lime. Over perlite or vermiculite base coat, add at least ½ cu ft fine silica sand or perlite fines per 100 lb gauging plaster.

Keene's cement-lime. putty

Medium hard: 100 lb dry Keene's cement, 50 lb dry hydrated lime

Extra hard: 100 lb dry Keene's cement, 25 lb dry hydrated lime

Add ½ cu ft silica sand or perlite fines per 100 lb Keene's cement if applied over perlite or vermiculite base coat.

Float

Keene's cement-lime

400–600 lb silica sand, 150 lb dry Keene's cement, 100 lb dry lime hydrate

Prepared finishes, and acoustical plasters, follow manufacturer's directions.

Source: *Manual of Gypsum Lathing and Plastering*, Gypsum Association, Chicago, Illinois

Wallboard. Lath and Plaster

kind of masonry wall or of condensation occurring, it is far preferable to fur such walls and provide an air space or insulation behind the lath and plaster.

e On hard dense masonry or concrete surfaces, it is preferable to apply metal lath. In any event, if plaster is applied to concrete, the first coat should be a bond plaster.

f Bituminous coatings are frequently applied over concrete or other masonry surfaces to exclude moisture. Such coatings do not provide a satisfactory base for plaster. Furring is recommended.

12. 27 Plaster

a Gypsum is the most commonly-used plaster. It is manufactured by calcining naturally-found gypsum to drive off most of the water of crystallization. When calcined gypsum is mixed with water, the water of crystallization is replaced and the material sets or hardens. Because this reaction can occur quickly (as in Plaster of Paris, a similar material), retarders are added to the gypsum to reduce the rate of set and to permit mixing and application before it stiffens too much to be worked. As normally employed, gypsum plaster hardens in a matter of several hours.

b For finish coats, lime is commonly employed in combination with gypsum or Keene's cement. Lime is made by burning limestone, or calcium carbonate, into the form known as quick lime, or calcium oxide. This, in turn, is mixed with water to form slaked lime, or calcium hydroxide, the form in which lime is used as plaster. When it is mixed with water and applied to a wall, the calcium hydroxide slowly reacts with carbon dioxide in the air to convert the material back to the original calcium carbonate. This is a slow reaction and, in thick layers, may never go to completion.

c Keene's cement is similar to gypsum, except that it is calcined at a higher temperature and the water of crystallization is virtually all driven off. The resulting material, when mixed with water, crystallizes as does ordinary gypsum plaster, but the final product is considerably harder and denser.

12.28 Plaster Aggregates

a A number of materials, or aggregates, can be added to gypsum to provide bulk, reduce shrinkage, increase strength, and for economy.

1. *Sand* is the most commonly-used plaster aggregate because of its economy and general availability. For best results, the particle size should

have the distribution recommended in Table 12.3. Proper gradation adds workability to the mix. Sand must be clean and usually has to be washed. Foreign matter, such as dirt, can easily reduce the strength of plaster by 50 percent. Commonly-employed mixes are shown in Table 12.4.

2. *Perlite* is frequently used as a plaster aggregate. Raw perlite is a volcanic glass which, when roasted at 1,400°–2,000° F, expands into a frothy mass of glass bubbles 4 to 20 times the original volume. Thus it weighs only 7½ to 15 lb per cu ft and effectively reduces the dead weight of plaster to approximately half that made with sand. Perlite plasters are effective fire barriers and are superior to sanded plasters. The strength, on the other hand, is somewhat less than that of sanded plaster. Recommended mixes are shown in Table 12.4.

3. *Vermiculite* is another lightweight aggregate. It is based on a form of mica which expands 6 to 20 times when heated at 1,600°–2,000° F. Several types are manufactured, of which Type 3 is used for plastering. Perlite and vermiculite, like sand, must be properly graded to form a satisfactory aggregate for plaster. Vermiculite weighs 6 to 10 lb per cu ft and, like perlite, therefore reduces the dead weight of plaster. As is true of perlite, vermiculite plasters are less strong than sand plasters and, therefore, the same precautions with respect to proportions should be followed as for perlite.

4. *Aggregates for Finish Coats.* For smooth trowel finishes, fine silica sand or perlite fines are added. They increase the resistance to cracking or crazing, especially over lightweight aggregate base coats. For float finishes where a textured surface is wanted, clean graded sand passing through number 12, 16, or 20 sieves is added. Other aggregates, such as vermiculite, may also be employed.

12.29 Application

a. Three-coat Plaster This consists of (1) a scratch coat applied to the plaster base and cross-raked after it has begun to stiffen, (2) a brown coat, which is brought out to the grounds, smoothed, and allowed to set and dry partially, and (3) the finish coat. Three-coat plaster is normally employed over metal lath, over gypsum lath on ceilings where the lath is attached by clips providing edge support only, over ⅜" perforated gypsum lath on ceilings, and over ½" gypsum lath attached to horizontal supports more than 16" on centers.

Wallboard. Lath and Plaster

b. Two-coat Work This is similar to three-coat except that the scratch coat is not raked and the brown coat is applied or "doubled back" within a few minutes to the unset scratch coat. This method is normally employed in applying plaster to masonry and gypsum lath. Three-coat work is considered to be somewhat superior to two-coat, but in the interest of speed, the two-coat method is widely employed.

c The thickness of the base-coat plaster, whether scratch plus brown or doubled up, is crucial to obtaining good work. Grounds are employed around openings, at baseboards, and at trim generally to control the thickness and give the plasterer a straight line with which to work. On large surfaces, intermediate screeds may also be employed so that the plasterer has a guide to work to. Grounds and screeds should be employed as follows:

1. $\frac{1}{2}''$ over gypsum lath and gypsum partition tile,
2. $\frac{5}{8}''$ over brick or other masonry,
3. $\frac{5}{8}''$ over the face of metal lath, that is, $\frac{3}{4}''$ from the back of metal lath.

12.30 Standard Base-Coat Plasters

a Base-coat plasters are applied directly to the lath or masonry substrates and support the finished coats. They are also truing coats; they are brought to an essentially level surface and take up any irregularities that may occur in the substrates. Since the base coat also provides the foundation for the finished coat, it must be compatible with and capable of supporting the finish coat without cracking, spalling, or otherwise deteriorating.

b Base-coat plasters come in several varieties:

1. *Neat plaster* must usually be mixed with an aggregate such as sand, perlite, or vermiculite, plus water. The neat plaster may be fibered or unfibered.
2. *Ready-mix plasters* contain all of the aggregates necessary, so that only water needs to be added at the job.
3. *Wood-fibered plasters* are neat plasters containing wood fibers, and may be used without other aggregate. The wood fibers give bulk and coverage to the plaster.
4. *Bond plasters* are especially formulated for use on interior monolithic concrete surfaces to provide a stronger bond than would be available with ordinary plasters. Only water needs to be added at the job.

c. Mixing Mechanical mixing is preferred to hand mixing, but both are feasible. In mechanical mixing, approximately 90 percent of the water is charged into the mixer, and, if sand is employed, half the sand, followed by all of the plaster and then the other half of the sand. Some water is held back so that overwatering will not occur. The moisture content of sand varies from dry to quite high. If perlite and vermiculite are used, they are all charged into the mixer with 90 percent of the water before the plaster is added. Mixing time is not less than a half minute nor more than three minutes, after which the additional water can be added, if necessary, to get the proper consistency. The entire batch is dumped at one time.

d In hand mixing, a plaster aggregate, whatever it is, should be mixed dry to a uniform color in a mixing box. The water is added, then the plaster is hoed in immediately and thoroughly mixed to the proper consistency.

e Once mixing is completed, the plaster should be used immediately. A batch of plaster that has started to set should not be employed.

f Plaster may be applied by machine or by hand. Machine application is preferred, especially on large jobs. There is a shorter time between mixing and actual application to the wall and greater uniformity and density can easily be obtained. One man handles the nozzle and applies the plaster. He is followed by other plasterers who darby the plaster. In hand work, the plasterer applies the plaster to the wall by hand, pressing it firmly to the surface and making certain that, in the case of metal lath, the plaster penetrates the metal lath and completely encloses it from the far side. He follows it with a brown coat.

g Drying is extremely important to the quality of the base plaster. It should neither be too fast before the plaster is set and thus leave insufficient water for the chemical reaction, nor should it be so slow after the plaster has set as to impair the strength. Rapid changes of temperature should be avoided while the plaster is setting and drying. The minimum temperature should be 55°F for a period of at least seven days before plastering, and during the plastering operation until the plaster is dry. There should be an even distribution of heat without local hot spots. In hot dry weather, openings should be closed with cloth or sash during the application of the plaster and until it has set. Afterward, ventilation should be provided until it is dry, and drying should be as quick as possible once the setting has occurred.

12.31 Finish-Coat Plasters

a The finish coat provides the visible surface of the plaster and must, therefore, not only be architecturally correct, but must also meet the user requirements and be compatible with the base coats and substrates.

Wallboard. Lath and Plaster

b There are six principal types of finish coats:
Smooth finishes:
 Gypsum-lime putty, trowel finish
 Keene's cement-lime putty, trowel finish
 Prepared gypsum, trowel finish
Float finishes:
 Keene's cement-lime sand float finish
 Gypsum-sand float finish
Acoustical Plaster

c The gypsum used in finish plaster is known as gauging plaster and is specially ground to a coarse texture with low consistency. It absorbs water readily and blends easily with lime and other constituents.

d In mixing, the lime is first mixed with water into a smooth putty which may be allowed to stand for the necessary length of time for complete wetting, or may be ready to use immediately, depending on the type of lime employed. To this is added the recommended volume of silica sand or perlite fines together with the gauging plaster or Keene's cement sifted into water. The mixture is then thoroughly blended.

e Trowel finishes are applied over the nearly-dry base coat with sufficient pressure to force the material into the slightly roughened surface. Very thin applications are made over the wall or ceiling, and suction by the base coat draws the finish into the pores of the base coat. After this has "drawn up," the second or leveling coat is applied and left smooth under the trowel. The plasterer wets the finished surface, at the same time applying considerable pressure with the trowel to densify the surface and provide the smooth, hard finish desired.

f For sand-float finishes, the base coat should be uniformly damp to avoid drying the finish before it can be floated; otherwise, the application is quite similar to that for the trowel finish. In place of using a trowel, a wood float is frequently employed first to give a rough surface which may be worked into final form with a rubber float.

g In addition to these two standard types of finishes, many special finishes can be applied such as swirl, spatter and so forth. Evidently, there are many variations.

12.32 Veneer Plaster

a The term "veneer plaster" covers a multitude of gypsum plasters, almost exclusively proprietary in nature, which vary widely in strength, hardness, and composition. With such wide variations in physical characteristics, no meaningful generic description is possible.

b While a veneer plaster generally can be used over a veneer base (lath) other than that for which it was designed, each veneer plaster–veneer base combination is designed as a system, to provide optimum cost and performance for each market's requirements.

c Surface finish desired by a market dictates whether the veneer plaster shall be mill-aggregated, job-aggregated or unaggregated, and whether it shall be a two-material (base coat and finish coat) or a single material (finish coat) system.

d To provide the optimum application and finishing characteristics, veneer plaster thicknesses must be held to not more than $3/32''$ to $1/8''$. At this thickness, in combination with the specified veneer base, it must yield adequate strength, hardness, impact resistance, and crack resistance to meet the design requirements of the assembly in which it is used.

e With proper design, veneer plaster–veneer base systems can offer fire retardancy, sound control, crack resistance, and appearance at least comparable to most conventional wallboard and plaster systems.

f Instead of using standard gypsum lath, the base for veneer plaster is generally gypsum board $48''$ wide and $6'$ to $16'$ long, the latter on special order. Thicknesses are $3/8''$, $1/2''$, and $5/8''$. Regular, Type X (fire-retardant), and foil-backed insulating base are available.

g Requirements for wood or metal framing and support members such as studs, joists, and furring are the same as for gypsum wallboard and lath and plaster. Spacing is not greater than $16''$ on centers except for $5/8''$ thick veneer base for which a $24''$ spacing is permissible for sidewalls, and for ceilings if the base is applied horizontally, i.e., with edges perpendicular to supporting members (joists or furring).

h Veneer base is applied in much the same manner as wallboard and backing board, with nails, screws, and staples. Nail penetration into wood should be $3/4''$ to $7/8''$. Spacing of fastenings is essentially the same as for wallboard and backing board. Floating interior corners are recommended, as for regular gypsum board (Section 12.15).

i Backing board is employed for double-ply construction, with veneer base applied over it.

j Joints and corners are reinforced with a mesh tape, commonly glass fiber. This is stapled to the veneer base. Special corner beads, casing beads, grounds, and other trim are applied.

k Temperature and ventilation requirements are similar to those for gypsum wallboards. It is recommended that the temperature be maintained between 55° and 70°F for at least a week before application of plaster, during the plastering operation, and at least a week afterward. Hot spots and drafts should be avoided. In hot, dry weather, windows should be opened only partially (3" top and bottom), to avoid excessive drying; unglazed openings can be protected with open-weave fabric such as muslin or cheesecloth. On the other hand, once the plaster has hardened, excess water must be allowed to evaporate, and sufficient ventilation, free of local drafts, is needed.

l Mixing must be in accordance with manufacturer's directions. These materials come ready-mixed and need only to have water added.

m Plaster is applied by hand or machine.

n In most cases, two coats are applied. Flat mesh-reinforced joints are first troweled over with sufficient pressure to embed the mesh and press the plaster against the base. The first or base coat is then immediately applied directly to the veneer base and straightened to a level surface between $\frac{1}{16}$" and less than $\frac{1}{8}$" thick.

o The finish coat can be applied immediately to green base coats, or it can be applied to partially dry or dry base coats. A smooth trowel finish is obtained by using a steel trowel to "scratch in" a thin layer of plaster securely to the base and doubling back to provide a true, hard surface. Textured finishes are obtained by trowelling and building up to a true surface and finishing by floating with the appropriate tool. Finish coat thickness is approximately $\frac{1}{16}$".

p Veneer plaster may be applied as a single coat by using a properly formulated single-coat mix and applying directly to the veneer base and building up to the required thickness.

q Veneer plaster may be applied over masonry surfaces and cast concrete. Such surfaces must be clean, free of dirt, oil, grease, or other contaminants. Mortar joints should be struck flush, but, if recessed, are first filled with base coat material firmly pressed in. Cast concrete surfaces should be treated with a bonding agent.

r Any marked irregularities should be smoothed, and depressions filled in with base coat, scratched in, levelled off, and allowed to set.

s A tight scratch coat is trowelled over the entire area and doubled back to a minimum thickness of $\frac{1}{16}''$ or more as needed to obtain a level surface. Finish coats are applied as described above.

t The foregoing procedures and thicknesses are approximately correct, but individual manufacturers have their own formulations, and their recommendations should be carefully followed.

u Under favorable conditions, if recommended procedures are followed, veneer plaster may be hard and dry enough to permit finishing operations to go forward in as little as 24 hours, but it is probably safer to permit a longer time in many situations.

STUCCO

12.33 General

a Stucco, or portland cement plaster, makes an excellent exterior and interior wall covering, as is attested by its long and honorable history. Care, patience, and meticulous attention to detail are required to obtain satisfactory stucco, particularly in rigorous climates.

b Lime has been used for centuries in stucco, but its slow hardening has led to the general adoption, in the United States, of more rapidly hardening cements. Portland cement is the most widely used. With white portland cement and lime, a wide range of colors is obtainable, particularly when colored sands and stone chips are employed with the colored pigments.

c If the surface to be stuccoed is masonry, two coats are applied; if the surface is metal lath, three coats are required. These are similar to the coats of plaster employed on interior wall surfaces.

12.34 Preparation of the Surface

a Because stucco, like plaster, is a thin, hard, unyielding, and brittle material, its support must be strong and rigid. If walls must be of frame construction, they should be well braced and so framed as to avoid shrinkage and expansion in the frame and sheathing which would cause cracks in the stucco. With any kind of surface—masonry, concrete, or frame—foundations must be firm and unyielding, with ample and carefully proportioned footings, to avoid differential settlement which would be reflected in cracked stucco.

Wallboard. Lath and Plaster

b. Masonry Brick, stone, concrete, tile, or any other masonry surface must be cleaned of all loose particles, dirt, and grease, and smooth surfaces hacked, sand-blasted, or otherwise roughened to improve the mechanical bond. Tile should be heavily scored, and all mortar joints rough struck. Cleaning, especially of greasy surfaces and of thin mortar streaks, may be accomplished with dilute commercial muriatic acid rinsed afterward with clean water. Form oil can be removed with strong soap and water, thoroughly rinsed. Walls should be thoroughly hosed and then allowed to dry only enough to leave the surfaces well dampened before applying the base coat.

c Wood-framed bases are of two types, open and sheathed. In open framing, the sheathing is omitted. Strands of soft annealed 18-ga. or heavier wire are strung tightly 6" on centers along the studs, first nailed or stapled every fifth stud, then pulled alternately up and down as nailed to the other studs to pull the wire taut. Waterproof building paper is attached over the wires. This combination of wire and paper forms a backing for the scratch coat of stucco. Metal lath, such as hexagonal stucco netting, is attached to the studs, and properly furred.

d On sheathed walls, the wire strands are not necessary. Sheathing is covered with building paper. Metal lath is applied over the building paper, and furred out at least $\frac{1}{4}$".

e. Building Paper Wood walls must first be covered with heavy, impregnated, waterproof building paper, lapped at least 3" horizontally with the upper layer lapping over the lower. Vertical laps should be at least 6".

f For furring on wood walls, self-furring lath (Section 12.25), furring nails, or furring strips are employed. On masonry walls, furring is not generally necessary but may be employed over unsuitable surfaces. Furring must allow at least $\frac{1}{4}$" between the surface of the wall and the under surface of the lath.

g. Metal Reinforcement Expanded metal lath, expanded stucco mesh, stucco netting, and wire fabric are most commonly employed. Paper-backed metal fabric is available. Expanded lath and mesh should be made of copper-bearing sheets. Reinforcement of all types should be galvanized or protected with rust-inhibiting paint. Openings should not exceed 4 sq in. in area, but must be large enough to permit the scratch coat to penetrate and to embed the metal completely to prevent corrosion.

h Expanded metal lath should weigh not less than 3.4 lb per sq yd for horizontal surfaces, and 2.5 lb per sq yd for vertical surfaces. Expanded stucco mesh should weigh not less than 1.8 lb per sq yd.

i Flat and self-furring expanded metal mesh are recommended, but not the V-rib lath because it is practically impossible to embed the latter to avoid rusting in outdoor and moist situations.

j Metal reinforcement is applied over wood or steel framing, flashing, masonry, and concrete if these do not provide satisfactory surfaces; chimneys; and unsound old stucco if it does not provide a satisfactory bond.

k If masonry or concrete does not provide a satisfactorily rough surface, a dash bond coat can be applied by mixing one part portland cement with one to two parts sand, mixed with water to a fairly stiff consistency. This is dashed and spattered against the wall with a straw broom or long-fibered brush, or by machine.

l Bond can also be improved by using one of the bonding agents, usually a water-based emulsion. This provides a strong bond between the surface and the stucco.

12.35 Mixes

a The first (scratch) and second (brown) coats have essentially the same mix. One part by volume of portland cement and 25 percent by volume of hydrated lime to act as a plasticizer, are mixed with three to five parts by volume of damp, loose sand. Sand proportions vary because of the wide variations in sand. Trial mixes are made until the proper workability is achieved. Short-fiber asbestos and diatomaceous earth are also used as plasticizers.

b The portland cement employed should conform to ASTM C150 Types I, II, and III. Type I is the general-purpose cement most widely employed; Type II generates heat a little more slowly; and Type III is high-early-strength. All may be modified with air-entraining ingredients, in which case, they are Types IA, IIA, and IIIA. Plastic cements may contain up to 12 percent plasticizing agents. Masonry cement, ASTM C91 Type II, is also employed. Plasticizers should not be added to plaster and masonry cements because they contain their own plasticizers. Otherwise, the mixes are 1 part cement to 3 to 5 parts sand.

c Finish coats are usually premixed by the manufacturers, and if true colors are wanted, it is best to use such ready-mixed materials. For job mixing, if white or colors are wanted, white portland or waterproof white portland cement is employed. The usual mix is one part cement, up to

Wallboard. Lath and Plaster

¼ part hydrated lime, and 2 to 3 parts fine-graded light-colored sand, plus mineral oxide pigments, the latter carefully weighed if uniformity from batch to batch is to be obtained.

12.36 Application (Figure 12.21)

a The first (scratch) coat should be applied with enough pressure (hand or machine) to make sure that it is pressed through the lath against the backing and completely embeds the lath. It should be approximately ½" thick. After application, it is horizontally scored with a scoring tool to provide a good mechanical bond to the second coat.

b The second (brown) coat is applied as soon as the first coat has set or hardened enough to carry the weight of both coats. This may be as little as four to five hours, perhaps less in hot, dry weather, more in cool weather. The first coat on open wood framing should be allowed to stand at least 48 hours to achieve enough strength to withstand the pressure of subsequent applications. The first coat should be kept damp, by fine spraying if necessary, until the second coat is applied. Just before the second coat is applied, the first coat should be evenly dampened.

c The second coat is applied evenly, to approximately ⅜" thickness, over an entire area such as a wall, without stopping, if possible. Stopping points, if needed, should be at natural breaks in the wall such as corners, pilasters, and windows. Otherwise, the starting and stopping joints may show through the finish coat.

d The brown coat is moist-cured for at least 48 hours and allowed to dry at least 5 days, or longer if possible, before applying the finish coat. More uniform suction for the finish coat is attained by the brown coat under these conditions and results in more uniform texture and color.

e The brown coat is uniformly dampened just before applying the finish coat.

f Many textures from smooth and hard troweled, through float, to pebbled may be attained in finish coats. They may be applied by hand or machine. Deep textures include "Spanish," "brocade," dashing, scoring, and combing. A pebbled or "rock dash" or "marblecrete" finish is attained by throwing or machine-blowing small pebbles or chips of marble or other stone against the freshly-applied finish.

g Moist-curing should begin the day after the finish coat is applied. Fog spraying should be used first, and moist-curing should be continued for at least a day.

Dwelling House Construction 308

Figure 12.21 Stucco. (Upper Left) Wire Mesh Attached over Building Paper with Self-Furring Nails (Above) and Nailed to Furring Strips (Below). (Upper Right): Wire Mesh, Scratch Coat, Brown Coat, and Finish Coat. (Lower Left) Close-Up of Scratch, Brown, and Finish Coats Showing Treatments and Thicknesses. (Lower Right) Close-Up of Finish Coat Showing One Type of Texture.

12.37 Epoxide-Based Stucco

a A recent development is the use of epoxies in place of portland cement. A liquid epoxy formulation is mixed with hardener and silica sand and troweled to the surface, such as masonry. Chips such as marble are immediately thrown or blown against the fresh mortar. Hardening occurs quickly, depending upon temperature, so the area of surface treated should be adjusted to the working time of the formulation.

TILE

12.38 General

a Tile floors and walls are widely employed in bathrooms and similar spaces where water is apt to be splashed about, and an impervious waterproof surface, easily cleaned, is required. Tile may be baked clay, terra cotta, glass, concrete, plastic, or cement-asbestos, but the baked varieties are usually employed for interiors.

b Glazed, ceramic, and vitreous varieties are commonly used for walls, ceramic and vitreous for floors. Colors may be surface colors or integral; the more glossy finishes usually contain surface colors; the duller-finished tiles usually have integral colors imparted by minerals added to the basic mixture and brought out by baking.

12.39 Wall Tile (Figure 12.22)

a Wall tile are set against a properly-prepared scratch coat of cement mortar applied to metal lath, special waterproof gypsum board lath, or to a masonry surface. Because of the weight of the tile finish and the need for a completely rigid backing, it is best to set studs not more than 16″ on centers. When metal lath is employed, lines of horizontal blocking 16″ apart should be solidly nailed between the studs.

b Metal lath must be carefully and firmly nailed to studs, with horizontal and vertical laps occurring at horizontal blocking or at studs.

c Walls of brick or terra cotta tile (heavily scored surfaces) should have rough-struck mortar joints. Cork, gypsum block, and similar walls should be covered with wire mesh, expanded metal lath, or special waterproof gypsum board lath.

d When special waterproof gypsum board lath is used, it should be at least ½″ thick. It should be applied with edges horizontal, leaving ¼″ space between an edge and the lip of a tub or shower pan. Horizontal blocking should occur 1″ above the edge of a tub or shower pan. Blocking should be employed to support soap dishes, towel bars, and other fixtures. Nails should be not more than 8″ on centers and screws not more than 12″ on centers, unless ceramic tile more than $5/16''$ thick is used, in which case, nails should be spaced 4″ and screws 8″ on centers.

e Cutouts for pipes and fixtures should have the cut edges protected by a waterproof compound or adhesive tape.

f Over gypsum board, ceramic tile is applied with organic adhesives. These must be water resistant and should conform to the requirements of the American National Standards for Organic Adhesives for Installation of Ceramic Tile. Each manufacturer has his own set of instructions for applying tile, and these should be carefully followed.

g On *metal lath,* two coats are applied to the wall: first, a scratch coat to form a base and second, a cement mortar in which the tile are embedded.

h The scratch coat is composed of 1 part portland cement, about $1/4$ part hydrated lime, and 2 parts clean sand. It is mixed with sufficient water to form a thick mortar, and is applied in a coat at least $1/2''$ thick, or thick enough to form a true, even surface $3/4''$ back of the finished surface of the tile, if tile $3/8''$ thick are employed, or to allow for a cement mortar bed $3/8''$ thick behind the tile. The scratch coat is thoroughly scratched and allowed to harden at least a day before tile are set.

i The cement mortar is commonly 1 part portland cement to 2 parts clean sand plus 25 percent lime putty.

j After tile (except vitreous types) have been well soaked in clean water, and after the scratch coat has thoroughly hardened and has been thoroughly dampened with clean water, the actual setting of tile may proceed either by "floating" or by "buttering" (Figure 12.21).

k When tile are "floated" the procedure is as follows:

1. Wood guide strips (lath are excellent) are placed vertically against the wall in a thin bed of mortar, and are carefully plumbed.
2. Mortar is spread between the guide strips and brought out flush with them by hawk, trowel, and float rod.
3. Guide strips are removed and the resulting grooves filled with mortar.
4. A thin grout of pure cement and water is spread on the back of each tile, as it is set.
5. Tile are placed against the wall and carefully beaten into place with block and hammer.

l Tile are "buttered" in the following manner:

Wallboard. Lath and Plaster

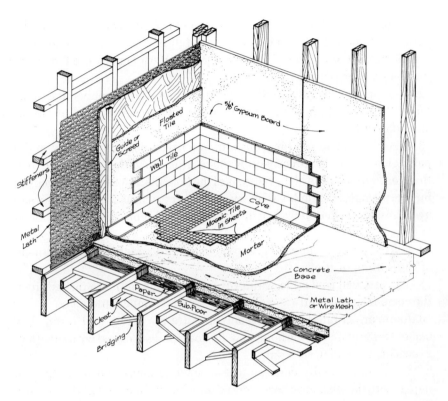

Figure 12.22 Tile Walls and Floors. Wall Tile Floated (Left), Buttered (Right). Mosaic Floor Tile. Special Cove.

1. Small pieces of tile, set in mortar, are spotted about 30 inches apart on the surface of the wall, and their faces carefully brought into a true even plane. These act as guides for the rest of the tile.

2. As each tile is set, a wash of neat cement and water is spread on the back followed by the proper thickness of mortar. Each tile is then set against the scratch coat and tapped into place until firmly set and plumb with the spotter tile.

m Whether tile are floated or buttered, the final steps are to fill the joints between tile with a thin grout of white or gray portland cement and water, and finally to wipe off all traces of cement on the surfaces.

12.40 Floor Tile (Figure 12.22)

a Preparation of floor framing and subflooring has already been discussed in Chapter 5. After the sub-flooring and framing have been covered with waterproof building paper, the base for the tile is prepared as follows:

1. A concrete bed of 1 part portland cement, 2 parts clean sand, and 4 parts small aggregate, is poured to a depth of $2\frac{1}{2}''$ and levelled. This brings the concrete $\frac{3}{4}''$ to $1''$ above the tops of the joists, where the chamfered joist is employed (Chapter 5).

2. Well-painted or galvanized wire mesh or metal lath is embedded in the top of the concrete to form reinforcement for the setting bed of mortar spread over it.

3. A setting bed of mortar, consisting of 1 part portland cement and 3 parts clean sand, is spread and levelled at an elevation $\frac{1}{16}''$ lower than the eventual elevation of the bottoms of the floor tile. Screeds are employed for this purpose. A thin layer of dry portland cement is frequently spread over the well-soaked surface of the concrete slab before the cement mortar is spread.

4. A $\frac{1}{16}''$ layer of dry cement is spread over the cement mortar just ahead of the setting of tile.

5. Tile are well soaked in water (except vitreous tile), are placed upon the cement mortar, and are well hammered and tamped into place.

6. Joints are grouted with a creamy mixture of cement and water. Dry cement is sprinkled on the floor, and the floor finally cleaned with sawdust, excelsior, or similar material.

b Mosaic tile, usually ceramic, are commonly cemented to a sheet of paper, with the mosaic pattern worked into the arrangement at the factory. If separate borders are required, some cutting of the central field, where it abuts the border, is required. Tile are bedded as noted above, but cannot be grouted until the paper has been removed.

c Where floors and walls join, special cove tile are placed, before either wall or floor tile are set. Cove tile provide smooth transition from floor to wall.

d If wall tile do not run to the ceiling, that is, if only a tile wainscot is wanted, special dado tile are provided to form a finished edge for the top of the tile wainscot. If walls are "returned" into window and door openings, special bull-nosed tile are provided to make the return. Similarly, at reentrant angles, special cove tile may be employed, although this is not always done.

e Towel bars, recessed soap and water-tumbler fixtures, and other special fixtures to match the tile, may be had. These are set at the same time as the rest of the tile.

13 Interior Finish

13.1 Specification Clauses

INTERIOR TRIM AND DOOR FRAMES

a All interior trim and door frames shall be of C grade Idaho white pine or ponderosa pine. All grounds shall be furnished and applied where directed by the architects or where demanded by the other trades.

INTERIOR DOORS

b Interior doors throughout the house, except in the basement and in the maid's wing, shall be stock white pine, ovolo molding, raised panel 2 sides. The doors in the other parts of the house above mentioned shall be stock 2-panel fir, bead and cove molding, flat panel 2 sides. All doors shall be $1\frac{3}{8}''$ thick.

KITCHEN AND PANTRY CABINETS

c The kitchen and pantry cabinets as indicated on the drawings shall be furnished and erected complete with hardware.

MAID'S SITTING ROOM CABINETS

d Furnish as indicated on the plans two cabinets of $\frac{3}{4}''$ white pine with $\frac{5}{8}''$ fir plywood backs, $\frac{3}{4}''$ birch or maple counters, and white pine doors.

CLOSETS

e All ordinary closets shall be equipped with a $\frac{3}{4}''$ white pine shelf supported on wall cleats, and a $1\frac{3}{4}''$ hardwood hanger rod. A hook strip of white pine on three sides of the closet shall be furnished.

SPECIAL CLOSETS, WARDROBES, DRAWER CASE WORK, BOOKCASES

f These shall be of white pine according to details furnished by the architect.

FIREPLACE IN LIVING ROOM

g The fireplace mantel and the panelling in the living room shall be built according to details furnished by the architect.

STAIRCASE WORK

h All staircases shall be supported on heavy carriages. The treads shall have molded nosings with moldings below. Treads and risers shall be tongued and grooved together, and both shall be housed into the wall stringers according to what is known as the best Boston practice. Treads shall be $1\frac{1}{8}''$ plain select oak, and risers $\frac{3}{4}''$ select oak for all staircases.

Install mahogany rail and white pine turned balusters in main staircase according to details furnished by the architect. Install stock 2⅛" birch or beech handrails where handrails are elsewhere indicated on the plans.

13.2 General

a Interior finish includes all exposed woodwork in the interior of the building. This consists mainly of stairs, doors, built-in features such as cupboards and bookshelves, trim of all kinds, and finish floors. All require great care in their installation to insure neat appearance, tight joints, and smooth surfaces whether the protective coatings (paint, enamel, varnish, etc.) are to be transparent or opaque.

b For painted (opaque) finishes the wood must be fine (close) grained and be free of knots, pitch streaks, resin pockets, or other imperfections which cannot be concealed by paint. It must have a surface which will not "raise," that is, parts of which will not pull away from the rest of the surface. It must be easy to work. Northern white pine is excellent, but less readily available than ponderosa pine and Idaho white pine.

c Transparent or "natural" finishes call for decorative woods free of blemishes. Hardwoods such as oak, birch, maple, walnut, mahogany, and lauan are excellent. Redwood, white pine, Idaho white pine, ponderosa pine, gum, and others are also employed.

13.3 Care of Wood Finish

a Wood finish is a valuable commodity and must be handled with care. The primary requisite is that it be kept dry and straight. It should be stored in dry warehouses at the supply house, should be covered when being transported in damp weather, should not be permitted in the building until the latter is completely dried after plastering, and should be piled carefully and straight so that surfaces are not damaged and individual pieces do not become crooked. The best practice is to stand the individual pieces on end if they are not too long, rather than piling them on top of each other. The precaution of keeping wood trim *dry* cannot be over-emphasized since most troubles with faulty trim can be traced to material which was damp when installed and which dried subsequently. In damp cool weather artificial heat must be installed in the building before the trim is brought on the job.

b If trim is to be painted or enamelled it is excellent practice to have it primed at the mill, i.e., to be given the first coat of paint, before it is shipped.

Interior Finish 315

This initial coat has some retardant effect on the penetration of moisture and slows swelling and shrinkage with varying humidity conditions. In any event it is good practice to back-prime all trim, i.e., to paint the back faces, before it is installed. Such treatment helps to retard the penetration of moisture from the back and therefore aids in slowing the swelling and shrinkage. Even if the front faces are to be finished "natural," it is wise to back-prime the material.

13.4 Joints (Figure 13.1)

a In ordinary framing no particular care is exercised to obtain neat, tightly fitting joints, and most members are simply butted or lapped. Strength and rigidity are important, but appearance is secondary because the frame, sheathing, rough flooring, and roof boards are subsequently covered. In trim and finish of all kinds, especially interior finish, appearance is of foremost importance, and the joints between members must receive close attention. Generally speaking, joints are either so made as to be permanently tight and completely inconspicuous, or they are worked into the design in such a way as to be inconspicuous if they do open.

b Speaking very broadly, four principal types of joints—butt, miter, shiplap, tongue-and-groove—are employed, each with numerous variations. Several of the typical applications of each are shown in Figure 13.1. These are merely representative and by no means exhaustive.

c Butt joints (Figure 13.1a), either longitudinal or corner, are easily made but are weak, open easily, and may show end wood, which is usually objectionable because it is difficult to finish well and is considered unsightly. Butt joints may be strengthened and reinforced by dowels, splines, dovetails, and tenons. When snugly fitted and glued, these devices aid materially in keeping joints tight. Generally speaking, butt joints, especially in the end-to-end directions of the boards, are avoided in finish of highest quality. The principal exception to this general rule is in re-entrant, or interior, corners where plain or molded members meet. One member is, in this instance, butted and cut to fit (coped) against the other.

d Miters (Figure 13.1b), particularly at exterior corners, are almost always preferable to butt joints in finish work of all kinds. Plain miters are frequently reinforced with miter brads, splines, and numerous types of patented devices designed to prevent the joints from opening. To provide a snug firm joint, the miter is sometimes shouldered, especially when the parts are made in a shop instead of on the job. The corner is occasionally

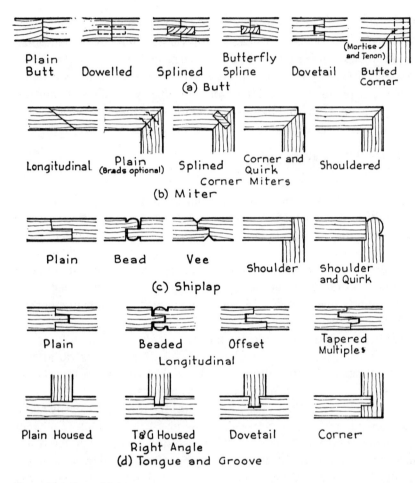

Figure 13.1 Typical Joints.

Interior Finish

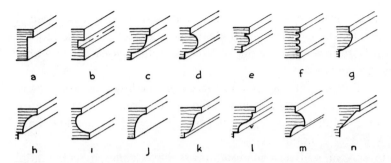

Figure 13.2 Basic Profiles Employed in Moldings. (a) Raised Fillet. (b) Sunk Fillet. (c) Quarter Round or Ovolo. (d) Torus. (e) Bead or Astragal. (f) Reeding. (g) Thumb or Ovolo. (h) Cavetto. (i) Scotia. (j) Conge. (k) Cyma Recta. (l) Cyma Reversa. (m) Beak. (n) Splay.

emphasized by a small re-entrant angle, called a "quirk." Longitudinal miters are employed in baseboards, moldings, and other linear members too long to be made in one piece. If these miters open slightly they do not show as unsightly cracks as do butt joints.

e Shiplap joints (Figure 13.1c) are quite common in both rough lumber and millwork. Although more common along the edges of adjoining members, they are also found at the ends and at corners. To avoid a corner joint involving both side grain and unadorned end grain, a bead and quirk or similar molding are sometimes formed at the corner. Various methods of emphasizing shiplap joints, such as beads or vees, are commonly employed.

f Tongue-and-groove (T and G) joints are widely employed for both "rough" boards and for finish (Figure 13.1d), either in longitudinal joints involving ends and edges of boards, or in joints involving angles such as re-entrant corners. Tongue-and-groove joints may be plain, beaded, or otherwise emphasized; they may be provided with offsets to increase the overlap, or formed in multiples to obtain large gluing areas; they may be made by "housing" the end of one member into the side of another; or they may be provided with dovetails, rabbets (rebates), or other means of tying the two parts together.

13.5 Moldings (Figure 13.2)

a Much of the woodwork employed for trim and finish of all kinds is shaped, or "molded," to various profiles, depending upon the ultimate use. Moldings are of two kinds, stock and special. Stock moldings are made in large quantities by millwork manufacturers and stocked by lumber dealers, hence the name "stock." Special moldings are made to the special profiles specified for a particular application by the architect, and require special knives to be made up for the job. Special moldings are considerably more costly than stock.

b Although the possible shapes and sizes of moldings are practically numberless, most are combinations of the basic shapes shown in Figure 13.2. Differences in terminology exist, but the ones here employed are those of Webster's Dictionary.

c Associations of millwork manufacturers have come to substantial agreement as to sizes and shapes of stock moldings. Many of these have also been standardized (see also Chapter 5 for standard sizes of framing lumber). In addition to these generally accepted stock moldings, many millwork manufacturers make certain stock items of their own design.

d It is beyond the scope of this book to discuss and illustrate the many stock moldings available, much less the specials which might be designed for a particular application.

STAIRS

13.6 General (Figures 13.3, 4, 6–8)

a The following discussion is based mainly upon conventional stair construction and the details shown are typical. The general principles are valid for conventional or for less traditional details. Figure 13.3 shows a typical conventional closed-riser stair of all-wood construction. Figure 13.4 shows a heavy-tread open-riser stair of wood and metal construction. Stairs are often installed by a separate craft. They may be plain, simple, and run in one flight between partition walls from floor to floor, or they may be ornate, open circular, or elliptical, with complex construction and ornamentation. Stairs have their own set of terms and parts, of which the principal ones follow.

b. Flight (Figure 13.3) Flight is a single set of stairs running as a unit from floor to floor, or from floor to an intermediate *landing* or *platform*. Landings are used to break up what would otherwise be long and fatiguing flights, or to provide for turns in the stairs.

c. Tread (Figures 13.3, 4, 5, 7–9) This is the horizontal member forming the top of a single step. It is associated with a *riser* or vertical member of a step. The front edge of a tread is the nosing.

d. Rise and Run (Figure 13.3) These terms each have two meanings. The *rise* is either the total height of a flight of stairs, that is, the height from finish floor to finish floor or from finish floor to top of landing; or the rise means the height of a single step, from top of one tread to top of the next tread. The latter definition is the more useful, and is more commonly employed. To distinguish the rise in this sense from the total height, the latter is sometimes called the *total rise* and this term will be used in this discussion. The *run*, similarly, may mean the horizontal projection of the flight of stairs, or it may mean the horizontal distance from the face of one riser to the face of the next. In this discussion the run has this latter meaning, and the term *total run* will be used for the horizontal projection of the entire flight.

e. Strings (Figures 13.3, 4, 7, 8) Strings are the sloping side boards of a flight, against which the ends of risers and treads are terminated. Three types of strings exist: *wall strings* which rest against partitions or walls adjacent to the stairs, *open strings* (free-standing, not adjacent to a wall), the tops of which are cut to the profiles of the stairs, and which allow the treads to project beyond the outer faces of the strings, and *closed or curb* strings which, like open strings, are not adjacent to walls, but do not have their upper edges cut to the profiles of the stairs. Treads and risers, consequently, are terminated against their inner sides. Partitions may be brought up under open and curb strings, but the upper edges are free.

f. Open and Closed or Enclosed Stairs An *open* stair is one which is open to a room or hallway on one side (occasionally both sides are open). It calls, consequently, for open or curb strings on the open side. *Closed* stairs are enclosed by partitions or walls on both sides, and consequently employ only wall strings.

g. Rail and Baluster (Figures 13.3, 4, 9) An ornamental or plain bar of wood or metal attached to an adjacent wall or attached directly to the stairs at a convenient distance above the level of the stairs, to be grasped by the hand, is called the rail or handrail. When the rail is attached directly to the stair treads, as in open stairs, the attachment is accomplished by means of balusters, which are vertical plain or ornamental sticks or bars fastened to the treads or strings and to the underside of the rail.

h. Newel and Angle Posts (Figure 13.3) At the foot of an open stair the rail usually terminates against a post of some kind and against a similar post at its head. This is called a newel post. Intermediate posts at landings or other breaks in the stair are called angle posts, and when an angle post projects down beyond the bottom of the strings, the ornamental detail at the bottom of the post is called the drop. Newel posts at heads of stairs

Dwelling House Construction

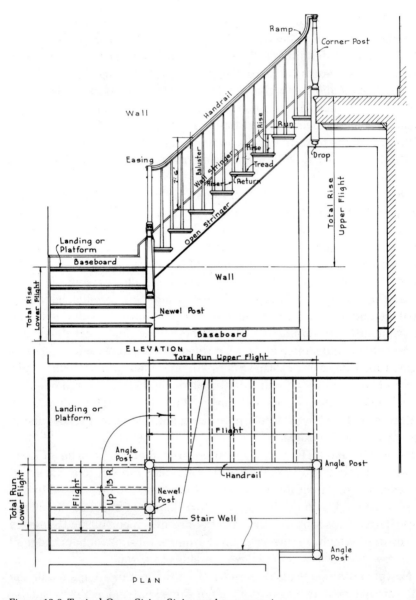

Figure 13.3 Typical Open-String Stairway Arrangement.

Interior Finish

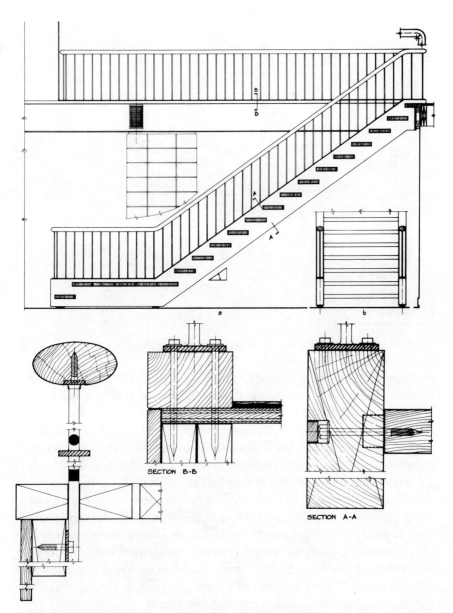

Figure 13.4 Open-Rise Stair. (a) Elevation. (b) View from Left Toward Stair. Section A-A, Detail of Tread and Stringer Connection, with Base of Rail. Section B-B, Detail of Upper Rail and Base. (c) Detail of Rail, Spindle, and Attachment.

Dwelling House Construction   322

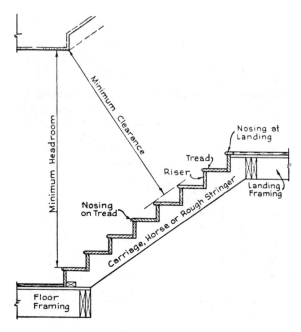

Figure 13.5 Minimum Headroom and Clearance Requirements.

are frequently halved because they are built directly against a wall. Sometimes the half newel is omitted and the rail is terminated by a small metal or wood plate fastened to the wall.

i. Head Room (Figure 13.5) The clear height from a tread to the overhead construction above (such as the upper floor) is called the head room. It is essential that ample head room be provided, not only to prevent actual collision but to give a feeling of spaciousness. Even when the head room provides clearance, if an individual using the stair has the impression that he is about to collide with the structure overhead, the head room is not sufficient. For a feeling of roominess, an average person should be able to extend his arm full length forward and upward without touching the ceiling of the stair. In any event, the absolute minimum allowable when measured from the nosing of a tread vertically to the overhead construction is 6′-6″. This is not a comfortable height.

j. Horse, Carriage, or Rough Stringer (Figures 13.5, 7, 8; Chapter 5) The finished stair must be supported on rough structural members underneath. These members go by the various names given above, although the term rough stringer is probably the most common (see Chapter 5).

Interior Finish

13.7 Rules for Laying Out Stairs

a Stairs must be neither too steep nor too shallow, since either extreme is fatiguing. Risers over 8" high are too steep for comfort, and less than 6" are too slow in their ascent. Best practice calls for risers slightly over 7" high, the exact height depending upon the total rise. Since the risers must all be the same height, the total rise must be divided among some whole number of risers which gives a rise in the neighborhood of 7". For example, if the total rise is 114", 17 risers provide a rise of 6.71" and 16 risers a rise of 7.13".

b For comfort, certain proportions between the magnitudes of rises and runs must be maintained. In general, the greater the rise the less the run and vice versa. For this reason, grand staircases frequently have shallow rises and wide runs, whereas secondary stairs for which a minimum of space is to be sacrificed may have high rises and narrow runs. A number of rules have been worked out for maintaining the proper proportions, of which the following are the most common:

1. Rise times run equals 70 to 75.
2. Rise plus run equals 17 to $17\frac{1}{2}$.
3. Twice the rise plus the run equals 24 to 25.

Any of these rules provides narrow runs for steep stairs and wide runs for gradually rising ones. In addition to these thumb rules a number of formulae have been devised to give comfortable proportions of treads and risers for various pitches.

c In any one flight of stairs there is always one less tread than riser. Hence, if a single flight runs from floor to floor and has 16 risers, 15 treads are required, an intermediate landing reduces the number to 14 and so forth. Knowing the total rise and the number of risers, the total run per flight can easily be computed and the size of stair well determined by adding to the space required by the stairs the amount of space needed for landings or for winders.

13.8 Stair Plans (Figure 13.6)

a Various arrangements of stairs within the stair well (opening in two adjacent floors within which the stair is located) are possible, some simple and others complex.

b The simplest is a single flight running from floor to floor. Because this type is fatiguing if the total rise is high, an intermediate landing may be built approximately halfway between floors. In buildings used by the public, laws often restrict the number of risers permissible per flight.

c A single flight or several flights with landings may require a longer stair well than can be obtained. The stair may therefore be turned 90° or 180° at a landing and be continued to the next floor. This construction can be varied still further by using several landings with short flights running from landing to landing. The short flights may consist of only one or two risers or may be quite long.

d Where space is at a premium, the landings are often omitted and the treads are carried around the angle by running their interior ends to a point at the turn of the stair (Figure 13.6c). At this point, then, the ends of risers are directly above each other. This kind of construction is called a winder, and should be avoided if at all possible. No adequate foothold is afforded at the angle and there is an almost vertical drop of several feet if a number of risers converge on the same point. The construction is dangerous and may easily lead to bad accidents. It is poor planning so to restrict the stair well that winders must be employed.

e The sharp drop which ordinary winders afford may be lessened in a stair with a 90° turn if the center of convergence for the various risers of the winder is brought out from the corner a foot or more (Figure 13.6c). The risers then do not converge directly at the corner but afford some foothold. The same holds true of a 180° turn if the corner is made broad, as would be the case in a stair with a short intermediate flight. The center of convergence in this instance is also brought out from the corner. The center of convergence should be so located as to provide treads of equal width along a line 1'-4" from the inner edge of the stair. This is the line of usual travel. The ordinary winder does not provide equal treads along this line.

f A helical stair, usually called a spiral stair, is a continuous winder from top to bottom. This type is rare in dwellings but is sometimes used for service stairs in buildings where space is at an absolute premium. Such a stair, of course, contains all the bad features of the winder multiplied several times.

g Ornamental stairs are frequently built in the shape of a portion of a circle or an ellipse (Figure 13.6d). This type of stair is almost always open on the underside, requiring the inner string to be open or curbed. Strings frequently have to be fairly massive, because the stair may be open underneath so that no carriages can be employed, and the curved section imposes torsion as well as bending in the inner string. Such strings may consist of a number of smaller pieces shaped to form and fitted, and then pegged, pinned, dowelled, and glued together to form a single unit; or they may consist of thin boards bent to shape and glued—laminated—together. For additional stiffness flat steel bands may be fastened to the inner side of such a string, or be concealed in its interior.

13.9 Stair Construction (Figures 13.7, 8)

a Stairs may be built in place or they may be built as units in the shop and set in place. Both methods have their advantages and disadvantages, and custom varies with locality.

b. Built-in-Place (Figure 13.7) Although exact details may differ, the procedure for building stairs in place is as follows:

1. Carriages are carefully cut to exact size and set in place, lower ends resting on the lower floor and upper ends framed against the header of the upper floor construction (Chapter 5).

2. The wall string is "housed" at least ½" deep to the exact profile of the risers and treads. The bottom of the string is cut to the profile of the stairs at the backs of risers and treads, and the string is set in place against the wall, so that the back of the housed-out profile coincides with the profile of the carriages. Treads and risers are firmly nailed to carriages. Rear edges of treads are best grooved to fit tongues in the lower portions of risers, and tops of risers are tongued to fit grooves in the bottoms of treads. The projecting portions of treads, or nosings, are usually finished underneath with cove or other moldings.

3. The wall string is either fitted over the treads and risers, or the ends of treads and risers are inserted into the housed-out portion of the string at the time they are placed and nailed.

4. The open string is carefully cut to the same profile as the treads and risers and mitered to fit corresponding miters cut in the ends of risers. The ends of treads project out over the string, and the nosing is returned along this projecting end.

5. If the outer string is curbed it is housed in the same manner as the wall string and is fitted over the ends of risers and treads. Setting the outer string, whether open or curbed, finishes the stair.

c The built-in-place method of stair building is not particularly satisfactory for enclosed stairs because the second string is not as easily fitted

Dwelling House Construction

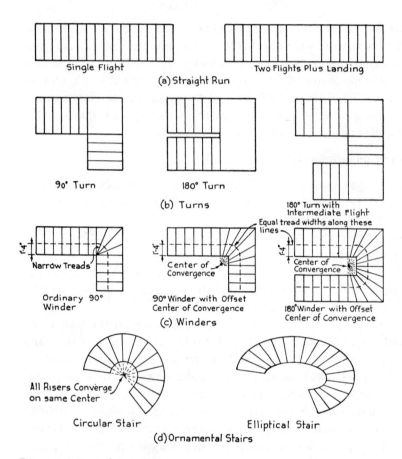

Figure 13.6 Typical Stair Plans.

Interior Finish

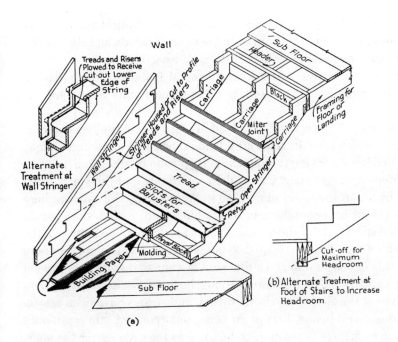

Figure 13.7 Built-In-Place Stairs.

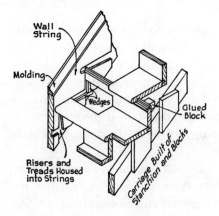

Figure 13.8 Detail of Shop-Built Stair.

into place as in the case of the open stair. When so built, however, both wall strings are ordinarily fitted over the ends of treads and risers after the latter have been nailed into position on the carriages.

d As figure 14.4 shows, many other means of fastening treads, risers, or both to strings can be devised. In this instance, only treads are employed.

e. Built-in-Shop (Figure 13.8) This kind of stair is essentially a self-contained unit which is set in place after being fabricated elsewhere. It may be set either before or after the application of plaster or other wall finishes. If set before, carriages may be installed separately and the stairs fitted to them by an arrangement illustrated in Figure 13.8. The carriage here consists simply of a stanchion to which blocks are nailed after having been fitted snugly under the treads. If, as is preferable, the bottom of the stair opening as well as the rest of the stair well is finished before the stairs are set, it is impossible to make this kind of stair rest snugly on pre-set carriages, and the carriages therefore must be incorporated into the stairs themselves at the shop after risers, treads, and strings have been assembled as described below. Except for being incorporated into a pre-built stair, such a carriage is essentially the same as the carriage for the stairway built in place. Ordinarily its lower edge is in the same plane as the lower edges of the strings, unlike the carriage shown in Figure 13.8. Construction details are:

1. Strings, instead of being cut or completely housed underneath to the profile of the stair, as are wall strings in the built-in-place type, are left solid their full depth to attain strength and are only housed to receive the ends of risers and treads. The housing is made wider than the actual thickness of treads and risers, and is flared on the underside. Inner strings in open stairs are either cut to the profile of the stair, if the string is an open one, or are housed in the same manner as the wall strings, if curbed. The curbed type is stronger and stiffer.

2. Risers and treads are best provided with tongue-and-groove joints in the same manner as the built-in-place type, and are firmly nailed together. Joints can, moreover, be glued together and blocks glued into the angles on the underside for additional reinforcement. The ends of risers and treads are inserted into the housed profiles in strings, and are tightened into place by wedges driven into the flared under-portion of the housing. Wedges are nailed and glued into place.

Interior Finish

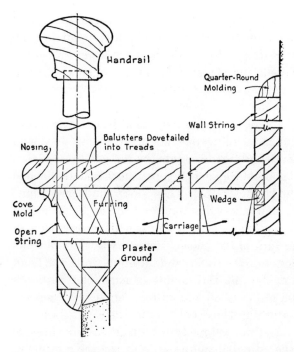

Figure 13.9 Details of Stairway Construction Showing Baluster and Handrail.

f When finished, these stairs are set as units in the stair well. Needless to say, they must be carefully made to fit the opening exactly, both in length and in width, and with proper rises and runs so that treads are level and risers plumb. A molding similar to a base mold (Section 13.14a) is usually necessary on top of the string to obtain a tight joint with the wall surface.

13.10 Rails and Balusters (Figures 13.3, 4, 7, 9)

a Rails are required on all open stairs for protection as well as to afford hand holds, and must be carried arounds the edges of open stair wells. Along the slope of a stair, rails are approximately 2'-6" above the treads, depending upon the pitch, when measured from the face of a riser. On landings and around open wells the height is the same as when measured from the center of a tread, namely 2'-8" to 2'-10" high.

b Rails are often ended against newel and angle posts at an oblique angle. If it is desired to have all rails meet the posts at the same elevation, the upper ends of rails may be flared upward in a curve called a "ramp" and then turned at right angles to meet the posts perpendicularly to their vertical faces. The lower ends of rails may be curved in an "easing" to meet the vertical faces of the newels at right angles. Angle posts may be omitted and the rail carried up in continuous curves from bottom to top. In such instances the newel post is frequently omitted and the lower end of the rail is finished in a spiral resting on a cluster of balusters or on a large single baluster. Rails may be carried around bends and corners in continuous curves as shown in Figure 13.4.

c Balusters may be plain square or turned rods, or may be ornately turned and twisted, depending upon the effect desired. In any event, they must be firmly fastened to the rail and treads so that the rail will provide firm support. The easiest and poorest way to fasten balusters is merely to nail them to rail and treads. Next easiest is to drill holes in rail and tread, round off the ends of balusters, and insert the rounded ends in the holes. Although this is better than nailing, it is not entirely satisfactory because balusters work loose and turn in the holes. The best way is to dovetail the lower ends of balusters into the treads and fit upper ends into holes bored in the lower side of the rail. Dovetailing is accomplished by removing the return nosing of the tread and cutting rectangular holes in the treads just back of the nosing. These holes, or mortises, are flared out at the bottoms, and the balusters are slipped into the mortises from the side and tightly glued into place. After balusters are firmly set, the return nosing is replaced, thereby concealing the joint. This dovetail joint is tight, rigid, and cannot work loose.

d The balusters of Figure 13.4 are welded to a continuous metal plate in turn firmly fastened down.

e Rails along enclosed stairs are commonly fastened to the walls with metal wall brackets. Such brackets must be substantial and should be not over 10' apart unless the rail is exceptionally rigid. Brackets must be securely screwed or lag-bolted to the frame of the wall or to blocking firmly spiked to the frame and set at the proper elevation to receive the brackets, since it is imperative that brackets be sturdy.

DOORS

13.11 Types

a Doors may be several different types, of which the commonest are batten, flush, and panelled. Each of these has several subdivisions.

b. Batten (Figure 13.10a) This consists of boards nailed together in various ways. The simplest is two layers nailed to each other at right angles,

Interior Finish

usually with each layer at 45° to the vertical. This construction is sometimes used for the cores of metal-clad fire doors. The second type of batten door consists of vertical boards nailed at right angles to several (two to four) cross strips called ledges or ledgers, with diagonal bracing members nailed between ledgers. When vertical members corresponding to the ledgers are added at the sides, the verticals are called frames.

c Batten doors are often found in places where appearance is not a factor and economy is important.

d. Flush (Figure 13.10b) Solid flush doors are perfectly flat, usually on both sides, although occasionally they are made flush on one side and panelled on the other. Flush doors sometimes are solid planking, particularly if authentic old details are required, but much more commonly they are veneered, and possess a core of small pieces of white pine or other

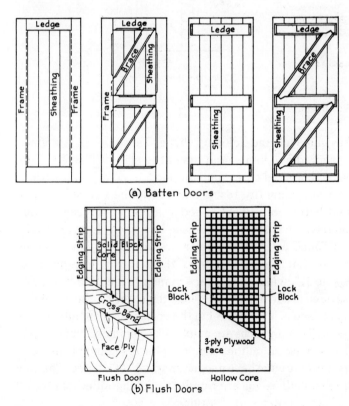

Figure 13.10 Batten and Flush Doors.

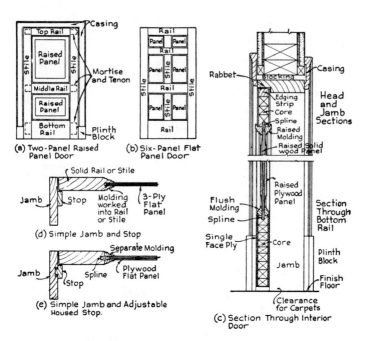

Figure 13.11 Panelled Doors.

wood which takes glue well and holds its shape. These pieces are glued together with staggered end joints, and the entire slab is dressed to the proper thickness. Other solid cores include particle board, and various mineral boards for maximum fire resistance. Along the two side edges, and along top and bottom are glued ¾" edge strips of the same wood (usually hardwood) which is to be used to face two sides of the door. The two sides are then faced with either one or two layers of veneer. If only one layer or ply is used, it is usually ⅛" to ¼" thick; if two plies, the inner is customarily ¹⁄₁₆" to ⅛", while the surface veneer is much thinner. The inner ply or cross band is laid at right angles to the core, i.e., horizontally, and the outer ply or face is laid parallel to the core, i.e., vertically, or the long way of the door. Two-ply faces are much more common than one-ply.
e For interior doors, good but not maximum water resistance is needed; for exterior doors, the glue must have maximum water resistance. Thermosetting synthetic resin adhesives are rapidly replacing all others for all veneered doors, both interior and exterior.

f. Hollow-Core Doors (Figure 13.10b) Like the solid flush doors, these are perfectly flat, but unlike the solid doors, the core consists mainly of a grid of crossed wooden slats or some other type of grid construction such as an impregnated paper honeycomb. Faces are 3-ply plywood instead of one or two plies of veneer, and the surface veneer may be any species of wood (usually hardwood) possessing any figure or pattern desired. The edges of the core are solid wood and are made wide enough at the appropriate places to accommodate mortise locks and butts. Doors of this kind are considerably lighter than solid flush doors.

g. Panelled (Figure 13.11) Many doors are panelled, with most panels consisting of solid wood or plywood, either "raised" or "flat," although exterior doors frequently have one or more panels of glass, or "lights." One or more panels may be employed, although the number seldom exceeds eight. They may be horizontal, vertical, or combinations of horizontal, vertical, and square, and different sizes may be employed in the same door.

h Horizontal members are called rails (Figure 13.11a, b). Every door has a top and bottom rail, and it may have one or more intermediate rails. Top and intermediate rails are usually the same width but bottom rails are wider than the other rails and stiles.

i Vertical members are called stiles (Figure 13.11 a, b). There are at least two stiles, one on each side, the full height of the door, and there may be one or more intermediate stiles, which may or may not run full height (i.e., from bottom rail to top rail). When intermediate rails occur, the rails are continuous and stiles are butted against them. Rails must be the stronger members in order to keep doors from sagging.

j The set-in thinner sheets are called panels (Figure 13.11). These are held in place by the rails and stiles, either directly or through the medium of moldings.

k Panelled doors may be either "solid" or veneered. Ordinarily exterior doors which are to be painted on the outside are commonly "solid" softwood, particularly white pine, and inexpensive interior doors are also "solid," with Douglas fir a favorite material (Figure 13.11d, e). The term "solid" refers primarily to the rails and stiles, which are run out of single pieces of wood.

l Where the panels are fitted into rails and stiles, the edges of the latter are worked down to an ornamental molding and are plowed (grooved) to receive the edges of the panels. Running the molding is called "sticking" or "stickering." Lights must be removable, so the molding on one side (exterior side of outside doors) is made removable and is held in place by brads.

m Veneered doors (Figure 13.11c) are used where better quality is desired and where faces are to be hardwoods. This makes for both economy and better doors, because thick hardwood stock is costly and, moreover, frequently has a bad tendency to twist. Veneering is usually considered advisable even for such woods as mahogany in which there is little tendency to twist. Rails and stiles of veneered doors are made in much the same way as solid flush doors (Section 13.11d). Cores of small pieces of softwood are first assembled, often by dovetailing, and glued. The edges of cores are provided with edging strips of hardwood, and the faces are covered with veneer.

n Panels are either flat (plane surface) or raised (bevelled edges), and may be solid wood or plywood. If plywood panels are to be raised, the surface plies must be thick enough to allow for the bevelling without showing the cross bands. By selecting and carefully matching sheets of figured veneer, various patterns are possible in the panels.

o Panels may be inserted directly into grooves plowed in the edge strips of the stiles and rails, or may be held in place by moldings. In the latter instance, splines are inserted into the plowed groove to fill in the space behind the edge of the panel. The splines are needed to prevent light from showing through the door between molding and edge of rail or stile.

p Three types of molding details are used. The simplest is the so-called *solid* sticking in which the molding is run directly in the edges of rails and stiles (Figure 13.11d). When heavier ornamentation is desired, separate pieces of molding are bradded and glued into the corner between panel and rail or stile (Figure 13.11c, e). Flush molding does not project beyond the surface of the door, raised molding protrudes beyond the faces of rails and stiles.

q Lights in exterior doors call for special details, because water running down the surface of the glass is likely to work its way behind the veneer and cause it to loosen. This may be prevented by inserting a piece of molding under the glass which extends through the door and is turned down over the face of the door on the outside to form a drip. The same result can be achieved by inserting a piece of sheet metal under the removable outer molding which holds the glass in place, and turning the metal up behind the glass and down slightly over the face of the door so that a narrow edge of metal shows (Figure 13.12c).

Interior Finish

13.12 Construction

a Doors may either be stock—made to manufacturers' details and stocked by dealers in their warehouses—or they may be made especially to architects' details. Stock doors commonly are 1⅛", 1⅜" and 1¾" thick. For interior work, 1⅜" doors are commonly employed, especially for ordinary medium-sized doors up to 2'-8" or 2'-10" wide and not over 6'-8" or 6'-10" high. Exterior doors and doors larger than those indicated, or doors which are to receive a great deal of use, should be at least 1¾" thick and very large exterior doors should be 2" or 2¼". The 1⅛" doors are found in closets and other minor rooms, particularly if the openings are quite small. Stock doors are made in 2" multiples in both width and height, and odd sizes must either be obtained by cutting down standard-sized doors, or must be made to order.

b. Frames (Figures 13.11a, c, d, e; 13.12a, b) The frame of a doorway is that portion of the ensemble to which the door is hung and against which it closes. It consists of two sides or jambs and a head, with an integral or attached stop against which the door closes.

c Exterior door frames (Figure 13.12a, b) are ordinarily of softwood plank, with jambs and head rabbeted to receive the door in exactly the same way as casement windows (Chapter 7). At the foot is a sill, almost always of hardwood to withstand the wear of traffic, and sloped down and out to shed water. A threshold is provided to prevent water from driving under the door. This often is integral with the sill and consists of a raised lip at the back whose width is the thickness of the door but which is slightly bevelled at the front to allow the front face of the door to project and form a drip. The back of the lip laps over the finish floor and conceals the joint.

d Thresholds may be separate pieces of hardwood (Figure 13.12b) placed over the joint between sill and floor, and hollowed on the bottom to fit snugly against sill and flooring. Such thresholds must be tightly fitted and securely nailed or screwed after the underside has been liberally coated with a sealant to seal the joint. Thresholds may also be metal, such as brass, but the same precautions must be followed when setting metal thresholds as when setting wood. Metal thresholds are often employed with stone sills, particularly if the adjacent floor is tile or some other masonry material.

e Interior frames may be set on the rough floor and the finish floor fitted around the bases of their jambs, or the finish floor may be laid first and the frames set on top. It is easier to obtain a tight joint by the latter method, but the former anchors the frame more firmly in place. Rough-framed openings must be wide enough to allow for the thickness of the jambs, and to provide an extra inch for blocking (Figure 13.11c) between the backs of the jambs and the rough framing (Chapter 5). The blocking is firmly wedged to bring the jambs and head to plumb and level straight lines, after which the frames are nailed solidly to the rough framing through the blocking. Finish nails are used and the nail heads are set (Chapter 14).

f If rabbeted interior frames (Figure 13.11c) are employed, the stock must be thick enough to allow the depth of the rabbet; if the stops are attached (Figure 13.11d, e), the frame is commonly $\frac{3}{4}''$ thick. Attached stops are small molded pieces firmly nailed to the faces of the frame so as to leave a space on the door side equal to the thickness of the door. The door is hung with its hinge or butt edge (Chapter 14) in line with the edge of the frame. Attached stops may simply be nailed to the frame or they may be tongued in back and set into a groove plowed in the frame. The simple nailed stop makes it easier to re-hang a door to swing the other way, because the stop need merely be reversed and nailed in its new position. Rabbeted and tongued stops provide lightproof joints which, moreover, make it impossible to open a door by inserting a thin strip between stop and frame and pushing back the latch.

g. Casings (Figure 13.10a, c; Figure 13.12a) Door casings have the same function as window casings (Chapter 7), namely, to set off the doorway and to cover and seal the space between frame and rough opening.

h Casings may be simple or ornate, depending upon the architectural design. The simplest treatment consists of two side pieces and one head piece. They are installed most simply by carrying the head casing across the full width of the opening plus the side casings and butting the side casings against the head casing. More commonly the joint is made by mitering the three pieces. The mitered joint is satisfactory as long as the wood does not shrink or swell, but with changes in width the joint opens at either the inner or the outer end. To guard against this, fine miter joints are reinforced in a variety of ways; by dowelling and gluing the joint, by mitering only the face and half-lapping and gluing the back, or by mitering the front and mortising, tenoning, and gluing the back. Sometimes special metal ties are employed.

i Next to the simple 3-piece casing is the casing plus backband. The backband is a narrow molding nailed to the outer or "back" edges of the casing. The joints in the backband are mitered, but the casings may either

Interior Finish

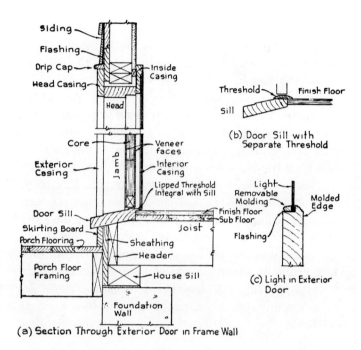

Figure 13.12 Exterior Door Details.

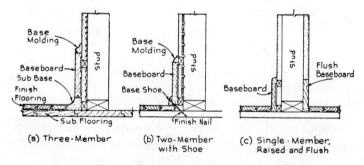

Figure 13.13 Types of Baseboards.

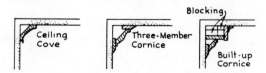

Figure 13.14 Cornices.

be butted or mitered. Backbands are used particularly to accentuate the outside edge of the casing without using heavy stock for the casing.

j Wide and ornate casings are built up of several pieces with moldings either run in the casing or attached to the casings as a part of the architectural profile. Frequently, ornamental pilasters are made a part of the casing in ornate entrances, and the head is then made in the form of a pediment. No matter what the detail may be, an essential part of the construction is plenty of blocking to which the parts may be attached.

k In thick walls, the door frames are augmented by additional casings called jamb casing. These may be panelled to match the panelling of the door or molded and ornamented to match the regular side casing. A refinement of this type of jamb casing in very thick walls is to have the sides splayed to give additional width and perspective to the opening. Again, no matter what the detail, it is essential to provide plenty of blocking, carefully set and true to line, to which the panelling and other ornament are attached. Setting the blocking is frequently the most important part of the work from a construction standpoint, because it must be accurately and sturdily built so that the finish will fit perfectly and be given adequate support.

l Casings are sometimes provided with short blocks called "plinth" blocks, at the bottoms of the side casings. They are thicker than the casing itself and are usually fairly plain. They form a stop for the baseboard as well as a base for the side casing. At the juncture of side and head casing, ornamental corner blocks are occasionally found.

MILLWORK

13.13 General

a Millwork in general comprises all trim or finish woodwork and therefore includes a number of items which, for convenience and completeness or because of their individual importance, are taken up elsewhere. Window trim is discussed in Chapter 7, stairs in Sections 13.6 to 13.10 of this chapter, and doors and door trim in Sections 13.11 to 13.12. Other customary items of interior trim are grouped in this section.

13.14 Baseboards (Figure 13.13)

a Baseboards, also variously called mopboards, skirting, or skirting boards, are found at the juncture of floor and wall. Ornate baseboards

Interior Finish

may consist of two or three parts, including a sub-base put down before the finish flooring, a center board or base proper, and an upper molded piece called the base molding, which might either be flush or raised, similar to the moldings employed in panelled doors. More commonly, baseboards are fairly plain and frequently quite narrow, with simply molded top edges, or with narrow simple molding nailed along the top. The backs of baseboards should be hollowed the same as are window and door casings so that they will hug the wall finish closely along the edges. Baseboards are nailed through plaster or wallboard to the studs with 8d or 10d finish nails. At internal corners one piece is butted and coped to the profile of the other, but at external corners the pieces are mitered and firmly nailed together. Where baseboards adjoin door openings, the ends of the baseboards are carefully fitted to the edges of the casing or to the plinth blocks (Figure 13.11a, c).

b Baseboards may be put in place either before or after the finish floor is laid. If before, the finish floor usually ends against the baseboard; if afterward, the finish floor is carried under it (Figure 13.13b, c).

c Often the joint between baseboard and flooring is covered with a small molding, either quarter-round or similar, called base shoe or carpet strip. The base shoe is best not nailed to finish floor or to baseboard because shrinkage in either causes it to move away from the other and open a crack. Nails should be driven through to the subfloor, which has little tendency to move.

13.15 Chair Rails, Cornices (Figure 13.14)

a. Chair Rails Chair rails are horizontal strips of wood applied to the walls at the height of chair tops to prevent the latter from marring the walls. When window sills are at the prcper height, the chair rail is continuous under the stool and takes the place of the apron.

b. Cornices Cornices consist of single-piece or multiple-piece ornamental details at the juncture of walls and ceilings. Generally no cornice is used at all. When employed, the cornice is often a wide crown mold set into the corner and nailed to studs and joists. It may be made a trifle heavier by attaching a narrow frieze board to the wall first, or a frieze may be employed on the wall and a similar piece on the ceiling, with the molding nailed to both. The frieze may be further ornamented with any detail the architect may desire. As far as construction is concerned, the cornice consists of a number of horizontal strips of plain or molded wood nailed either to each other or to the studs and joists. Dentils and similar details have to be nailed on separately. In order to obtain a snug fit between plaster or wallboard and wood, the wall finish must be applied straight and without any waves. This can be accomplished with grounds. Heavy cornices require blocking.

13.16 Mantels (Figure 13.15)

a Little can be said in general about the construction of mantels, since almost all are different when built to the architect's details. They may consist merely of a shelf, called the mantelpiece, at the top of the masonry, or the entire front of the fireplace may be faced with wood, marble, limestone, or other ornamental material, leaving only a narrow edge of masonry showing around the fireplace opening. The mantelpiece is often omitted entirely.

b No matter what the facing of the fireplace may be, the opening should be edged with masonry material such as brick, tile, or stone, because the heat of the fire causes materials such as wood to char if they are brought to the very edge of the opening. If brick or tile are used, the fireplace opening must be designed to fit the sizes of these units, in order that a full number may be carried up the sides and across the top without requiring any to be cut. Stone slabs and rubble can be fitted to desired dimensions.

c The principles underlying the construction of panelled fireplace fronts are the same as for any panelled work, and call for rails and stiles to hold the panels in place. If the chimney juts into the room the panelling is returned at the sides and may be continued as part of the wainscot. Jutting chimneys may be panelled up to the ceiling to emphasize the importance of the fireplace in the room. In any event, as far as construction is concerned, it is essential to provide plenty of solid blocking, carefully set to line, plumb and level, to which the finished woodwork is applied. Furthermore, since this woodwork is likely to be warmed a great deal of the time and consequently to become very dry, it is advisable to use only very dry wood in its original construction because otherwise joints are apt to open.

d In Figure 13.15 is illustrated a typical mantel installation, in which simple molded trim surrounding a brick-faced opening is surmounted by a one-piece mantelpiece with supporting molding, face, and small pilasters. The wall is finished with random-width vertical boarding tongued and grooved into molded separating strips. All woodwork is simply nailed

Interior Finish 341

to blocking inserted among the studs of the partition and to furring strips built around the brickwork.

e Figure 6.12 shows another type of fireplace detail. There are many variations.

13.17 Cabinets (Figure 13.16)

a Probably no part of the house has received more intensive study recently than has the kitchen. Equipment makers have employed engineers to study the work of the housewife in the kitchen, to arrange the needed equipment in the most efficient manner, and to work out floor plans to give the maximum of convenience in the minimum of space. Various plans have been evolved; and both equipment—refrigerators, sinks, and stoves—and cabinets have been worked into them for greater efficiency.

b Kitchen cabinets (Figure 13.16) ordinarily are in two parts, a rather wide lower portion surmounted by a counter, and a shallower upper portion, starting approximately 16″ above the counter and extending to the ceiling. The cabinets contain drawers and shelves, often have special compartments, and usually possess bread boards and chopping boards which slide into slots under the counter.

c. **Toe Strip** (Figure 13.16) To prevent toes from being stubbed at the bottoms of the cabinets and to allow closer approach to the counters, the bottoms of cabinets are recessed approximately 3″ and the bottom shelf is raised off the floor approximately 4″. This provides a toe-space and greatly increases comfort when working at the counter.

d. **Shelves** (Figure 13.16) Shelves in the best-constructed cabinets are housed into the side of the cupboard with dovetail joints and do not, consequently, have to be nailed (Figure 13.1d). Shelves and cupboard sides are firmly joined and cannot separate. In less expensive work, shelves are supported on cleats fastened to the sides and backs of cupboards. In kitchens, shelves are usually fixed in place, but they may be made adjustable by movable brackets which fit into slotted standards fastened to the sides and backs of cupboards. Ordinarily $\frac{3}{4}''$ stock is used for shelves and sound knots are considered no detriment inasmuch as shelves are concealed behind cupboard doors.

e. **Drawers** (Figure 13.16) To prevent sagging, sticking, and loss of shape, drawers must be carefully and sturdily built. Sides should be dovetailed to the front, and the back should be housed into the sides. Bottoms should be rigid material (plywood is excellent) and should be inserted into slots

Dwelling House Construction

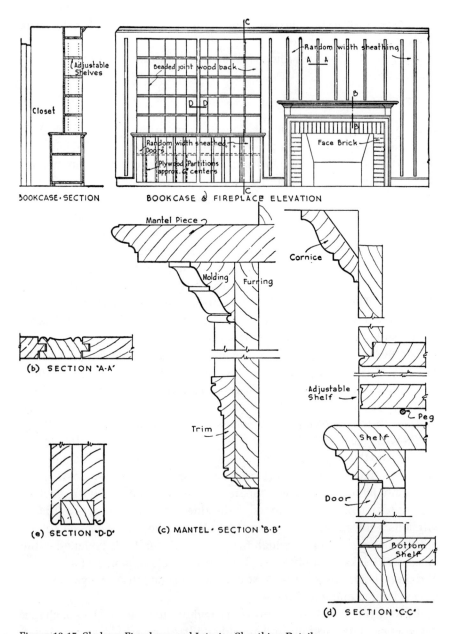

Figure 13.15 Shelves, Fireplace, and Interior Sheathing Details.

Interior Finish

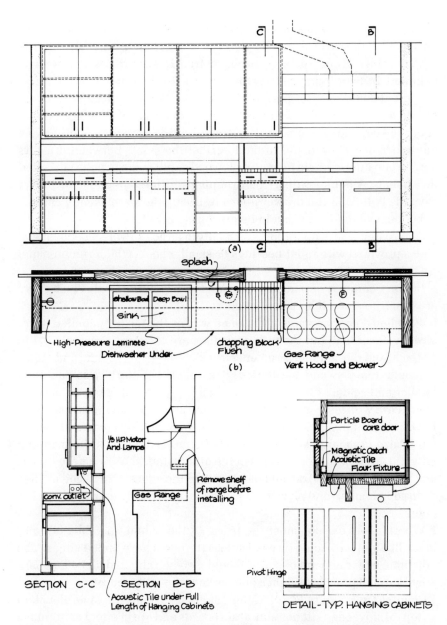

Figure 13.16 Kitchen Cabinets. (a) Elevation. (b) Plan. Sections B-B and C-C Show Shelves, Drawers, Counters, Toe Space, and Exhaust. Details Are of Hanging Cabinets.

plowed in the front and sides, and not merely be nailed to the bottom. For best results, the parts should be glued as well as nailed together, and the drawers must, of course, be squared and true. Drawer fronts may be flush or lipped. Drawers must rest on sturdy slides firmly fastened to the frame of the cupboard and set perfectly level. The bottoms of large drawers should have intermediate reinforcing strips which should run in auxiliary slides to keep the drawers moving straight. Wide drawers are otherwise apt to become skewed and to bind against the sides. Between drawers may be laid intermediate panels of plywood or other sheet material, called dust panels, which keep dust from dropping into the contens of one drawer from the bottom of the drawer above it. Dust panels are usually omitted in kitchen cabinets, but should be present in furniture.

f For greater ease of operation, especially of large drawers, special metal slides provided with roller bearings or small wheels which run on metal tracks are employed.

g. Doors Doors may be panelled, provided with lights, or flush. Flush doors are easier to clean than panelled doors.

h Panelled doors are made in the same way as the "solid" panelled doors already described, but the rails and stiles are thinner, commonly $3/4''$ to $1''$ thick, and the molding is run or "struck" directly to the rails and stiles. Panels may be plywood, hardboard, glass, or other materials, seldom solid lumber and seldom raised. Flush doors are plywood, lumber or particle board core with veneer or other hard facing and solid edging strips, hardboard, or combinations. Plywood doors are more easily made and can be cut to size from large sheets, whereas the other type must be made up either to order or in stock sizes. On the other hand, screws for the hinges and butts (Chapter 14) do not hold so well in the edges of plywood as they do in solid lumber, and plywood may twist more readily than lumber-core doors.

i Where doors close against the frame of the cabinet a rabbeted joint much like shiplap should be used unless the doors have lipped edges. The adjoining edges of pairs of doors should also be rabbeted since a square edge does not provide as tight a joint as does the rabbeted type.

j. Counter Tops (Figure 13.16) Counter tops are usually the same elevation as tops of stoves and kitchen sink drain boards and are in effect continuous with them. Although the tops may be exposed wood which is enamelled, painted, or oiled, the majority are covered with some sheet material, most commonly decorative high-pressure laminate, carried at least part way up the wall over a coved corner and molded down over the front edge. Other water-resistant sheet materials are suitable for covering counters. Commonly the edges are finished with metal strips such as stainless steel.

k. Metal Cabinets Factory-made metal cabinets for kitchens have come into general use. Doors, drawers, shelves, backs, sides, and fronts are enamelled pressed steel sheets. Doors are almost always hollow flush doors with concealed hinges. Shelves are frequently adjustable. Drawers in the better cabinets move on roller slides, others move on friction slides similar to those in wood cupboards.

l Metal cabinets are generally made in unit sizes, frequently two doors wide, which are put together in multiples for any particular plan. It is generally necessary, therefore, to plan the cabinet arrangement to fit standard units. The same holds true of standardized wood cabinets, or cabinets made of a combination of materials.

m Figure 13.17 shows details of built-in lavatories and cabinet work. Shelves, doors, toe space, and counters are typical.

13.18 Bookshelves (Figure 13.15, 17)

a Bookshelves are much the same as cabinets in their construction, except that doors are usually omitted and fronts are left open. Glass doors, either swinging or sliding on rollers resting on metal tracks, are occasionally found.

b Bookshelves are commonly adjustable rather than housed into the ends. A simple adjustable shelf may be obtained by providing the ends of cases with vertical wood strips pierced with a series of holes bored at short intervals (Figure 13.15a). Wood pegs or small metal brackets inserted in these holes support the shelves. Various metal standards with adjustable brackets are common.

13.19 Closets

a Hook strips and shelves are found in clothes closets. The hook strips are high enough to be within convenient reaching distance and yet provide sufficient height to permit garments to hang free of the floor. Clothes rods fastened to opposite hook strips permit clothing to be hung free of the wall and allow many garments to be hung in a small space. Hook strips must be firmly nailed to the framing of the walls and if the studs are not properly spaced to provide nailing, blocking must be provided. Shelves rest on top of the hook strips or, if hook strips are omitted, rest on narrow cleats nailed to the wall framing or to blocking.

Dwelling House Construction

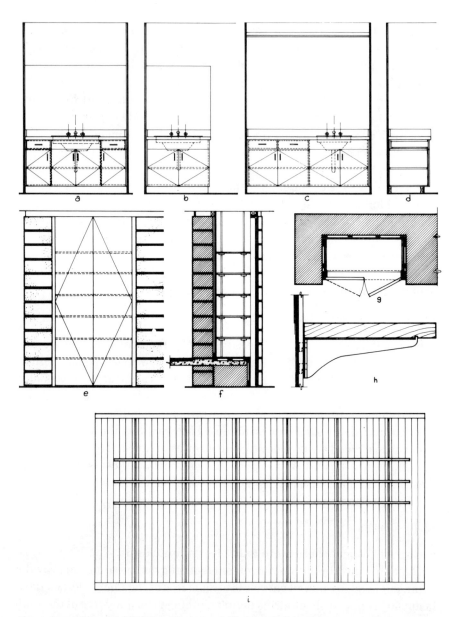

Figure 13.17 Details of Cabinet Work. (a–c) Elevations of Built-In Lavatory and Cabinets Below. (d) Cross-Section of a–c Showing Shelves, Counter, and Toe Space. (e, f, g) Elevation, Section, and Plan of Cabinet Built into Masonry Wall. Recessed Carpet in Concrete Floor. (h.i) Open Shelves on Vertically Boarded Wall with Shelves and Adjustable Shelf Support.

Interior Finish

b Clothes closets may be provided with sets of shelves and drawers to accommodate the various articles of wearing apparel. For instance, at the bottom may be special inclined racks with heel strips upon which shoes are placed. Above this may be a set of drawers with ventilated fronts to receive garments which can be laid away flat. Above this and full length to one side may be a space provided with a clothes rod to accommodate other garments which must be hung on clothes hangers. Shelves may be provided with compartments instead of being open, and may be covered to exclude dust.

c Linen closets are commonly supplied with both shelves and drawers. Shelves may be slatted instead of being solid so that linens, woolens, and other folded materials may be ventilated from below and thereby be kept dry and fresh. Drawer fronts are frequently ventilated for the same reason.

d Together with other built-in features, increasing attention is being given to the design of closets for maximum efficiency and convenience. Closets are made wide and shallow rather than narrow and deep, doors frequently run the full width, and whole walls are sometimes built as compartment closets with separate doors for the individual compartments. In small houses, where space is at a premium, closet design is especially important.

FLOORS

13.20 Materials

a Both hardwoods and softwoods are regularly used for finish floors, but the hardwoods are generally considered to be better and more attractive. Hardwoods, e.g., oak (both red and white), maple, beech, birch, pecan; and softwoods, e.g., southern yellow pine, Douglas fir, western larch, western hemlock, and redwood, are commonly used for finish flooring. The last possesses superior durability when the heartwood is used, and is mainly intended for exposed places such as porches.

b All species are manufactured in various grades, but not all species are graded the same. Oak is classified as quarter-sawed and plain-sawed and has two sub-grades in the quartered: clear and select; and four in the plain: clear, select, No. 1 and No. 2 common, plus $1\frac{1}{4}'$ shorts. Beech, birch, and maple have first, second, and third grades plus special combinations. Pecan has first grade red and white, second grade and second grade red, and third grade. All of these species have "nested" flooring bundles consisting of mixtures of grades and lengths.

Table 13.1 Hardwood Flooring

Nominal Size (in.)	Actual Size (in.)	Nails
$25/32 \times 3\frac{1}{4}$	Same	7d or 8d screw nail or cut nail
$25/32 \times 2\frac{1}{4}$	"	"
$25/32 \times 2$	"	"
$25/32 \times 1\frac{1}{2}$	"	"
$3/8 \times 2$	$11/32 \times 2$	Always on subfloor. 4d bright wire casing nail
$3/8 \times 1\frac{1}{2}$	$11/32 \times 1\frac{1}{2}$	
$1/2 \times 2$	$15/32 \times 2$	5d screw, cut, or casing nail
$1/2 \times 1\frac{1}{2}$	$15/32 \times 1\frac{1}{2}$	"

c Finish flooring (Figure 13.18a) is made in strips, tongued and grooved and end-matched so that strips can be driven tightly sidewise and lengthwise to provide tight joints, with each strip held in place by its neighbors. Thicknesses of hardwood flooring and recommended nailing are given in Table 13.1. The lower faces of flooring strips are hollowed so that they will bear firmly along both edges and so that difficulties in laying caused by irregularities in subflooring are minimized.

d Resilient flooring includes plastic (PVC), asphalt, linoleum, rubber, and cork. These are applied over a variety of substrates, and the principles of installation are discussed below (Section 13.26ff).

13.21 Preparing the Subfloor

a It has already been pointed out (Chapter 5) that subflooring, if wood boards, should preferably be laid diagonally. Before the finish flooring is laid the subfloor should be carefully inspected, any loose boards nailed down tightly; buckled, warped, or twisted pieces levelled by nailing down or replacing, and any raised edges levelled. Openings should be filled and broken pieces patched. The floor should be cleaned of plaster droppings, dirt, and nails, and be swept clean. Finally, it should be covered with a good grade of building paper lapped at ends and edges. Probably the best is a 15 lb asphalt-impregnated felt, although rosin-sized, black-sized, and slaters' felt are common. The paper helps to make the floor draft-tight so that dust does not work up from below, particularly from basements. If it is a vapor-seal type of paper, it also helps to exclude moisture from below which might otherwise cause the finish floor to warp and twist.

b If the subflooring is plywood and if it is not difficult to clean (e.g., no plaster droppings), the building paper may be omitted.

Interior Finish

13.22 Laying

a Flooring strips are blind nailed or toe nailed (Figure 13.17b), that is, the nails are driven into the angle between tongue and front edge of strip. Nails are 8d wire flooring nails, or cut steel flooring nails (Chapter 14). Nails are driven at an angle of 45° to 50° to the horizontal, and are driven flush either with a nail set or by a blow of the edge of the hammerhead. Very hard woods or those that split easily may have to be drilled for nail holes, particularly if wire nails are employed, because these have a splitting action whereas the blunt-ended cut nails punch through the wood. Nails should be from 10" to 16" apart in thick flooring, and not over 8" apart in the thinner varieties.

b The first strip of flooring (Figure 13.17c) is selected for straightness and is nailed down firmly parallel to and close to the side of the room, held just far enough away from the baseboard to be covered by the base shoe which is nailed down later. If no base shoe is employed, the flooring is carried under the baseboard just far enough to be covered or is butted against the baseboard. This first strip is face nailed along the back edge and is frequently also blind nailed. Subsequent strips are driven up tightly against the first strips and all are blind nailed. Flooring strips are made in random lengths ranging from about 2' to 12' or more, with the longer lengths predominating in the higher grades and shorter length in the lower grades. The bundles should be sorted over and the longer pieces reserved for the larger and more important rooms, leaving the shorter pieces for smaller rooms, and for closets or small halls. Strips are laid end to end, and when the end of the room is reached, the last piece is cut to length and the cut-off piece used to start the next stretch.

c Various patterns (Figure 13.17c, d) can be worked out in the flooring. The simplest method is to run all the strips in the same direction, which should be the long dimension of the principal rooms. The direction can be varied if the long dimensions of rooms do not run in the same direction. In such cases the change is usually made under doors to conceal the joint when doors are closed.

d A second method is to lay the flooring parallel to all four sides of the room, which makes changes in direction from room to room unnecessary. The ends of strips parallel to one side butt against the sides of strips parallel to the adjacent sides, which makes them appear to lap each other. This kind of pattern is often called "log cabin."

Dwelling House Construction

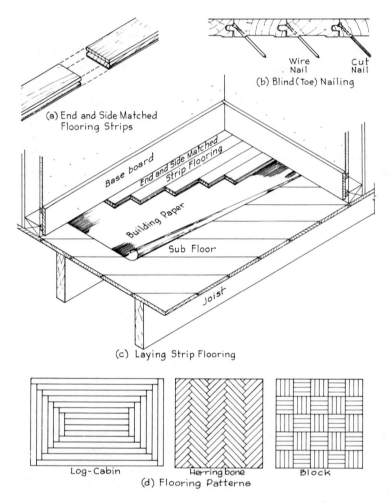

Figure 13.18 Wood Flooring.

Interior Finish 351

e Parquetry is a term applied to a floor in which the strips are cut and laid to a geometrical pattern. The most common are the so-called herringbone and block flooring. In herringbone, short pieces, perhaps 1' to 2' long, are laid zig-zag so that they are all at 45 degrees to the side walls and at right angles to each other. Block floors consist of squares of short lengths of flooring usually 8" to 12" square, which are usually first assembled into units in the factory and are then laid as blocks on the subfloor. When laid on wood subfloors they are blind nailed along the tongued edges; when laid on concrete they are usually cemented down with stiff mastic.

f Parquetry may be ornate and, when made up to some pattern as laid out by the designer, and with various colors and figures provided by different species and cuts of wood, gives the intricate patterns sometimes found in pretentious buildings. In such instances the individual pieces of flooring must be cut to pattern and then be completely laid out on the floor, and each piece must be individually fitted into its proper place.

g End-matched wood-strip flooring $25/32$" thick may be laid directly across joists without subflooring. Joists should not be more than 16" on centers, and both ends of a single piece of flooring should not be in the same joist space.

13.23 Precautions in Handling Flooring Material

a Finish flooring is a carefully-manufactured product, cut and shaped to precise sizes and cross-section, and dried to a low moisture content. It is a valuable commodity, and should be treated as such. The principal precaution, and one which cannot be stressed too much, is that it must at all times be kept *dry*. Flooring is commonly kiln-dried to a moisture content of 6 to 8 percent, which content it will maintain if kept in dry storage and not permitted to become damp. If it is exposed to dampness it quickly absorbs moisture, swells, twists, and otherwise gets out of shape, so that strips no longer fit well together. Furthermore, dwellings are commonly dry in winter because of artificial heat and lack of sufficient humidity, and flooring laid down in a moist condition shrinks, leaving unsightly open joints between adjacent strips. Flooring should *(a)* be stored in dry sheds at the supply yard, *(b)* be covered during transportation from yard to house during damp weather, *(c)* be stored in dry conditions in the house, *(d)* not be brought onto the premises until plaster is thoroughly dry and concrete floors have completely dried, *(e)* be laid as soon as possible after delivery, and *(f)* be finished by the painter as soon after laying as possible.

Dwelling House Construction 352

Flooring should preferably be one of the last items in the construction.

13.24 Sanding

a Formerly all floors were scraped by hand with special floor scrapers. Today, sanded finishes are achieved by motor-driven floor-sanding machines which consist of a revolving drum on which is mounted sandpaper of varying degrees of fineness depending upon the stage of the sanding operation. Sanding is begun with coarse paper to break down any irregularities in the floor and to provide a plane surface but one which looks coarse. Subsequent sanding with finer papers brings out the figure of the wood and provides a smoother surface, the degree to which the polishing is carried depending upon the fineness of the paper in the final sanding operation.

13.25 Specialties

a A few special types of wood finish floors may be considered here.
b. Random-Width Flooring To simulate plank floors, flooring is obtainable in random widths and lengths. Widths usually vary from 4" to 10" or 12" in multiples of 1" or 2", so that two or three narrow boards can be made to match the width of a wide one. Frequently this flooring is cut from specially selected knotty material such as oak. Such floors may be blind nailed or may be nailed and pegged to simulate the appearance of old plank floors.
c. Ready-Finished Flooring When it is desired to lay a new floor over an old one in an occupied house or to avoid the delay caused by applying finish coatings, ready-finished flooring may be used. This is commonly provided with a vee joint instead of tongue and groove; already stained, sealed, or otherwise finished; and ready for use as soon as laid. Because great care must be exercised in laying such flooring to avoid marring, the edges are frequently factory-drilled for nails, and the strips are carefully wrapped in packages for delivery.

13.26 Resilient Flooring

a The principal types of resilient flooring used in house construction are vinyl (PVC, Chapter 16), asphalt, linoleum, and cork, with vinyl and asphalt the most common. Vinyl flooring is made as both sheet and tile, asphalt as tile, linoleum as sheet, and cork as tile. Vinyl, in addition, is made as vinyl-asbestos and as "homogeneous" sheet. As the name implies, asbestos

Interior Finish 353

fibers, as well as other ingredients and pigments, are added to PVC in vinyl-asbestos, resulting in a harder but somewhat more brittle formulation than the homogeneous. Of all the forms of vinyl, vinyl-asbestos tile are probably the most widely used.

b Asphalt tile are based on various resins, asphaltic compounds, and asbestos fibers, plus other ingredients and pigments. Linoleum uses an oleoresinous base incorporating linseed oil, cork, and wood flour formed on a burlap backing or rag fiber.

c In all of these materials the wearing layer may go through the entire thickness, or it may be only partial. Common overall thicknesses range from $\frac{1}{16}''$ to $0.090''$.

d Some vinyl sheet materials have a felt back, e.g., asbestos fiber, for adhesion to a substrate. Others have a thin foamed plastic base to add a soft, resilient cushion, especially useful where hard substrates might otherwise lead to fatigue.

e Although most resilient flooring has its surface improved by periodic waxing, some of the denser, harder formulations can dispense with waxing.

f A wide variety of colors, textures, and figures is possible. Mottled, plain, chips in a matrix, embossed to simulate brick and tile, smooth, and other surface characteristics are available.

g The term resilient flooring is applied to these materials to distinguish them from hard materials such as concrete and ceramic tile (Chapter 12). They are softer underfoot, with homogeneous softer than vinyl-asbestos, but they should not be so soft as to show permanent indentation under long-continued or impact loads. Such indentation is more likely to occur at elevated temperatures because the vinyls and asphaltic compositions are thermoplastic. At lower temperatures they become harder.

13.27 Installation

a Resilient flooring may be installed over a variety of substrates, but concrete and wood are the most common in house construction.

b Concrete may be suspended, that is, have an air space below it; it may be at grade, or it may be below grade. The latter two cases are most common in dwelling houses, as exemplified by slab on grade and by basement construction.

c Suspended concrete floors do not pose any particular problem. Fresh concrete must be dry enough so that a good bond can be achieved, and there must be enough ventilation under it to allow the water in the concrete to escape from the lower surface.

d Concrete on soil may present a problem. A bed of gravel or other porous material should be placed under the concrete, and it should be well drained. A vapor barrier should be placed under the slab but, even so, it is entirely likely that dampness will exist in the concrete which will tend to migrate to the upper surface. Such slabs should be allowed to dry, after curing, for the longest possible time before resilient flooring is applied. Some types, such as linoleum, probably should not be applied to such floors in any event.

e Alkaline water below the slab, or even the alkalies in the concrete itself, may cause difficulties with some cements or adhesives. Only those adhesives and cements formulated for such conditions should be employed, but when they are, satisfactory results can readily be obtained with vinyl and asphalt floors.

f Rough surfaces, such as are likely to be found on rough-cast concrete, are almost certain to show or "telegraph" through a resilient floor. Consequently, such floors must be smoothed, usually with a topping layer. This may be a concrete mortar but, if it is, it must be thick enough not to crack later under traffic. Such material cannot be feathered successfully. If the topping must be thin, it should be based on a latex rather than straight portland cement, and the formulation must be alkaline-moisture resistant if on a slab on soil.

g Wood makes a satisfactory substrate for resilient flooring but the problem of telegraphing defects exists here as it does with concrete. This is particularly true of the joints between rough subflooring, especially if the boards are wide (more than 3″–4″) and not tongue and groove. In such instances, it is best to apply an underlayment of underlayment-grade plywood or hardboard (Chapter 5). Particle board is also employed if of the proper grade for that purpose. Underlayment should be thoroughly fastened by close nailing, and joints filled if necessary to provide a smooth surface.

h Once the substrate is prepared, resilient flooring may be installed. Because of differences among proprietary materials and cements or adhesives, and different substrates, it is essential that manufacturers' directions be closely followed. For tile, the cement is applied over the entire surface of the substrate, in small enough areas at a time to permit the tile to be placed and pressed down firmly before the adhesive becomes too

Interior Finish

dry for a good bond. For sheet materials, adhesive may be spread over the entire surface, or in strips, e.g., 6″ along perimeters and 8″ (4″ each side) along intermediate joints. Any such adhesive must be resistant to water that may seep down into the joint.

i Special cove strips are made for use with resilient flooring. These provide a smooth curved transition between the foor and the wall, and provide the same function as a baseboard. In some instances, sheet materials are warmed and bent into a gradual curved cove at the floor-wall interface, but the specially formed cove provides a surer detail less prone to difficulty.

14 Hardware

14.1 General

a Builders' hardware is divided into two general categories—rough and finish. Under rough hardware are lumped such items as anchor bolts and other bolts, rough screws, nails, hangers, strapping, and similar miscellaneous iron and steel which is not of a heavy structural nature. Finish hardware includes butts and hinges, locks, door knobs, window fasts, drawer pulls, and similar items.

14.2 Nails (Figure 14.1)

a Nails are practically the universal means of fastening the wood members of a house together. Wood and metal pins are occasionally found, screws and bolts have some use, some adhesives are employed and their use is growing, but nails are easily the most important of all fastening means. Two principal types are employed, cut and wire, with wire nails far overshadowing cut nails in importance. In the following discussions, only those nails are included which are at all likely to be found in building operations. In other chapters, reference is made to the types of nails employed for the particular operation under consideration.

14.3 Cut Nails (Figure 14.1)

a Historically these are the older variety. They are cut or sheared from flat iron nail plate of the proper thickness, and heads of various types are formed at the wide end. The shanks taper along the two edges but the two sides are flat. Points are blunt.

14.4 Wire Nails (Figure 14.1)

a Wire nails, first introduced during the second half of the last century, have practically superseded all others. Most are manufactured from mild steel wire of the proper gauge, but aluminum and copper or copper alloy are employed for corrosion resistance. They may have any one of a number of different kinds of heads or no head at all, and may be pointed in a variety of different ways. The shank is cylindrical, and of uniform diameter. The holding power of the nail depends to a considerable extent upon friction caused by the pressure of wood fibers pried apart by the point of the nail as it penetrates the wood. Because of the point, wire nails have a greater tendency to split the wood than do cut nails.

The section on Finish Hardware was prepared by W. R. Haverkampt, of Sargent and Company.

Hardware

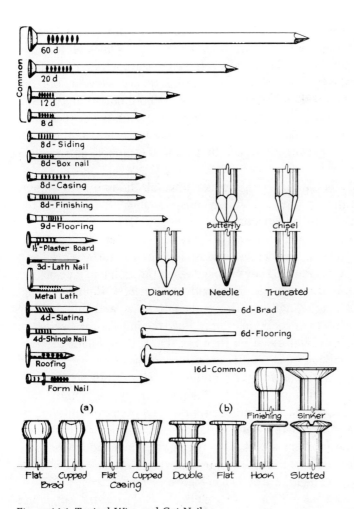

Figure 14.1 Typical Wire and Cut Nails.

b. Sizes Lengths of nails are in general designated by the "penny," an expression whose origin, although known to be English, is difficult to trace. One theory is that the word originally referred to the cost, for example, a 6-penny nail cost 6 pence per hundred. Another theory is based upon the fact that the symbol "d" was originally used for both pound and penny, and a 6d nail therefore was one which weighed 6 pounds per thousand. Whatever the origin, the symbol "d" is still used to designate length. In general, the range is from 2d to 60d, or 1 inch to 6 inches, as shown in Table 14.1.

c Diameters of shanks vary with the use to which the nail is to be put. For heavy work they are large (low wire-gauge number); for fine work the shank is thin (high wire-gauge number) (see Table 14.1).

d. Heads Depending upon the use, nails may have flat, tapered (sinkers), or countersunk heads. Flat heads provide the greatest surface area and the greatest gripping power. They are used when the nailhead may be permitted to show, and are driven flush with the surface. Common nails, spikes, box nails, siding nails, roofing nails, lath nails, and shingle nails are examples. With the exception of brad heads, countersunk heads are conical and are only slightly larger than the shank. Brad heads are partially spherical and are even smaller than the conical. Both are designed to be driven below the surface (countersunk) with a nail set, and almost always are concealed by putty. Common brads, nails and brads for flooring, casing nails, and finishing nails have countersunk heads. Special types of heads are round or oval; hooked as in metal lath nails; and double-headed, as found in concrete form nails which must be driven home firmly for holding power, but which must subsequently be easy to withdraw. A few types of heads are shown in Figure 14.1.

e. Points Practically all nails used in building have the ordinary diamond point shown in Figure 14.1. Blunt-pointed nails may be employed with refractory woods which split easily (an ordinary diamond point when blunted by striking with a hammer is a satisfactory substitute). A few other types of points are shown in Figure 14.1.

f Shanks for most building applications are cylindrical, but many other types are found. These include barbed, fluted, grooved, knurled, threaded, and twisted types. Twisted shanks are square in cross section. Threaded nails, in particular, are used for superior gripping power. Threads may be annular or helical, and result in alternate ridges and depressions in the shank. Annular-threaded nails are often called ring nails. Drive screws

Hardware

have helically-threaded shanks and are likely to have slotted heads. They are made to turn when driven with a hammer and can be retracted with a screwdriver.

g. Surface Treatment Most nails are made of wire cold-drawn in the final stages and therefore are bright, smooth, and covered with a thin film of lubricating oil. Lath nails turn blue when they are cleaned of oil and sterilized. A coating derived from resins or shellac provides a temporary bond between nail and wood and protects against corrosion in storage. Etching increases holding power. Shingle and roofing nails, and others destined for exposure to severe atmospheric conditions, are usually zinc coated.

h Nails may be of metal other than mild steel. Aluminum is used for its corrosion resistance. Copper is especially desirable for roofing. Zinc, brass, Muntz (yellow) metal, and copper-bearing steel have increased corrosion resistance. Stainless steel and monel metal have special uses. Coatings, in addition to the widely-used zinc, may be tin, copper, cadmium, or brass. Nickel and chromium plated nails are used in conjunction with trim of the same metal. "Parkerizing" increases the paint-holding power.

i Where greater hardness is needed, as in driving nails into masonry or concrete, the steel may be medium to high carbon, heat treated to produce greater hardness and resistance to bending than usual.

j Many special types of nails are made for special purposes. Some of these are described in other chapters of this book in connection with those topics (e.g., gypsum board). A few others are listed in Table 14.1.

14.5 Screws and Bolts

a. Wood Screws Screws are made with various shapes of heads and threads, depending upon the uses to which they are to be put. The types usually employed in building are (Figure 14.2):

Flat head
Round head
Oval head
Clove head
Bung head
Winged head
Pinched head
Headless
Drive
Dowel

Table 14.1 Wire Nails

Size	Common[a]				Box,[b] Casing				Head Diam. In.		Finish				Flooring[c]			
	Length (in.)	Gauge No.	Shank Diam. (in.)	Head Diam. (in.)	Gauge No.	Shank Diam. (in.)			Box	Casing	Gauge No.	Shank Diam. (in.)	Head Diam. (in.)		Gauge No.	Shank Diam. (in.)	Head Diam. (in.)
2d	1	15	0.072	11/64	15½	0.067			3/16	0.098	16½	0.058	0.086		15	0.072	9/64
3d	1¼	14	0.080	13/64	14½	0.076			7/32	0.113	15½	0.067	0.098		15	0.072	9/64
4d	1½	12½	0.098	¼	14	0.080			7/32	0.120	15	0.072	0.105		13	0.091	5/32
5d	1¾	12½	0.098	¼	14	0.080			7/32	0.120	15	0.072	0.105		13	0.091	5/32
6d	2	11½	0.113	17/64	12½	0.098			17/64	0.142	13	0.092	0.135		11½	0.115	13/64
7d	2¼	11½	0.113	17/64	12½	0.098			17/64	0.142	13	0.092	0.135		11½	0.115	13/64
8d	2½	10¼	0.131	9/32	11½	0.113			19/64	0.155	12½	0.098	0.142		11½	0.115	13/64
9d	2¾	10¼	0.131	9/32	11½	0.113			19/64	0.155	12½	0.098	0.142				
10d	3	9	0.148	5/16	10½	0.127			5/16	0.169	11½	0.113	0.155		10	0.135	¼
12d	3¼	9	0.148	5/16	10½	0.127			5/16	0.169	11½	0.113	0.155		10	0.135	¼
16d	3½	8	0.162	11/32	10	0.135			11/32	0.177	11	0.120	0.162		9	0.148	9/32
20d	4	6	0.192	13/32							10	0.135	0.177				
30d	4½	5	0.207	7/16													
40d	5	4	0.225	15/32													
50d	5½	3	0.244	½													
60d	6	2	0.263	17/32													

[a] Aluminum and copper shank diameters are slightly different.
[b] Commonly cement-coated. Shanks are generally 1 ga. smaller. Also supplied galv.
[c] Also flooring brads, brad head, slightly different shank diameters.

Asbestosboard. Galv. steel or alum. Helical. 1⅛″–1½″. Flat or casing head.
Brad. Common. ⅜″ to 60d, brad head 0.050″ to 0.331″.
Clinch. 2d to 20d, duckbill or clinch point. Oval head.
Concrete. Hardened steel, ½″ × 0.135″ to 3½″ × 0.207″, countersunk head.
Dating. Galv. copper, brass. ⅜″ to ½″ flat numeral head.

Hardware

Double-headed. Concrete forms. $1\frac{3}{4}''$ x 0.113 to 4'' x 0.207'', $\frac{3}{16}''$ to $\frac{7}{16}''$ double heads.

Fiberboard. Low-carbon or hardened. 1'' x 0.054'' to 2'' x 0.062'', needle point.

Gypsum-lath. Bright or blued. 1'' to $1\frac{3}{4}''$ x 0.092'', $\frac{19}{64}''$ to $\frac{3}{8}''$ head. 1'' x 0.120'' to $1\frac{1}{2}''$ x 0.148'', $\frac{1}{2}''$ head. Alum. alloy $1\frac{1}{8}''$ x 0.099'' to $1\frac{1}{2}''$ x 0.105'', $\frac{19}{64}''$ or $\frac{5}{16}''$ head.

Gypsum-wallboard. Smooth or annular thread. $1\frac{1}{8}''$ x 0.062'' to 2'' x 0.105'', $\frac{1}{4}''$ to $\frac{19}{64}''$ head.

Hardboard. Bright, colored lacquer, or galvanized. Medium carbon or hardened. Annular or helical thread. 1'' x 0.058'' to 3'' x 0.115''. Small flat to countersunk head.

Hinge. $1\frac{1}{4}''$ x $\frac{3}{16}''$ to 4'' x $\frac{3}{8}''$, flat or oval countersunk, long diamond or chisel pt.

Insulation board. Zinc, nickel, or cadmium plate. $1\frac{1}{4}''$ and $1\frac{3}{4}''$ x 0.054'', $\frac{3}{32}''$ head, needle pt.

Masonry. Hardened, knurled, vertically threaded, fluted, plain or zinc coated. $\frac{1}{2}''$ to 4'' x 0.148'' to 0.250'', flat or checkered $\frac{5}{16}''$ to $\frac{9}{16}''$ head.

Roofing. Plain, galvanized, copper-coated steel; aluminum. Flat, checkered, small to large, reinforced, lead and cast-lead heads plus lead, neoprene, plastic washers. $\frac{3}{4}''$ x 0.092'' to 2'' x 0.150'', heads $\frac{1}{4}''$ to $\frac{5}{8}''$.

Shake. Galv. steel and alum. $1\frac{1}{4}''$ x 0.086'' to $2\frac{1}{2}''$ x 0.092'', heads $\frac{1}{8}''$ to $\frac{5}{32}''$.

Shingle. wood. Bright or galv. steel, alum., plain or threaded shank, heads $\frac{7}{32}''$ to $\frac{3}{32}''$, med. or blunt diamond pt.

Shingle. asbestos. Bright or galv. steel, alum.. threaded, 1'' x 0.118'' to $2\frac{1}{2}''$ x 0.113'', $\frac{3}{16}''$ to $\frac{5}{16}''$ head, diamond or needle pt.

Siding, aluminum. Plain or helical thread. 1'' x 0.099'' to $2\frac{1}{2}''$ x 0.135'', $\frac{1}{4}''$ to $\frac{5}{16}''$ hd.

Siding, asbestos. Bright, colored lacquer, galv. hardened steel; alum., bronze, stainless steel. Annular, file-grip, or screw thread. 1'' x 0.080'' to $2\frac{1}{2}''$ x 0.099'', $\frac{3}{16}''$ striated flat or button head.

Siding, wood. Bright, colored lacquer, steel alum., or stainless steel. Plain or threaded. $1\frac{3}{4}''$ x 0.080'' to 3'' x 0.148''. Flat, casing, or sinker head. $\frac{5}{32}''$ to $1\frac{3}{32}''$.

Slating. Galv. steel, alum., copper. 1'' x 0.106'' to 2'' x 0.148'', $\frac{5}{16}''$ to $\frac{7}{16}''$ head.

Underlayment. Low, medium-carbon or hardened steel. Annular thread. 1'' x 0.080'' to 3'' x 0.148'', flat or countersunk head.

Wallboard. Low, medium-carbon or hardened steel. Colored lacquer. Plain or annular thread. $1\frac{1}{8}''$ x 0.062'' to 2'' x 0.083'', countersunk head.

Table 14.2 Wood Screws

No.	Diam. (in.)	No.	Diam. (in.)	No.	Diam. (in.)
0	0.060	6	0.138	12	0.216
1	0.073	7	0.151	14	0.242
2	0.086	8	0.164	16	0.268
3	0.099	9	0.177	18	0.294
4	0.112	10	0.190	20	0.320
5	0.125	11	0.203	24	0.372

Of these, flat-head screws are used much more than all the others combined, round-head screws rank second, and the first three listed above fulfill almost all building requirements. The others are used for limited specialized applications.

b Most items of hardware (see Sections 14.6 ff.) are applied with flat-head screws. Round-head screws are used for surface hinges and for other applications where appearance is a factor. Oval-head screws find some use for applying interior trim which must be removed periodically.

c Standard metals for wood screws are steel and brass, although bronze, monel metal, and other special metals may be employed, and the screws may be plated for special purposes. Steel screws are ordinarily made in lengths from $\frac{1}{4}''$ to 5 ", and brass screws from $\frac{1}{4}''$ to $3\frac{1}{2}''$. Sizes are given in Table 14.2.

d. Lag Screws These are large-sized, gimlet-pointed steel screws provided with square or hexagonal heads to be turned by wrench instead of by screw driver. Lag screws are used for fastening heavy framing members and for attaching structural iron such as angles, channels, and strap iron to wooden members.

e Holes for all types of screws should be prebored. In soft woods, holes for the smaller sizes of screws may be approximately equal to the diameter at the base of the thread, but in hard refractory woods it is often necessary to bore two holes—one for the threaded portion and a larger one—slightly smaller than the shank diameter, for the unthreaded portion of the screw. Lubricating with wax or paraffin often helps to drive the screw home in hard refractory woods.

f. Bolts Bolts are used to fasten the sills to foundation walls, to join heavy wood framing members, and to attach metal framing members to each other or to wood.

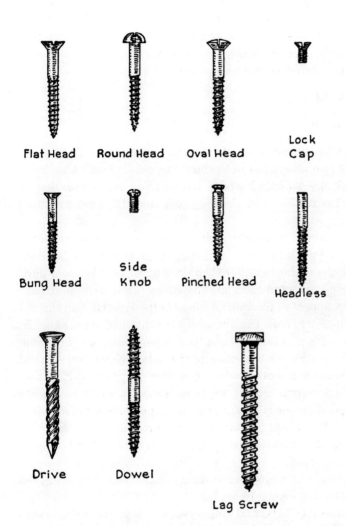

Figure 14.2 Types of Screws.

g Holes for bolts are bored $\frac{1}{16}$ inch larger than the diameter of the bolt unless an absolutely snug or "driving" fit is required, in which instance the bolt hole is the same diameter as the bolt.

FINISH HARDWARE

14.6 Butts and Hinges

a A hinge is, strictly speaking, a pair of straps joined together by a pin which allows the two straps to rotate about the pin (Figure 14.3a, b). A butt, often mistakenly called a hinge, is the butt end of a hinge, that is, the portion near the pin but with the long portion of the straps omitted (Figure 14.4a, b).

b Today the term "hinge" is usually applied to a member screwed to the surface of a door and the term "butt" to a member mortised into the edge. Most doors are hung on butts rather than on hinges. For the most part, hinges are employed on batten doors and similar doors not easily mortised.

c The round central part of the butt or hinge is the knuckle and the flat portions are the flaps or leaves. The pin, which is contained in the knuckle, when made removable is called a "loose" pin, otherwise it is a "fast" pin. Loose pins sometimes have a tendency to rise when leaves are worked back and forth because the pin binds and rotates only one way. Consequently, specially designed "nonrising" pins, forced to rotate both ways may be substituted. If the pin is so made that it cannot be withdrawn when the door is closed, it is a self-locking pin. This prevents doors from being tampered with and removed from their hinges when closed. (Figure 14.5c—Non-rising hinge)

d The proper location of a hinge on a door is determined by using the formula as illustrated in Figure 14.5a, b.

e In addition to ordinary butts and hinges, several types of butts, hinges and pivots are made for special purposes. Open or enclosed springs may be incorporated at the pin to provide a self-closing feature. Such hinges are especially common on screen doors which must be kept closed as much as possible (Figure 14.6).

f Invisible hinges consist of a number of small flat plates rotating about a central pin and provided with shoulders which are mortised into the edges of doors and frames. When closed, the hinges are completly out of sight (Figure 14.7).

Hardware

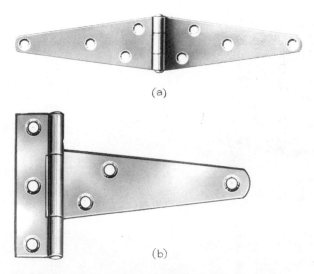

Figure 14.3 (a) Strap Hinge. (b) Half-Strap Hinge.

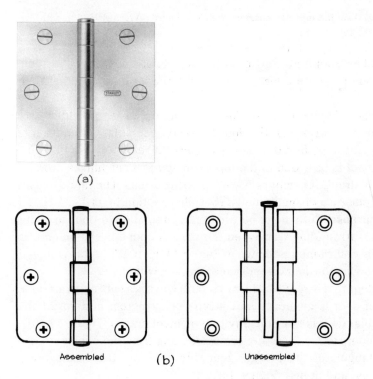

Figure 14.4 (a) Butt. (b) Assembled and Unassembled Butt Showing Pin and Leaves.

Dwelling House Construction   366

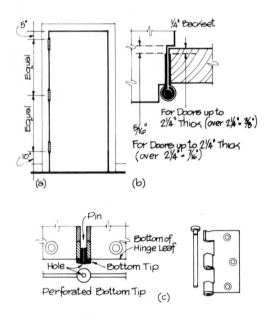

Figure 14.5 (a) Door and Butt Arrangement. (b) Clearances. (c) Non-rising Pin with Perforated Bottom Tip.

g Doors that are required to swing both ways (in and out) are called double-acting doors. These doors are usually hung on double-acting floor hinges.

h Floor hinges (Figure 14.8) for double-acting doors are provided with springs which either permit the door to stand open at 90° or cause it to close to its central point if the door is released when open less that 90° A mortise pivot is provided at the top. "Checking" floor hinges are provided with hydraulic chambers as well as with springs. The liquid causes the door to close slowly and quietly. (See also Section 14.11.)

i. Special Purpose Butts must be wide enough to allow a door to swing clear of the surrounding trim when the door is opened. The manner in which this is determined is shown in Figure 14.9a, b. Wide-throw hinges are used to swing the door clear of exceptionally wide trim.

j Another door that is required to swing both ways is a cafe or dwarf door. This type of door is generally louver-type construction and very light. This particular door is hung on gravity-type pivot hinges (Figure 14.10).

k Cabinet doors today generally come pre-hung from the cabinet manufacturer. Hanging of these doors generally falls into three categories: flush, overlay, and lipped (Figure 14.11).

Hardware

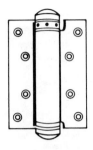

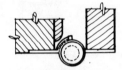

Figure 14.6 Spring Hinge.

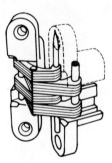

Figure 14.7 Invisible Hinge.

Figure 14.8 Floor Hinge for Double-Acting Door.

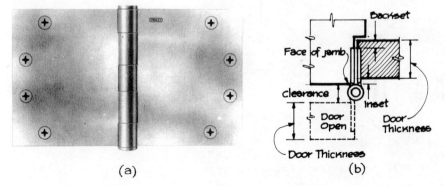

Figure 14.9 (a) Wide Butt. (b) Determination of Inset and Clearance for Wide-Throw Door.

Dwelling House Construction

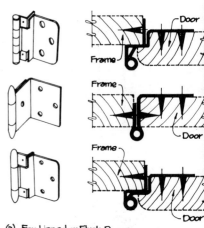

(a) For Lipped or Flush Doors

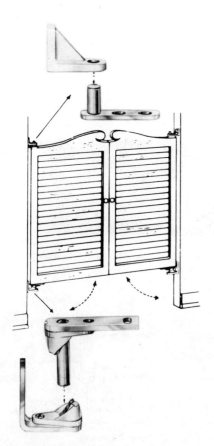

Figure 14.10 Cafe Door with Gravity Pivot Hinge.

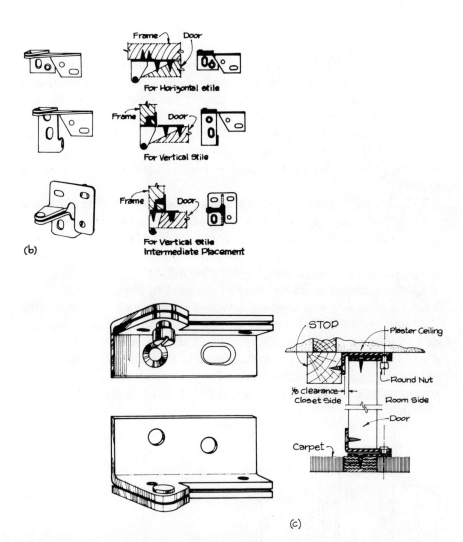

Figure 14.11 (a) Hinges for Lipped or Flush Doors. (b) Pivot Hinges for Vertical and Horizontal Stile. (c) Pivot Hinges for Full-Height Wardrobe Door.

14.7 Locks and Latches

a Both locks and latches are devices for holding doors in the closed position. Generally speaking, the latch consists of a beveled or otherwise shaped bar, called a latch bolt, which slides into position when the door is closed. Generally, also, the device for opening the latch is a knob or lever, connected with the latch. These are called operating trim. In some instances the latch is so arranged or fitted that it can be opened by such a device from one side only and must be opened from the other side with a key.

b Dead bolts are rectangular-shaped bolts which do not slide into place automatically but must be thrown into place by key or turn knob. It is common practice in builders' hardware to combine latch bolts and dead bolts into the same unit and such combinations are simply called locks. When a lock has a dead bolt only, it is called a dead lock.

c Various types of locks and latches are used in today's construction. They are generally named after the type of construction and installation they require. Basically there are mortise, unit and integral, cylindrical and tubular, and special-application locks and latches.

d Mortise locks and latches are designed to fill a cavity or hole placed in the edge of a door (Figure 14.12). They are concealed except for the face or portion showing at the edge, the knob or lever, the cylinder, and the operating trim.

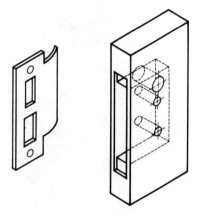

Figure 14.12 Mortise for Mortise Lock. Strike.

Hardware

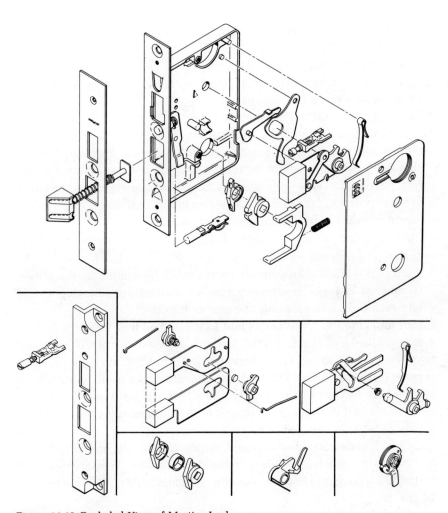

Figure 14.13 Exploded View of Mortise Lock.

e A mortise lock is made up of a lock body that contains the following features (Figure 14.13):
1. Latch bolt—(described earlier).
2. Guard bolt—Prevents the latch bolt from being retracted by surreptitious means when it is depressed.
3. Dead bolt—(described earlier).
4. Cylinder—See Section 14.18.
5. Stop Works—Slide stops operating on split hubs which engage the outside knob hub so it cannot be turned and the latchbolt cannot be retracted from that side without a key. Stop works are operated by a pair of buttons on the face of the lock. When one is depressed, the other is raised and the knob is stationary on one side.
6. Operating Trim—Can be either a knob or lever that comes in various design and finishes.
7. Functions—Mortise lock functions are made by combining the above-mentioned features in various ways. A manufacturer's catalog should be consulted when selecting the proper function.

f. Unit and Integral Type Locks and Latches The unit lock is a factory pre-assembled lock that fits into a door cutout (Figure 14.14). The integral type lock is also a factory pre-assembled lock that fits into a door mortise (Figure 14.15a, b). These similar locks offer the same basic features as a mortise lock, i.e., latch bolt, guard bolt, dead bolt, stop works and operating trim except that the cylinder is placed in the knob, and they have limited functions when compared to mortise locks.

g. Cylindrical and Bored Locks and Latches Cylindrical and bored locks feature a key-in-knob principle with rapid installation. They are designed to fit into a hole that is bored in the edge and lock stile of the door (Figure 14.16a, b).

Following are the components that make up a cylindrical lock (Figure 14.17):
1. Latch Tube—Contains both the latch bolt and guard bolt.
2. Aligning Tube—A cylindrical housing which receives the latch tube, knobs, and provides threads for the roses to fasten to.
3. Knob Assemblies—An outside and inside knob assembly that snaps into the aligning tube or cylindrical housing.
4. Cylinder—Placed in the knob, generally the outside knob.
5. Button Stops—Also placed in the knob. Generally the inside knob contains the buttons that lock the outside knob.

Hardware

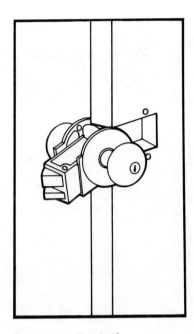

Figure 14.14 Unit Lock.

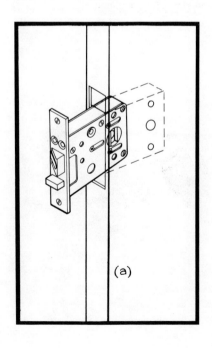

(a)

(b)

Figure 14.15 (a) Mortise Lock Inserted in Mortise. (b) Integral Lock.

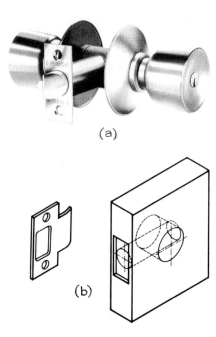

Figure 14.16 (a) Bored-in Lock. (b) Cutting and Boring for Bored-in Lock.

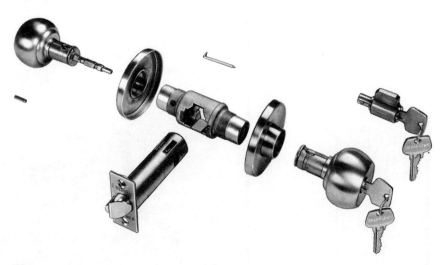

Figure 14.17 Exploded View of Cylindrical Lock.

Hardware

Cylindrical and bored locks are used today mostly in dwelling house construction because they are easy to install and inexpensive.

h. Special-Application Locks and Latches The more common special-application locks and latches found in dwelling house construction are rim night latches, rim deadlocks, jimmy-proof rim locks and sliding door locks (Figure 14.18). Rim locks are used to provide additional security when required. Sliding-door locks generally come furnished with the sliding door assembly, directly from the manufacturer.

14.8 Cylinders

a The locks previously mentioned accomplish their security by cylinders having keyways, cylinder pins and drivers. Normally five or six pins are used for locks keyed alike or differently (Figure 14.19). The mechanism of a cylinder for a mortise cylinder lock consists of a cylinder or shell containing a cylindrical barrel. The barrel is slotted lengthwise and requires corresponding slots or keyways in the side of the flat key. At right angles to the barrel and cylinder shell are holes containing the cylinder pins, drivers and cylinder springs. The key is notched along one edge in such a way that when it is inserted it raises the pins. If the notches are correctly cut the breaks between the cylinder pins and the drivers line up where the barrel and shell meet, allowing the barrel to be rotated. Otherwise, the cylinder pins would obstruct the rotation of the barrel. Rotation of the barrel turns a cam which engages the deadbolt lever, the latch bolt lever, or both, and withdraws the bolt or bolts, allowing entry. Key-in-the-knob cylinders are substantially the same as the above except that the door knob acts as the cylinder shell.

14.9 Operating Trim

a In conventional mortise and bored-in type locks the operating trim or combination of knob, spindle, and hub makes it possible to retract the latch bolt when the door is closed. The operating trim comes in various designs and finishes to complement the various types of dwelling-house construction seen today.

b Operating trim varies so widely that the manufacturer's catalog should be consulted when selecting an appropriate design.

Dwelling House Construction

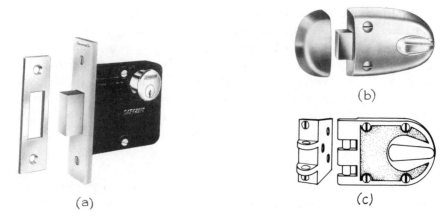

Figure 14.18 (a) Rim Dead Lock. (b) Rim Night Latch. (c) Rim Lock.

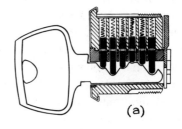

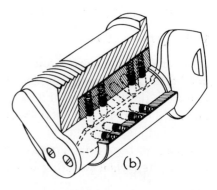

Figure 14.19 (a) Cylinder Showing Pins Lined Up in Barrel by Notched Key. (b) Key Turning Barrel and Cam, Maximum Security Cylinder.

Hardware

14.10 Handing the Door

a In order for the various locks and latches and hinges previously described to function properly on the door they must be handed first. The manner in which a door swings determines its "hand" and the hand of the lock required (Figure 14.20).

b The hand of a door is determined from the "outside", i.e., the street side of an entrance door, the corridor side of a room door, the space between twin doors, and the side of the door against which it is to be locked if the door communicates from one room to another. In general, the outside of a door is the side which must be opened by key, if the door requires locking. The "outside," of closet doors and cabinet doors of all kinds are the room sides.

1. A right hand is a door that swings away from you to the right. You cannot see the hinges from the outside when the door is closed. Usually this is abbreviated, "R.H.".
2. A left hand is a door that swings away from you to the left. You cannot see the hinges from the outside when the door is closed. Usually this is abbreviated, "L.H.".
3. A right hand reverse bevel door is a door that swings toward you to the right. You do see the hinges when the door is closed. Usually this is abbreviated, "R.H.R.B.".
4. A left hand reverse bevel door is a door that swings toward you to the left. You do see the hinges when the door is closed. Usually abbreviated, "L.H.R.B.".

c The four hands apply to descriptions of hands for door locks. This is necessary because the swing of the door must be known, also, the outside hand for the key when keyed locks are required and the bevel of the front of the lock where beveled doors occur. In the case of hinges only, two hands occur, right or left. This also applies to the handing of cabinet doors where again only two hands occur, right or left.

14.11 Door Closers

a Door closers or "checks" are designed to close a door quickly but to do it without the slamming that occurs with ordinary springs and spring hinges. In dwelling-house construction, two general types are employed, liquid and pneumatic. The liquid or hydraulic type is used to control larger doors than the pneumatic which is almost exclusively used on screen doors, or gates (Figure 14.21).

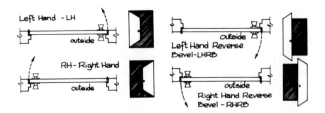

Figure 14.20 Hands of Doors Showing Left Hand (LH), Right Hand (RH), Left Hand Reverse Bevel (LHRB), and Right Hand Reverse Bevel (RHRB).

Figure 14.21 Pneumatic Door Closer.

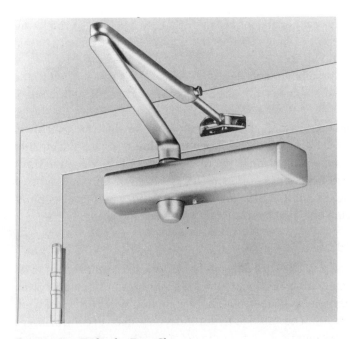

Figure 14.22 Hydraulic Door Closer.

Hardware

b The hydraulic closer is generally used on garage doors that enter directly into a house as well as front and rear entrance doors. Fire regulations today call for a self-closing "C" label fire door between the garage and house. These doors are generally of the larger size requiring the hydraulic type closer (Figure 14.22).

14.12 Miscellaneous Hardware

a In this category may be lumped the many small items of commonly-found hardware where function is more or less self-evident. Cabinet door catches and locks, tracks and sliding sheaves for sliding cabinet doors, overhead track and hangers for heavier sliding doors, drawer pulls, door stops, coat and hat hooks, supports for adjustable shelves, and many others too numerous to mention, fall into this group.

14.13 Window Hardware

a Window hardware is no longer the worrisome problem that it was years ago. The reason is that the manufacturer supplies the windows complete with all hardware. In days gone by windows required sash weights, ropes, and pulleys that have been replaced by today's pre-hung windows.

14.14 Materials and Finishes (Table 14.3)

a The metals commonly employed in finishing hardware are brass, bronze, aluminum, iron, steel, and stainless steel. These metals can be cast, forged, extruded, or wrought. The cast, forged, or extruded metals are a minimum of 0.080" thick and are generally much too expensive for use in dwelling-house construction. The less-expensive wrought materials 0.050" thick are more commonly employed in dwelling-house hardware. Of the wrought materials, brass and bronze finishes are used most often.

b All hardware finishes are obtained by the careful processing of the above-mentioned metals. When required, a clear protective coating is electrostatically applied and cured in a high-temperature oven.

c To insure satisfaction, many factors should be considered when selecting a finish, e.g., the design surface, whether sculptured, etched, or smooth, matching decor, interior or exterior exposure, and climate. In areas where the finish is to be subjected to strong corrosive vapors, humid climate, sea air, or salt spray, aluminum or stainless steel finishes are recommended for durability and minimal maintenance.

d Finishes of one color tone are suggested for smooth surface designs while the ornamental designs with irregular surfaces are usually more

Table 14.3 Finishes

BHMA Code Symbol	Finish Description	Nearest US Equivalent
600	Primed for painting	USP
605	Bright brass, clear coated	US3
606	Satin brass, clear coated	US4
609	Satin brass, blackened, satin relieved, clear coated	US5
610	Satin brass, blackened, bright relieved, clear coated	US7
611	Bright bronze, clear coated	US9
612	Satin bronze, clear coated	US10
613	Oxidized satin bronze, oil rubbed	US10B
616	Satin bronze, blackened, satin relieved, clear coated	US11
617	Dark oxidized satin bronze, bright relieved, clear coated	US13
618	Bright nickel plated, clear coated	US14
619	Satin nickel plated, clear coated	US15
620	Satin nickel plated, blackened, satin relieved	US15A
621	Nickel plated, blackened, matte, relieved, clear coated	US17a
622	Flat black coated	US19
623	Light oxidized bright bronze, clear coated	US20
624	Dark oxidized statuary bronze, clear coated	US20D
625	Bright chromium plated	US26
626	Satin chromium plated	US26D
627	Satin aluminum, clear coated	US27
628	Satin aluminum, clear anodized	US28
629	Bright stainless steel	US32
630	Satin stainless steel	US32D

attractive when the two-tone oxidized finishes are specified. The darker oxidation remains in the lower surface areas, to accent the high-lighted upper portions where the oxidation has been removed.

e Some ornamental finishes are actually the result of "handmade" processes to achieve an effect such as aging. Finishes are many and varied; a few of the more common are given in Table 14.3, following the listing of the Builders Hardware Manufacturers Association. In the table, the nearest US equivalents are also given.

15 Coatings

15.1 General

a The traditional classes of protective and decorative coatings or finishes for building have become blurred with the advent of new classes of coatings, but the traditional classes still make a useful starting point for discussion. The more important traditional classes are:
1. *Oil Paint.* A finish consisting of a drying oil acting as a vehicle or carrier for various kinds of pigments in suspension.
2. *Water Paint.* Interior finish in which the vehicle is based on water rather than oil. Ingredients in the water coalesce into a film as the water evaporates. In calcimine, glue or casein is added to the water and the principal pigment is powdered calcium carbonate to which others, such as colored pigments, may be added. Whitewash is lime in water.
3. *Varnish.* A transparent finish consisting of fossil or synthetic resins or esters dissolved in a drying oil. *Enamel* is pigment in varnish.
4. *Shellac.* A transparent finish consisting of the exudations of the lac insect dissolved in alcohol. It may be the natural color (orange) or bleached (white).
5. *Lacquer.* Lacquers used in building are completely unlike oriental lacquers, and consist of solutions of nitrocellulose combined with resins and plasticizers dissolved in volatile solvents. As the solvent evaporates, a film of the lacquer is left behind.
6. *Aluminum Paint.* Unlike usual oil paints, paints containing aluminum flake pigments usually have a varnish-like vehicle of oil and resin, nitrocellulose, or of pitch or asphalt.

15.2 Coating Systems

a Among the principal coating systems in general use are those based upon the following (see Chapter 16 for discussion of polymeric materials).
1. *Natural Drying Oils.* Most paints employ drying oils alone or in combination. These oils do not dry; they harden by oxidation. The outstanding natural drying oil has for years been linseed. When used raw as pressed from flax seed, it hardens relatively slowly, but when heat-treated with chemicals, or "boiled," it hardens more quickly. Boiled linseed oil is also treated to have higher viscosity than the raw oil. Among the other important natural drying oils are tung oil, which dries faster, has better compatibility with varnish resins, and greater water and weather resistance than most oils; oiticica oil, similar to and used in place of tung oil; safflower, one of the best non-yellowing oils; soybean, forming good flexible films

with excellent non-yellowing properties; dehydrated castor oil, with good hardening characteristics, and fish oil, considered to be an inferior oil, but low in cost. Natural oils may be subject to mildew unless specially treated, and may bleach in sunlight while yellowing in the dark.

2. *Alkyds.* These are synthetic resins modified with various vegetable oils (linseed, soya, tung, etc.) to produce clear resins much harder than ordinary oils. Because of their versatility, they are used in greater volume than any other paint vehicle, the properties of the resultant film depending upon the relative volumes of resin and oil present. Hardening is by both evaporation of solvent and oxidation. Other ingredients, such as chlorinated paraffin, may be added for such features as increased color retention, resistance to blistering and dirt collection, and gloss retention.

3. *Latices.* These "latex paints" are also called rubber-based paints, although rubber is not the only water emulsion used. The most widely-used emulsions are styrene-butadiene, polyvinyl acetate, acrylic, and water-thinned PVC. The use of these paints has increased rapidly because of their ease of application, quick drying, freedom from solvent odor, minimum fire hazard, and ease of cleanup with soap and water only. They adhere well to many surfaces, have good color retention, and are of varying degrees of flexibility.

4. *Epoxy and epoxy-polyester.* Catalyzed two-part epoxy coatings are mixed by adding a catalyst to the enamel or primer just prior to application (pot life a few minutes to a day). A chemical reaction produces a film as hard as most baked-on coatings, resistant to solvents, abrasion, traffic, and cleaning agents. Epoxy esters, produced by modifying epoxies with oils, harden upon oxidation without catalyst, and, therefore, have no pot-life restrictions. They are less hard and chemically resistant than catalyzed epoxies but are fast drying, easily applied, and produce hard, tough films commonly used as single-component paints. Polyester-epoxies produce a smooth tile-like finish applicable to many firm interior surfaces. The two components are mixed prior to use, with a pot life up to a full working day. They have good adhesion and chemical resistance with color retention and good stain resistance.

5. *Polyurethane.* These fairly recent additions to the coatings family can produce especially abrasion-resistant fast-hardening coatings. They may be made as two-component formulations that are mixed just prior to use and have variable pot life depending upon formulation, or they may be

one-component formulations that cure by evaporation and reaction with moisture in the air (30 to 90 percent relative humidity). Some formulations are modified with oils and alkyds.

6. *Vinyl solution coatings.* These are solutions of PVC and vinyl esters which dry rapidly by evaporation of the solvent, so spraying is the best method of application. Individual coats are thin, but to build up thickness, multiple coats can be applied in rapid succession because of the quick drying. Vinyl coats characteristically have low gloss, high flexibility, and inertness to water, but are sensitive to some solvents. They should not be used at temperatures above 150 °F. Weather resistance is excellent, but adhesion can be a problem unless proper precautions are taken.

7. *Oleoresinous coatings.* These are primarily oil-based but contain resins that make them harder, with higher gloss and better durability than straight oil coatings. At the other end of the scale are resins, modified with oil in varying amounts, that harden by evaporation and oxidation. Phenolics are in this class; they have good resistance to water and most chemicals but are attacked by strong solvents. They age well, become very hard, and tend to turn darker with time, especially upon exposure to sunlight.

15.3 Pigments

a White pigments are by far the most important of all, and generally form the base of colored or tinted paints as well. Pigments are held in the vehicle and fill the many pores and interstices which otherwise would be present in the hardened film formed by the vehicle. Opaque white pigments impart color and hide the surface.

b. White Lead Basic lead carbonate—white lead—is the oldest and formerly one of the most widely-employed white pigments. It mixes well with drying oils, may be used alone or mixed with other pigments, and is an excellent base for tints. With good quality drying oils it forms an excellent paint film of high durability. Generally it is ground in oil to a thick paste (hard paste), or in oil plus thinner (soft paste). As is true of lead pigments generally, its use is restricted because of possible toxic effects.

c. Basic Lead Sulfate This lead-base pigment is often considered interchangeable with white lead in paints employing mixed pigments.

d. Zinc Oxide Zinc oxide is widely used by itself or in combination with other pigments, upon which it has no visible effect. It is very fine in texture and is often considered to be whiter than white lead. Its color is unaffected by atmospheric gases, and it is therefore often used around seashores and in chemical plants or other areas where hydrogen sulfide would discolor other pigments such as white lead.

e. Zinc Sulfide This is a widely used, highly opaque white pigment, particularly in mixture with other pigments.

f. Titanium Dioxide Titanium pigments have the highest hiding power per pound of any of the white pigments and because of this, plus their other generally good qualities, have rapidly become some of the most widely used of all the white pigments, especially in mixtures.

g. Red Lead Chemically this is Pb_3O_4. Red lead has been found especially useful as a protective pigment on metal surfaces, particularly ferrous materials.

h. Transparent Pigments The most commonly employed pigments of this type are barium sulfate, magnesium silicate, silica, clay, and calcium carbonate. Transparent pigments, often called fillers, fill the pores and interstices, but have no marked coloring or hiding power.

i. Aluminum Metallic aluminum is used as a pigment in two forms: (1) finely ground metal in dry powder form, or (2) very thin metal foil broken into extremely small flakes and dispersed in a volatile paint thinner to form a paste. For most applications aluminum flake paste is preferred to the powder.

j. Colored Pigments For durability under atmospheric conditions, pigments must be mineral rather than organic dyes, but where brilliance is required and durability is not important, the various organic dyes may be employed. Of the mineral pigments, the following are commonly employed:

1. *Red.* Iron oxide is most common, either by itself or in combination with some of the "earth" colors.
2. *Dull browns, reds, yellows.* Raw and burnt sienna, raw and burnt umber — often called "earth" colors — are used by themselves or in combination with other pigments.
3. *Bright yellow and orange.* Chrome yellows (lead chromate and basic lead chromate).
4. *Green.* Chrome green (mixture of chrome yellow and iron ferrocyanide, or Chinese blue).
5. *Blue.* Iron ferrocyanide, or Chinese blue.
6. *Black.* Carbon pigments such as lampblack are best. Asphalt paints may be used if durability upon exposure to atmospheric conditions is not important.

Coatings

k. Zinc Chromate This yellow pigment has been found to be a good rust inhibitor and, like red lead, is used widely as a primer or complete paint system for corrosion resistance.

15.4 Mixed Pigments

a Mixed pigments are often favored over pigments of only one type. At one time, such formulations were based on white lead to which combinations of zinc and titanium pigments were added. Various formulations of differing mixes are available with different white pigments such as zinc and titanium, colored pigments, and transparent pigments combined to meet the requirements of particular applications. By mixing pigments, an attempt is made to obtain some of the best characteristics of each while minimizing deficiencies.

15.5 Fillers

a Certain hardwoods have large pores which must be filled if a smooth finish is to be attained with paints and varnishes. The more commonly-found "open-grained" woods are ash, butternut, chestnut, elm, hickory, mahogany and African mahogany, Philippine hardwoods, oak, and walnut. The more commonly-found hardwoods which do not have large pores (close-grained woods) and therefore do not need filling are alder, aspen, basswood, beech, birch, cherry, cottonwood, gum, maple, poplar, and sycamore. None of the softwoods requires filling. Birch does not have to be filled to obtain a smooth surface but has pores large enough to be accentuated by colored fillers if desired.

b Fillers ordinarily consist of ground silica paste which is thinned to the desired thick creamy consistency before application. They are spread on the surface, thoroughly rubbed in, and the excess wiped off, leaving the large pores filled with the paste. Colors may be added.

15.6 Formulations

a With so many coatings, bases, and combinations to choose from, a great multiplicity of combinations can be formulated to meet the needs of exterior and interior protective and decorative coatings for many different materials under diverse climatic and other environmental conditions. It is manifestly impossible to cover them all, and only a few indicative types can be examined briefly.

b For wood siding and trim, depending upon degree of gloss, whether only white or tints or deep colors are wanted, when fume resistance is a factor, and when there is danger of staining masonry below, systems may be based upon linseed or similar drying oil, linseed plus alkyd, acrylic latex, PVC latex, alkyd enamel, or alkyd plus chlorinated paraffin. For staining redwood and red cedar shingles, shakes and siding, stains may be based upon alkyds. If clear, tough finishes are wanted, as on siding or hardwood exterior doors, varnishes and urethanes may be chosen.

c Brick, stucco, and concrete walls free of chalky deposits, and concrete block and cinder block walls may be coated with PVA, PVC, and acrylic emulsions and alkyd-epoxy combinations, some for white only and others for white and colors.

d For ferrous metals that must be protected against corrosion, reliance is often placed upon a rust-inhibitive primer (see below) followed by finish coat or coats. Depending upon the color and degree of gloss wanted, systems may be based upon linseed or other drying oil, alkyd, and varnishes (including oleoresinous) and varnishes plus aluminum flake pigment. Galvanized metal is a special case. It must be free of the oil normally left as a result of processing. Systems may be based upon cement-oil primers and oil-based finishes, PVA latex emulsion, alkyd enamel finish on cement-oil primer, and alkyd-linseed oil-resin combinations.

e Aluminum may be coated with systems based upon drying oil, alkyd enamel, or PVC latex, depending upon the desired degree of gloss.

f Dry-wall construction may be coated, e.g., with alkyd resin, acrylic latex, PVA latex, and two-component polyester-epoxy. Wallboards with paper surfaces, e.g., gypsum board, must be sealed first.

g Plaster, depending upon color, texture, hardness, and the wear-resistance wanted, may be coated with alkyd-resin, alkyd or alkyd-resin enamels, acrylic latex, PVA latex, and two-component polyester-epoxy.

h Painted interior woodwork and trim may be coated with alkyd enamels, acrylic latex, drying-oil paints, alkyd-drying oil paints, and two-component polyester-epoxy, depending upon the color, degree of gloss, and hardness wanted.

i Natural and stained finishes on interior woodwork and trim may be achieved with shellac, lacquers, and varnishes based on alkyds, urethane, vinyls, and copolymers. Latex, oil, and water-based stains may be employed for colored transparent finishes.

Coatings

j Wood floors may be finished with shellac, standard varnishes, and coatings based on alkyds, epoxies, and polyurethanes.

k For durable finishes on concrete floors, PVA latex, alkyd, and two-component polyurethane may be used. Where concrete is below grade and dampness is a problem, alkyd resin–chlorinated rubber is a useful combination.

l For fire resistance, intumescent paints based on modified PVC latex may be applied. When heated, as by the approach of flames, these paints evolve an inert gas that causes the paint to bubble into a froth that insulates the surface below.

15.7 Application

a For successful application, the surface to be coated must be properly prepared. Most coating systems require clean, dry surfaces, especially free of grease, oil, dust, and moisture, although some formulations can be applied to damp or even wet surfaces. Strongly alkaline surfaces such as fresh concrete and plaster may have to be neutralized if oil-based paints, for example, are to be applied. Resinous wood, such as knots and pitch pockets, are likely to strike through ordinary paints, and should be coated with shellac over the prime coat, before finish coats are applied. With the many formulations available, manufacturers' directions must be followed.

b Most coating systems require a primer or prime coat, followed by one or two finish coats, although some systems can be applied in one coat, especially if sprayed. Application is usually by brush, roller, or spray.

c Thicknesses of the various coats are critical. They must be sufficient to accomplish the objective. Recommended thicknesses for prime and subsequent coats are often given in terms of square feet per gallon to be covered, or, more accurately, in wet film thickness. This latter thickness is usually expressed in mils (thousandths of an inch). Wet film thickness can be converted to square feet per gallon by noting that one gallon is nearly 230 cu in. One square foot is 144 sq in., from which it follows that one gallon, or 230 cu in., will cover nearly 1,600 sq ft, one mil thick. If a thickness of 4 mils is called for, the spreading rate is 400 sq ft per gallon.

15.8 Paint Deterioration and Defects

a. Deterioration Paint films are expected to deteriorate with time and to require renewal at intervals. All other things being equal, the most

rapid deterioration takes place where exposure to sunlight is most intense and prolonged. Several stages are observed:
1. *Soiling*, in which dirt collects on the surface.
2. *Flatting*, in which gloss disappears.
3. *Chalking*, in which the surface becomes powdery and the accumulated dirt is at least partly thrown off, leaving the surface dull but fairly clean.
4. *Fissuring*, either checking, in which small superficial cracks eventually penetrate to the wood; or cracking, in which the fissures pass through to the wood almost immediately.
5. *Disintegration*, either crumbling, which develops from checking and causes small fragments cut off by the checking to fall away; or flaking, in which the small scales caused by cracking curl up at the edges, and finally fall away.

b From the time that deterioration in the form of cracks reaches the wood surface, the nature of the wood controls the speed with which further deterioration occurs. The density of the surface, especially the proportion and width of summerwood bands, has marked influence upon the rate of deterioration after this stage.

c Pure white lead and other "chalking" pigments often tend to soil more rapidly than others in the first stages, especially in urban areas, but chalking usually sets in rather early and the dirt is thrown off readily and evenly, leaving a clean, if faded and chalky, surface. Incorporation of zinc oxide generally reduces excessive soiling and retards both chalking and checking. Appearance during the early stages may therefore be better than with chalking pigment, but ultimate disintegration is likely to be caused by curling and flaking, which may require the old paint to be removed before new paint can be applied.

d No matter what the pigment, if surfaces are repainted before deterioration reaches the wood, it is seldom necessary to remove the old paint; if disintegration reaches the wood, removal of old paint often cannot be avoided.

e. Defects These may be listed as of two types, defects caused by poor paint, and defects caused by improper preparation and maintenance of the surface of the wood or by poor workmanship.

1. *Widespread checking and alligatoring.* The surface is unusually heavily checked or broken up into a larger and more prominent alligator pattern than is normal. Usually the cause is too-soft undercoats, either because

the formula was wrong or because not sufficient time elapsed between application of coats.

2. *Heavy cracking and scaling.* Pigments which promote hardness and brittleness in the film are generally responsible for this condition.

3. *Blistering and peeling.* Moisture in the wood behind the paint film almost always causes this condition, either because the wood was too wet when it was painted, or because moisture has found its way behind the paint film.

4. *Spotting, or loss of gloss.* This is almost always caused by too thin paint films, which cause excessive absorption of oil by the wood and permit early loss of gloss, fading, and chalking.

5. *Washing.* Soluble pigments in the paint film may be washed out by rains and run down over foundations or other surfaces, causing streaks and causing the film to lose its hiding power.

6. *Wrinkling.* Excessively thick coats of paint may skin over quickly without hardening underneath until a later time, when wrinkles develop.

7. *Running and sagging.* These are associated with wrinkling in that they are also caused by too thick coats of paint. A paint high in oil content and applied too freely is apt to run down the wall, or to form a surface skin and then to sag under its own weight.

8. *Excessive soiling.* If the surface film contains too much oil it is apt to be soft and tacky, and to collect dirt in excessive amounts.

9. *Mildew.* Mildew may be mistaken for soiling. It is apt to occur in warm, damp, shaded areas in which ventilation is poor.

16 Plastics

16.1 General

a Plastics have entered into house construction in numerous ways and are referred to in various places in this book. Because of the lack of information and the confusion about plastics, this brief chapter is inserted in an attempt to clarify to some degree the nature of these materials.

16.2 Classes, Properties

a Plastics are a family of some 20 to 30 distinct materials, rather than a single material. They are used in many ways in a variety of industries including building. Without becoming involved in the chemistry of these materials, certain of their attributes of interest in house construction may be listed.

b The term "plastic" is, in some ways, a misnomer; but, in other ways, it is not. All plastic materials at some stage in their manufacture are plastic; e.g., they can be formed into whatever shape is desired, usually by pressure, heat, or both. Some, however, need merely to be cast into molds where they harden into their final shapes. Some plastics are plastic only once; and, having hardened, cannot be softened. To this extent, they are similar to concrete, which hardens once and remains hard; or like an egg, which becomes harder the more it is fried or boiled. Other plastics can be softened by the application of heat and hardened by cooling, and re-softened and re-hardened any number of times by heating and cooling. These materials are analogous to glass or to butter. All plastics, however, are organic, and are the product of synthetic chemical processes. The chemist sees them as giant molecules, or synthetic organic high polymers.

c The plastics that harden irreversibly and cannot again be softened are called "thermosetting" because the first plastics in this category had to be hardened by heat. Today, there are plastics which harden at normal temperatures, but the term "thermosetting" has persisted. Plastics in the second category, which can be softened any number of times by heating and hardened by cooling are, quite appropriately, called "thermoplastic." There are, then, two great classes of plastics—thermosetting and thermoplastic. In each class there are numerous types, and these types cover a wide range of properties from soft and flexible to hard and brittle, fully transparent to fully opaque, highly weather-resistant to rapidly deteriorating when exposed outdoors, infinitely colorable to only moderately colorable, non-flammable to highly flammable.

Plastics

d These materials are organic, and their properties are in general similar to those of wood, fabric, and other organic materials. Thus, all plastics can be destroyed by fire, but some do not support their own combustion, and in the presence of a fire, are slow-enough burning to be considered noncombustible under code provisions. Others burn rapidly. As is true of other organic materials, plastics may burn with a clear flame and give off nothing but carbon dioxide and water or, depending on their composition and depending upon the conditions, they may give off a good deal of smoke and noxious, or possibly toxic, gases. Here again, they are similar to materials such as wood and fabric, which can also give off smoke and noxious or toxic gases, depending on their composition and the burning conditions.

e The outdoor durability or weathering resistance of plastics varies with the materials and their compositions. Some have had histories of exposure outdoors of as much as 15 to 25 years and have stood up well. Others, on the other hand, have deteriorated badly in less than a year, depending on the composition and exposure conditions. None of the plastics, at this writing, have been used in buildings over the long periods of time that have characterized wood, stone, masonry, concrete, and glass.

f The strength of any particular plastic is highly dependent upon its composition. This is particularly true when reinforcing materials such as fibers are mixed with the plastic. (see Sections 16.5–10). Plastics unmodified by fillers or reinforcements have strength properties comparable to those of good quality wood parallel to the grain and to the compressive strength of concrete. When reinforced with high-strength filament such as glass, the strength approaches that of the best high-strength steel at a fraction of the weight, and, indeed, the resulting materials can be so light and strong that they are widely employed in space vehicles because of their high strength-to-weight ratio.

g The stiffness of plastic materials, by and large, is low. This is particularly true of the unmodified plastics. Some of them are soft and flexible; others are hard and rigid, but even these materials have stiffnesses which are less than that of wood parallel to the grain. When modified with high-strength fibers, such as glass, the stiffness is considerably increased, but is still only comparable to that of wood parallel to the grain, or to unreinforced concrete in compression. For many applications in housing, however, strength or stiffness is of less importance than other properties. The toughness of some plastics, for example, is outstanding, and their resistance to wear and abrasion may also be excellent. Flexibility may be extremely high, and resistance to cracking or fatigue caused by repeated bending can also be extremely high for some plastics.

16.3 Thermoplastic Materials

Among the most important of the thermoplastics used in house construction are the following.

a. Polyvinyl Chloride (PVC) This is usually called simply "vinyl." In its unmodified state, it is a fairly hard, rigid material, but when it is combined with plasticizers, it becomes soft and flexible, or when it is combined by copolymerization with other plastics such as polyvinyl acetate, it also forms a soft and flexible material. PVC may be transparent, or colored with dyes and pigments.

b. Acrylics These provide the clear, transparent materials widely used for tough, break-resistant skylights, glazing where resistance to breakage is important, and for lighting fixtures. They have infinite colorability with either dyes or pigments, and may range from fully transparent to fully opaque.

c. Polystyrene This material has excellent electrical characteristics and is commonly used for some electrical parts. As a foam, it is used for thermal insulation. It ranges from fully transparent to fully opaque and is infinitely colorable with dyes or pigments. Normally it is brittle, but when combined into a copolymer with acrylonitrile and butadiene, it gives the tough, impact-resistant material known as ABS.

d. Nylon Nylon is tough and wear-resistant and is easily molded into intricate shapes. It is not fully transparent and has moderate colorability.

e. Polyethylene This is a soft, flexible plastic. It is most often used for making film for packaging, and it has many other packaging applications. In its unmodified state, it is a waxy, light-gray colored material, easily molded into a great variety of shapes. Often it is modified with carbon black to increase its weather resistance.

f. Cellulosics The cellulosics include such materials as cellulose nitrate, cellulose acetate, and cellulose acetate butyrate. These materials are used for a variety of moldings and similar molded articles. Cellulose nitrate, in particular, is often used for impact-resistant handles for chisels and other tools. It is highly flammable, especially when in thin sheet or film form; but cellulose acetate and cellulose acetate butyrate burn relatively slowly. Cellulose nitrate is often used as a tough lacquer.

g. Polyvinyl Butyral This is mainly used as an exceptionally tough upholstering material, either with or without fabric as a backing. It also forms the interlayer for safety glass.

h. Polypropylene Polypropylene is similar to polyethylene in some of its properties, but is somewhat harder and more temperature-resistant than ordinary polyethylene. It has the peculiar property of withstanding repeated bending without cracking and is, therefore, used for lightweight hinges. It is also used for piping and for intricate molded parts generally.

i. Polycarbonate This tough transparent material is used mainly for applications where the breakage hazard is high.

j. Fluorocarbons The fluorocarbons, as exemplified by the commercial material Teflon, have outstanding resistance to high and low temperatures and to weathering generally. They also have extremely low coefficients of friction and are, therefore, used as supporting pads for steam lines and other structural parts that move back and forth. Other fluorocarbons are used as protective films on building boards and other surfaces against weathering and attack by sunlight.

16.4 Thermosetting Plastics

a. Phenolics The phenolics are the oldest of the thermosetting plastics. For housebuilding applications, they are mainly used for making a great variety of knobs, handles, switchplates, small electrical parts, and so forth. Dark colors ranging from green through purple, red, blue, and black are available. For ordinary molding, they are modified with wood flour; for electrical purposes, with mica; and for heat resistance, with asbestos.

b. Urea Formaldehyde, Melamine Formaldehyde These are used for making light-colored molded cases of various kinds. In building, however, the major use of melamine is as a constituent of the decorative high-pressure laminates widely employed for counter tops, table tops, and furniture generally (Section 16.11d).

c. Polyesters Polyesters of the unsaturated variety are mainly used in connection with reinforced plastics described in the following sections. In the electrical industries, they are sometimes used as castings in which are potted (embedded) delicate electronic parts. The same is true of the *epoxies;* but perhaps the most widespread use of epoxies in building applications is as adhesives, because they bond strongly to a great many different kinds of materials including hard, impermeable materials such as metal and glass. They provide surface finishes.

d. Polyurethanes The polyurethanes are mainly employed in the form of foams for insulation and as soft, flexible upholstery materials. They also provide surface finishes.

e. Silicones These are mainly used in building in the form of sealants for joints and for difficult glazing applications. They are also used in the form of liquid sprays to be applied to masonry walls for increased moisture repellency. Their weathering resistance is generally outstanding, as is their resistance to higher temperatures than are normally allowable with plastics. As silicone rubbers, they are highly resistant to exposure and to high and low temperatures.

PLASTIC-BASED COMPOSITE MATERIALS

16.5 General

a Plastics are often combined with other materials to form composites that have properties none of the constituents could provide by itself. The most common composites are particulate, fibrous, and laminar.

16.6 Particulates

a In particulate composites, particles are embedded in a matrix. The most important particulate composite, of course, is portland-cement concrete, already described elsewhere in this book. If unsaturated polyesters are substituted for portland cement and suitable aggregates are used, a "polyester concrete" results. It has higher tensile strength, substantially equal compressive strength, and greater toughness than portland-cement concrete, but less fire resistance, although it can be made to conform to Class A or non-flammable rating. When wood particles are bonded together with urea or phenolic binders, interior grade or exterior grade particle board results. Fine inorganic solids such as marble dust can be combined with epoxide or other binders to produce materials for molded lavatories and similar fixtures.

16.7 Fibrous Composites

a When fibers are incorporated into a plastics matrix the class of fibrous composites results. The most commonly used fiber is glass. The strength of ordinary window glass is not particularly high, and the material is quite brittle. When glass is drawn into extremely fine fibers, much finer than human hair, the resulting filaments have a strength that is the equal of the highest strength steel, at one-third the weight. Such fibers may be incorporated into plastic matrices, of which the most common is unsaturated polyester. The resulting material is a fiber-reinforced plastic often called fiberglass, although glass fiber is only one of the two constituents. The glass fiber may be incorporated in the form of continuous filaments, or as chopped fibers varying from $\frac{1}{2}''$ to $2''$ in length in a random matted configuration, or in the form of woven fabrics. For building purposes, the most common is chopped fiber. This is formed either into a random mat and incorporated into the polyester resin or it is chopped and sprayed simultaneously with the liquid resin onto a mold and consolidated by roller or other pressure.

16.8 Laminates

a Laminar composites consist of sheets or layers of material bonded together into a laminar structure. The layers may adhere merely by virtue of interface adhesives, or the bonding material may interpenetrate the various layers.

16.9 High-Pressure Laminates

a Sheet material such as paper, fabric, and wood veneers may be impregnated with a variety of plastics of which the most common are the phenolics and melamine formaldehyde. The usual procedure is to start with a decorative overlay sheet, usually of printed paper, but possibly fabric or wood veneer. This is saturated with the melamine formaldehyde and over that is laid a thin film of melamine formaldehyde in a cellulosic veil. This decorative sheet is backed with layers of high-strength kraft paper impregnated with a phenolic resin. The combination is placed on a hot-plate press and pressed and fused together at one-half to one ton per sq in. and at a temperature of 350 °F. The resulting decorative sheet is applied to a variety of substrates such as plywood and particle board.

16.10 Sandwiches.

a. Structural sandwiches consist of two relatively thin facings of hard, dense, strong material bonded to a relatively thick core of softer, weaker, less stiff material. The geometry of the sandwich results in stiffness and strength combined with lightness. In addition, the facings provide the appearance and resistance to weathering and wear and tear; the core provides most of the thermal insulation, and supports the facings against buckling under load; the adhesive must hold them together and resist any tendency to delaminate; and the combination, in addition to strength and stiffness, provides whatever acoustical value there may be.

b Common facing materials for dwelling-house sandwiches are plywood, hardboard, cement-asbestos board, high-pressure laminates, and metals such as aluminum and plain or coated steel. Cores include foamed plastics, phenolic-impregnated kraft paper honeycomb, egg-crate construction, wood strips, plywood, particle board, and fiber board.

16.11 Applications in House Construction

a A few building applications are briefly listed below and are referred to in other parts of this book. Figure 16.1 shows a hypothetical house with possible plastics applications.

b. Floor Covering The resilient vinyl floor coverings commonly found today are based on polyvinyl chloride with admixtures such as asbestos, pigments, and other materials to provide a great variety of colors, patterns, and textures which may simulate other materials. Asphalt floor coverings also contain a large amount of plastic-based material.

c. Wall Coverings Exterior wall coverings of extruded polyvinyl chloride made to simulate lap siding or clapboards are commonly employed. They must be properly formulated for weather resistance, and allowance must be made for expansion and contraction with changes in temperature when they are applied to the wall. Interior wall coverings may also employ polyvinyl chloride in the form of film or sheet, frequently backed with fabric, flock, paper, or other material for application to plaster, wallboard, and other surfaces by means of adhesives.

d. Counter Tops, Table Tops, and Furniture The high-pressure laminates previously described are commonly bonded to plywood, particle board, or other substrates to form counter tops and other surfaces. They may also be bent into moderate curves to form molded counter edges and backs. These sheet materials may be employed as wall covering, as facings for doors, and for furniture parts generally.

e. Skylights Heat-softened sheets of acrylics such as polymethyl methacrylate (MMA) are placed over a vacuum chamber and a vacuum is drawn, which causes them to take a bubble shape. When they cool, they harden in this form and are employed as skylights or top lights generally (Figure 16.1; See also Figure 7.20).

Plastics

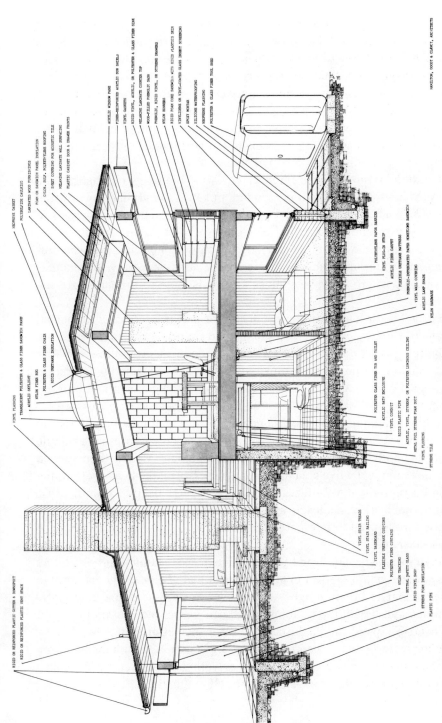

Figure 16.1 Hypothetical House Showing Where Plastics May Be Applicable.

f Where high breakage hazards exist, the acrylics may be used for glazing in place of ordinary glass. Similarly, they are commonly employed for lighting fixtures of many shapes, sizes, and forms because of their high transparence, their colorability, and their resistance to breakage. They are much more easily scratched than glass and abrasion can cause fogging and loss of transparence. Like all plastics, they have high coefficients of expansion and must, therefore, be supported in such a way as to allow movement with changes in temperature or they may buckle, crack, craze, or otherwise go out of shape.

g. Lighting Fixtures These employ a considerable range of plastics. The acrylics have already been mentioned, but PVC (vinyl) and polystyrene are also commonly employed, as are urea and melamine. All degrees of transparence and opacity and a wide variety of colors are available, depending on the plastic employed. For outdoor lighting, the weather-resistant materials, such as the acrylics, are best; polycarbonate is employed, where great toughness and resistance to breakage are important.

h. Foam Plastic foams for thermal insulation may be prefoamed or foamed in place. The two most commonly employed materials are polystyrene and polyurethane. Prefoamed polystyrene is made by incorporating a volatile material into the polystyrene and extruding it at an elevated temperature through a nozzle. As it emerges, it expands into a long, continuous lightweight "log" which can be cut into slabs and boards as necessary for building insulation. The most common density is approximately two lb per cu ft. For foaming in place, polystyrene beads incorporating a volatile ingredient are poured into the space to be insulated. Heat is then applied as, for example, by live steam. At this temperature, the beads soften, the volatile ingredient volatilizes, and the softened beads expand, press against each other, coalesce, and fill the space to be insulated. The same procedure is used to fabricate molded foams, by utilizing a mold of the desired shape.

Polyurethane is employed by mixing two or more liquid ingredients. As they are mixed, they react to give off a gas which causes the material to rise, similar to bread dough. At the same time, it hardens. If properly handled and timed, the liquid material can be poured into a cavity and allowed to expand and fill the cavity with an insulating foam.

Other materials may also be foamed and include the phenolics, urea formaldehyde, the cellulosics, vinyls, polyethylene, and so on. They may be formulated to be flame resistant and self-extinguishing or, on the other hand, the particular formulation may burn readily. Their rates of heat transmission are low and make them good insulating materials. When the polyurethanes are blown with heavy gasses, such as Freon, their heat transmission is exceptionally low. Such special formulations must be sealed so as to avoid the escape of the gas and replacement with ordinary air. For this reason, they are most commonly used in refrigerators and other places where they can be sealed in. Prefoamed polystyrene is composed of cells which are not interconnected and, because the polystyrene itself has extremely low vapor transmission, this material can provide its own vapor barriers. Other foams may or may not provide their own vapor barriers, depending on whether passages occur among or between the cells of the foam.

i. Vapor Barriers Where a barrier against the passage of water vapor is needed, polyethylene film is usually employed. This is particularly common under concrete slabs cast on soil. Other films that may be used for this purpose include the vinyls. Where exceptionally low vapor permeability is necessary, polyvinylidene chloride may be employed. Some plastic foams such as prefoamed polystyrene, mentioned above, form their own vapor barriers.

j. Hardware Many plastics materials may be used for hardware. Among them are nylon, which is readily molded into intricate shapes, is tough, and has good wear resistance. PVC is also employed for some hardware applications, as are the phenolics and other plastics as they meet the particular requirements.

k. Piping For waste, vent, and drain lines, in particular, two types of plastics are commonly employed. These are PVC and ABS, although others, such as polyethylene, are also in the picture. For domestic water lines variants of the vinyls, such as the divinyl and other copolymers, are employed; and other plastic materials are coming into the picture. Here the crucial point is the ability to withstand the prolonged pressures and temperatures involved in hot-water lines. For outside lines such as irrigation, water supply, and so forth, polyethylene is commonly used. It is usually modified with carbon black to enhance its resistance to outdoor exposure.

l. Flashing, Gutters, and Leaders PVC is finding use for gutters and leaders. It must be formulated to withstand both the very low temperatures found in the winter and the high temperatures which are likely to occur upon exposure to the sun in the summer. Gutters and leaders must be installed in such a way as to allow for expansion and contraction with changes in temperature and to avoid sagging between supports. On the other hand, their resistance to corrosion and decay is attractive. For flashing, the softer varieties of PVC and copolymers with such materials as polyvinylidene chloride are employed. They must be formulated for resistance to exposure to sunlight and the weather in general. Such formulations frequently call for carbon black inclusions as well as other stabilizers.

m. Plumbing Fixtures Glass fiber reinforced polyesters are used in molded bathtubs and similar plumbing fixtures. Tubs may have integral surrounding walls to provide seamless joints. The same is true of shower stalls. In some instances, thermoformed acrylic materials are shaped into the desired configuration and then glass-fiber polyester is sprayed on the back to increase the strength and rigidity of the unit. Such units are lighter in weight than standard cast iron, porcelain, or steel fixtures, and the gel coat or acrylic finishes are more resistant to cracking under impact. On the other hand, they scratch more readily and may stain more easily then porcelain, and they are susceptible to softening at elevated temperatures such as might occur under a burning cigarette. Such blemishes, however, may be more readily repaired than cracks in porcelain.

n. Sealants Caulking compounds and glazing materials may be formulated of a variety of polymeric materials including polysulfide, butyl rubber, neoprene, polyurethane, the acrylics, and the silicones. Some of these have to be mixed shortly before they are used, whereas others are single components that may already be packaged as cartridges ready for gun application. The two-component materials harden by a chemical reaction between the two components; the single-component types harden upon exposure to the atmosphere, especially moisture or water vapor in the atmosphere.

o. High Strength Mortar When vinyl and vinylidene combinations are added in the form of latex to standard mortars, their strength and adhesion to masonry units such as brick and concrete block are greatly increased. High-strength masonry can be achieved in this manner and sometimes 4″ thick brick walls can be substituted for standard 8″ walls. In some instances, walls are laid up in panels and then are hoisted as units into place. Polymeric materials may be added to plasters and stucco to increase their strength and toughness, as well as to increase their adhesion to the substrates.

17 Manufactured Housing

17.1 General

a The preceding chapters have discussed the principles of house construction in the field, which is the customary procedure by which houses are built. Manufactured housing is an alternative of growing importance. Some of the principles and details differ from standard construction, and the differences are dealt with in this chapter.

b Various degrees of factory fabrication of houses are practiced. "Precutting" is the oldest and simplest form. This method involves factory cutting of the shell elements, marking, bundling, and shipping the pieces to the building site. Precut packages usually include components such as doors and windows and all other elements, such as nails and shingles, necessary to complete the shell.

c "Frame and infill" (Figure 17.1a) is the next step in the sequence of industrialization. It consists of a structural frame comprising the carrying elements of floor, wall, and roof, e.g., girders, posts, trusses, to which are attached infill elements, e.g., small wall, floor, and roof panels, capable of transmitting imposed loads such as wind and live loads to the structural elements. Post and beam construction (Chapter 5) lends itself to this approach.

d "Panelized houses" (Figure 17.1b) involve precutting all shell elements and assembling the wall panels and, sometimes, floor and roof panels. Windows and doors are usually inserted into the wall panels. The panelized package includes the many precut pieces and elements needed to complete the shell at the site.

e "Volume elements" or big boxes (Figure 17.1c), frequently referred to as modules, are completely factory-fabricated three-dimensional dwellings or segments of dwellings. They are finished as completely as possible at the factory. Bathrooms, kitchens, and interior finishes are all factory installed. The foundation and the utility hook-ups comprise most of the work needed to complete the dwellings at the site. The materials and methods employed vary from system to system and from manufacturer to manufacturer.

f Many systems combine aspects of frame and infill, panels, and volumetric elements (Figure 17.1d).

g In this chapter, it will be neither possible nor desirable to discuss factory prefabrication or fabrication in detail. The essential features differentiating manufactured homes from standard on-site construction will

This chapter was prepared with the major assistance of Frances Fleetwood, graduate student in the Department of Architecture, M.I.T.

Dwelling House Construction

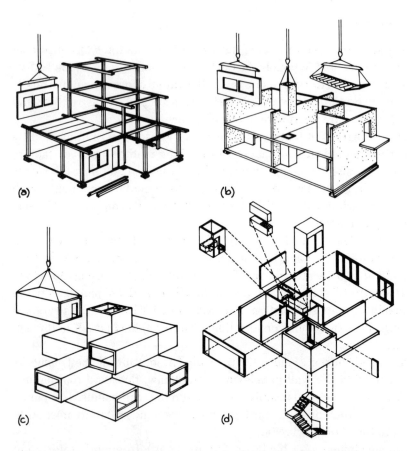

Figure 17.1 Elements of Manufactured Home Systems. (a) Frame and Infill. Post and Beam Frame Supports Infilling Wall, Floor, and Roof Panels. (b) Panel. Load-Bearing Panels Form a Complete Shell. (c) Volumetric Elements (Big Boxes, Modules) Are Stacked to Form the Building. (d) Combination of Various Types of Elements.

be presented with typical examples illustrating the range of practices and details.

17.2 Precut Construction

a Precut houses differ little from standard construction in detail and theory. They may be assembled more quickly than conventional houses since most of the cutting is done at the factory. The principle advantages are in: (1) simplified purchasing of materials, and (2) step-by-step directions for erecting a selected house. Frequently, this type of manufactured home appeals to the novice owner-builder.

17.3 Panel Construction

a Most panel construction is adapted from standard platform construction in order to meet differing local codes. The typical panel consists of studs 16″ or 24″ on centers and an exterior grade of plywood, hardboard, or factory-applied siding, with a down lap to cover the exposed floor construction at the edge (Figure 17.2). Insulation and interior finishing material are generally applied in the field, after wiring has been installed in the wall, thereby conforming with standard practices.

b Other panelized homes are being produced from stressed-skin panels (Chapter 5), which are the most efficient structural wood systems for walls, floors, and roofs. Stressed-skin panels are constructed from ribs to which plywood is bonded by gluing. Under load, the plywood sheathing acts integrally with the ribs, consequently requiring fewer and smaller ribs. Stressed-skin panels have been in use for over 35 years and tested by the US Forest Products Laboratory. Many building codes do not recognize this form of construction, and the use of stressed-skin panels has been limited in some areas.

c Wall panels are classified by their length. "Small" panels range from 2′ to 8′ in length. Any panels larger than 8′ are called "large." These panels may range from 9′ to 40′ in length. Panels longer than 40′ require special road permits for their transport and are, therefore, unusual. Panels smaller than 16′ do not require cranes or special mechanical equipment. Small panels may be lifted by a standard building crew of four men.

d Panel height and length are controlled by the economies available through the use of standard material sizes. Plywood is available in 8′ and 9′ lengths and panels are seldom taller, except at the gable end. Plywood is available in 4' and 5' widths. Panel length is designed, where possible, to utilize fully the materials. In other words, panels tend to be close to 4', 5', 8', 10', etc., in length.

e Panel systems may be either loadbearing or post and beam. Loadbearing panel systems are most common and economical, especially when large panels are used, but small loadbearing panel systems offer more design flexibility. Post and beam panel systems are gaining in popularity. Frequently, they are being used with stressed-skin or sandwich infill thereby avoiding code problems, since the posts and beams support the structure.

f To be successful, stressed-skin construction must fulfill the following requirements (Figure 17.3):

1. The skins or covering materials must be strong and rigid, available in continuous sheets, with stiffnesses approximately equal to or greater than the framing members. Plywood is a favorite material for use with wood framing members for this reason.

2. The skins or covers must be securely and continuously attached to the framing members, and to both edges for maximum efficiency. Nailing alone is not enough to provide strong and stiff stressed-skin panels. The covers or skins are therefore glued to the framing members. Adequate pressure for the glue bonds can be achieved by nailing the covers securely at close nail spacing, or glue bonds can be achieved by pressing the covers down firmly and continuously with presses or clamps.

3. The thicknesses of the covers and the spacing of the framing members must be such as to avoid buckling of the covers under the design load. Covers on joists, for example, must not only act integrally in the longitudinal direction with the joists, but must also act transversely to support the floor load between joists.

4. When framing members, such as joists, are relatively deep and thin, headers should be provided to help support the joists and skins against twisting and buckling.

17.4 Structural Sandwich Constructions

a Closely related to the stressed-skin panels are the composite constructions or materials known as structural sandwiches. (See Section 16.10).

b Structural sandwich panels may be used as the total loadbearing wall, floor, and roof construction, or may be employed as the infill units in frame-and-infill construction. In the latter case, the units are fitted into a wood or metal frame. Two methods are feasible. The first, or horizontal-rail

Manufactured Housing

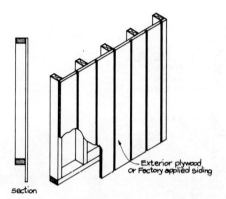

Figure 17.2 Typical Wall Panel Element Based on Platform Framing: Studs, Sole Plate, and Plywood with Downlap.

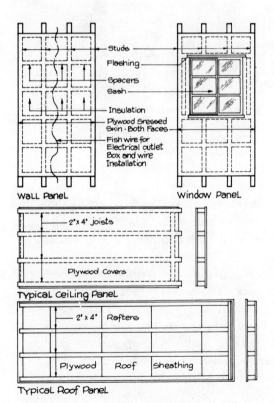

Figure 17.3 Typical Wall, Ceiling, and Roof Panels of Stressed-Skin Panel House.

Dwelling House Construction   406

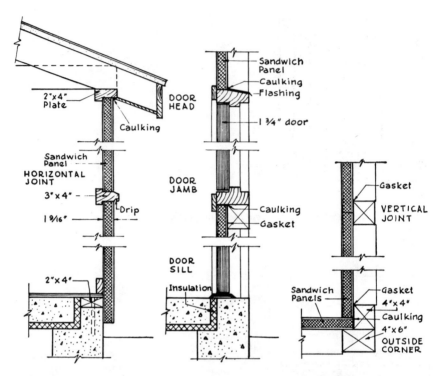

Figure 17.4 Cross-Section of Horizontal-Rail House Showing Sandwich Infill. Details at Floor, Rail, and Roof.

method, uses posts at corners and intermediate points, with horizontal rails running from post to post (Figure 17.4). The sandwich panels are fitted into the frame and screwed or nailed into place with caulking to provide watertight joints. In the construction shown, the long dimensions of the panels are horizontal.

c The second, or post and lintel, type of construction uses wall-height panels, places the long dimensions of the panels vertically, and eliminates the horizontal rails but uses more posts.

d Several variants of the post and lintel type of construction may be employed. The framing may be exposed on the exterior, and the sandwich panels applied to the inner faces of framing members. As usual, gaskets or caulking materials are depended upon to provide weathertight joints.

e An obvious variant of this construction is to place the framing members on the inside instead of the outside and to make the frame a feature of the interior of the house. A third variant is to use rabbeted or grooved framing

Manufactured Housing

members into which the panels are fitted in a manner similar to the horizontal rail construction described above.

f Partitions may be built of the same sandwiches as the outside walls and the construction is similar.

17.5 Panel Construction Details

a Panel construction differs from traditional construction in panel connection details. Panel joints are either edge or right-angle connections. They can be considered part of the traditional family of wood joints, except that panel joints involve two and sometimes three elements: the exterior skin, the frame, and occasionally, the interior skin. Frequently, one

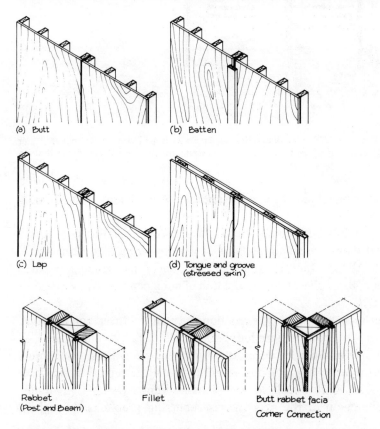

Figure 17.5 Various Types of Panel Connections. (a) Plain Butt. (b) Butt Covered with Batten. (c) Lapped Joint. (d) Tongue and Groove with Covers Bonded to Ribs for Stressed Skin. Variants with Post, Fillet, and Corner.

Dwelling House Construction

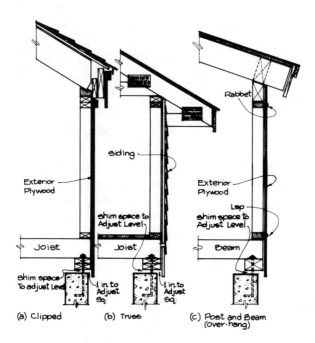

Figure 17.6 Typical Wall Section, Panel Houses. (a) Roof Clipped to Wall, Downlap Exterior Plywood. (b) Trussed Roof, Downlap with Bevel Siding. (c) Post and Beam, Rabbetted Exterior Joints.

element in the panel may butt while another element may pass, forming a lap joint.

b Several possible joint configurations are shown in Figure 17.5. It will be seen that these are all adaptations of the types of joints discussed in Chapter 13. All have their uses, advantages, and disadvantages in panel construction. Examples of the application of such connections are shown in Figure 17.6.

c The most important step in the erection of a manufactured home is starting out with a level and square foundation. Half-inch tolerances are usually acceptable and can be accounted for in the sill details. Time spent starting out right by aligning and levelling the sill is repaid many times in the simplicity with which the rest of the building goes up. Conversely, careless alignment of the sills causes recurring problems throughout erection.

d Steps in the erection of a typical panel system are shown in Figure 17.7. The floor consists of precut joists and headers and precut plywood rapidly

Manufactured Housing

assembled on the foundation. Wall panels and roof trusses all come in one truck load. Wall panels are placed and roof trusses upended into position. Roof plywood is nailed into place and the house is closed in in one day. Interior partitions are all non-bearing and are assembled later from precut stock.

17.6 Volumetric Elements

a Also called modules, these "big boxes" may consist of factory-assembled three-dimensional units employing standard framing, stressed-skin, sandwiches, or combinations. The obvious advantage is that exteriors and, especially, interiors, can be more nearly completely finished in the factory than is true of the other systems. Disadvantages are the bulk of the units, transportation restrictions on size, possible greater difficulty of making field connections, and lessened flexibility of design.

b Figure 17.8 shows large volumetric elements being hoisted and stacked to form two-story multiple housing. To a large extent, such elements are constructed in a conventional manner, although, in many instances, greater use of glued construction is employed, because of good shop control, than is normally possible in the field.

c Figure 17.9 shows details of a structural sandwich system, and a completely assembled two-story house. The basic sandwich starts with a phenolic-impregnated kraft paper honeycomb core with unsaturated polyester impregnated woven glass fiber roving fabric facings. Over this is applied ½" gypsum board, in turn faced with suitable interior and exterior facings, such as sprayed-on chopped glass fiber and polyester with, possibly, a finish painted surface. Full floor, wall, and ceiling or roof panels are assembled into the complete house shell. Non-bearing partitions, conventional or sandwich, are inserted, finish exterior surfaces and cabinetry are applied, exterior surfaces are caulked, flashed, and finished as necessary and the house is moved, set on the foundation, and field utility connections made.

17.7 Flexibility in Design

a One of the goals of panelized housing is the development of highly prefinished panels of a minimum number of standard sizes which can be put together in practically unlimited arrangements to meet the requirements of an individual family building on a specific site at the lowest possible cost in a minimum amount of time. The panelized house adapted from

Dwelling House Construction 410

(a)

(b)

Figure 17.7 Steps in Erection of Panel House. (a) Floor of Precut Joists, Headers, and Plywood Assembled on Foundation. (b) Truckload of Wall Panels and Roof Trusses. (c) Erecting Wall Panels. (d) Placing Roof Trusses.

Manufactured Housing

(c)

(d)

Figure 17.8 Volumetric Units Being Stacked.

Manufactured Housing

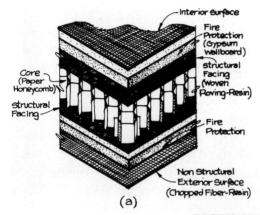

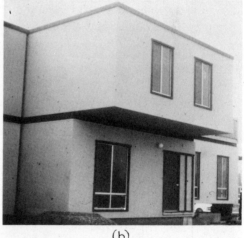

Figure 17.9 (a) Details of Sandwich With Honeycomb Core Faced With Woven Glass Fiber Roving Impregnated With Polyester Resin, and Covered With Gypsum Board Protected With Sprayed-On Chopped Glass Fiber and Polyester Resin. (b) Two-Story House of Stacked Big Boxes Built of Structural Sandwiches Such As Shown at (a).

standard framing practice can be made to meet this goal to a greater or lesser degree, with varying amounts of field labor required. Most builders of panelized houses have tended to move in the direction of a standardized plan, or a series of plans and variants. The goal of practically complete flexibility of arrangement, utilizing standard panels but not standardized plans, is yet to be attained.

b In manufactured housing, customers are able to choose one plan or another. Design flexibility is generally limited to (1) placement of doors and windows in the exterior shell and (2) the movement of a few interior partitions. Basic house configuration changes and floor-to-ceiling height changes are generally not possible. Manufacturing and material handling processes have not reached the level of sophistication necessary to allow complete flexibility. However, recent developments indicate that the use of computers will lead to greater design flexibility.

18 Mobile Homes

18.1 General

a Mobile homes represent the ultimate in completely prefabricated and largely completely furnished dwellings, ready to be transported on wheels and connected to utilities at a permanent or temporary site, ready for occupancy.

b Since its inception in the early 1920s, the mobile home industry has developed as a major producer of housing. In the last ten years, the industry's annual output in terms of unit shipments increased by more than 300 percent, producing in 1973 more than 600,000 units per year. A homogeneous production and distribution network covers the United States from coast to coast.

c One of the most important characteristics of the industry is its cost performance. Completely finished and furnished dwelling units have typically been produced at average F.O.B. factory prices one-third to one-half the equivalent per square foot costs for on-site home building. Consequently, the mobile home industry dominates the lower-cost spectrum of housing. Equally significant, during a period of generally rising costs, per square foot costs of mobile homes have slightly declined, although the product has been much improved.

d The volume of production by individual companies greatly exceeds that of the fabricators of manufactured, e.g., panel-constructed, homes. Because of the high volume, it has been possible to design for maximum use of the minimum amount of material, preassembly of components, mass purchasing, and efficient utilization of labor.

e The high degree of factory fabrication has been accompanied by a correspondingly high degree of organization and management, not only in the plant itself, with main assembly lines and feeders, but throughout the process from procuring of raw materials and components to distribution and sales.

f The industry has designed and implemented an entirely new production and delivery system not based upon the traditional housing delivery process. This system integrates the internal functions of production, transportation, and distribution, first, with the external support functions of financing and materials and land supply, and second, with the regulatory functions of highway and building code regulation, taxation and land use

This Chapter has been prepared by Arthur D. Bernhardt, Assistant Professor of Architecture and Director, Project Mobile Home Industry, M.I.T., and by Norman Y. Quon, with assistance from Jeffrey Ng.

controls. Thus, while this chapter is limited to the characteristics of mobile home construction, it must be kept in mind that product-design efficiency is but one of many factors explaining the performance of this industry.

g Road restrictions limit dimensions and weights. There is essentially no flexibility; the purchaser takes the product as it is, especially the spatial arrangement. The same is true of equipment and such attributes as thermal and acoustical adequacy.

18.2 Product Rationale

a The mobile home is a product that originated within the automobile manufacturing industry. Its objective was to provide an easily standardizable, low-cost living unit that would fully exploit the mass-production capabilities of the automobile industry. Moreover, the unit had to be transportable from the factory to its eventual site.

b Aside from the requirement for transportability, many of the design criteria of the mobile home are similar to those of conventional homes. It must be a suitable shelter of dimensions that would be usable by human beings. It must accommodate various generally needed facilities—lighting and electricity, waste and water transport, and heating or full air-conditioning facilities. It must be adequately insulated from the outside. Finally, it must be structurally sound, able to withstand effectively the loading expected during use.

c Although the design criteria are similar to those of conventional homes, the physical details of the mobile home are radically different. This is due primarily to a difference in approach to design. The mobile home industry has adopted a "systems approach," whereby it meets its design criteria by a solution that attempts maximally to utilize all its resources: material, labor, and capital. This attitude is reflected in the physical design of mobile homes, in which the properties of materials are utilized to the fullest practicable extent, avoiding many structural redundancies. Standardization and interchangeability of components allow for fast assembly-line construction.

d The mobile home industry commonly utilizes wood frame construction, but has used aluminum and steel framing when it has proven economical to do so.

e In meeting the requirements of a transportable living unit, the mobile home design consists of four major assemblies: chassis frame, floor assembly, wall assembly, and roof assembly, plus cabinetry and window and door details.

Mobile Homes

f The chassis frame serves as the structural base for the home, receiving all the vertical loads and transferring them either to the wheels when in transit or to the foundation at the stationary site. By contrast, in traditional homes, the vertical loads are transferred directly to the foundation. Floor, wall, and roof components serve the same function as in traditional homes: enclosing and insulating the living space and producing a fully rigid structure against wind forces and other loads. Plumbing, electricity, and heating ducts are housed within these assemblies. Figure 18.1 shows an exploded view of a mobile home structure as will next be described.

18.3 Chassis

a The basic chassis frame differs from that of an automobile, say, in terms of loading capabilities and size. Common dimensions for the chassis are approximately $12' \times 60'$, increasing from former $8'$ limitation. If permissible road widths are increased, the width will increase. Lengths are both smaller and greater than $60'$. The basic chassis consists of two steel I-beams running the full length of the chassis. These are reinforced by steel cross members. Outriggers are cantilevered from the sides of the two beams, allowing more floor area as well as making more efficient use of the chassis loading capabilities. In front of the chassis frame is an A-frame comprised of two tongue members and the coupling mechanism, forming a hitch assembly.

b The two steel I-beams are $8''$ or $10''$ in height. (Tubular steel beams are also used.) The beams have built-in camber to ensure that the mobile home will be level when the weight of cabinet walls and other assemblies is added at points not supported by the axle assembly. The cross-bracing members are $6''$ high, Z-, I-, or open-web joists spaced $48''$ apart and spanning the distance between the beams. The outriggers may be either open-web steel joists or steel beams. They start at $6''$ on the inside and gradually taper toward the outside. Additional $6''$ to $10''$ longitudinal beams reinforce the axle assembly, giving added protection at the area of load concentration.

c The tongue members of the A-frame are usually $6''$ high steel beams or tubular steel. Each starts at the hitch plate and extends through the front cross member to the main longitudinal beams. The coupling mechanism is a socket through which the mobile home is engaged to the towing vehicle. The hitch plate where this socket is connected is formed of die-pressed steel. It is reinforced around the socket and is equipped with a positive

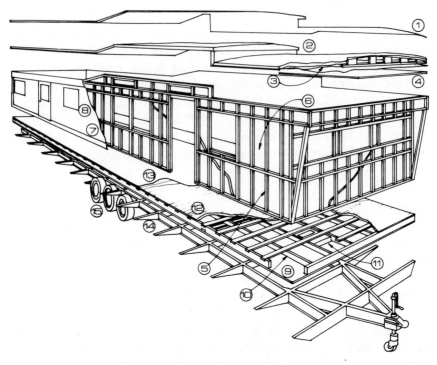

Figure 18.1 Details of Mobile Home. (1) Galvanized Roof. (2) Insulation Board over Roof Truss. (3) Fiberglass Blanket Insulation. (4) Polyethylene Vapor Barrier. (5) 2″ x 3″ Studs 16″ o.c. (6) Natural Wood Interior Panels. (7) Aluminum Exterior Sheet, Baked-On Enamel Finish. (8) Awning-Type Windows. (9) Asphalt Fiber Board Plus Fiberglass Blanket. (10) 6″ Floor Joists. (11) Aluminum Heating Ducts. (12) ⅝″ Flooring Panels. (13) Nylon Carpet Throughout Except Bath, Kitchen, Dining. (14) 10″ I-Beam. (15) Running Gear.

locking device for retaining the ball in the socket. The hitch is usually flush detachable. The running-gear assembly includes springs, spring hangers, axles, bearings, wheels, brakes, rims, and tires. These axle assemblies vary with the weight placed on them. In most cases either tandem or triple axles are formed. Springs are heavy-duty elliptical, mounted to the frame. Axles are heavy-duty one-piece cast steel drums. Wheels are 6.00" x 14.5" heavy-duty cast-steel. Brakes are operated by a 12-volt electrical system. Tires vary from 24" to 30" depending on the ply rating calculated to support the weight of the unit.

18.4 Floor Systems (Figure 18.2)

a. Standard The standard floor system is attached directly to the chassis frame. In contrast with traditional homes, where 2" x 8" floor joists are generally used, the standard mobile home floor system uses steel-spliced 2" x 6" floor joists. As in traditional homes, they are spaced 16" on centers, running parallel to the length of the chassis. Dadoed 1" x 4" cross members spaced 48" on centers act as cross-bracing for the floor joists and serve as nailers for the subfloor on top and undersiding on the bottom. Standard decking material is 4' x 8' sheets of ⅝" plywood or particle board, screwed, nailed, and glued to the floor joists.

b Some variations in the sizes of floor joists exist, varying with the width of the floor sections: 10' wides use 2" x 4"'s; 12' wides use 2" x 6"'s; 14's generally used 2" x 8"'s. The spacing of the 1" x 4" crossbracing varies. Spacings of 12", 16", or 24" also occur.

c Ductwork and piping can be laid easily in the longitudinal direction. For transverse distribution, openings in the floor joists must be made. Careful attention must be paid to the size and location of the opening to be sure no structural strength is lost. The sizes of the horizontal distribution heating ducts are limited. Openings in the decking must also be made to accommodate heating ducts, vents, furnace, and plumbing passing through the floor deck.

d. Insulation Fiberglass blankets 1½" thick are placed between the floor joists and heating ducts to insure insulation within the system. Polyethylene vapor barrier (0.004") is used beneath the fiberglass to protect it from outside moisture. Often, an aluminum-foil vapor barrier is placed on the inside floor structure to protect the fiberglass from condensation resulting from the colder winter air in contact with the warmer, moist interior air. Next, a ⅜" asphalt-impregnated, rigid insulation board is

Dwelling House Construction 420

Figure 18.2 Floor Section Mated with the Chassis.

nailed in place with galvanized nails through metal nailing plates. The floor system is thus sealed at the bottom against moisture and rodents.

e. Floor Finish The floor covering is carpeting or vinyl, or both. Carpeting is nylon. Usually it is installed in one piece over a pad. The vinyl flooring may come as separate tile or in rolls the width of the floor section. Rolled material predominates. A glued-on $\frac{1}{4}''$ overlay is often used over the $\frac{5}{8}''$ plywood subflooring with the vinyl to give added wear and provide a cushion when walking.

f. Cavity One other floor system, known as the cavity floor system, exists besides the standard floor system already discussed. Unlike the standard system where floor and chassis are independent units attached together, in the cavity floor system, the floor and chassis become an integrated whole. Instead of enclosing the bottom of the floor joists with rigid insulation board, the bottom is left open. A rigid insulation board is laid on the top of the chassis cross-members. This serves as the bottom of both the chassis and the floor system, extending the floor system area $4''$ to $6''$ in depth. A basement is thus formed within the chassis framework between the longitudinal I-beams. This area automatically becomes a storage space for numerous items. The primary items placed here are longitudinal plumbing pipes and heating ducts. Two levels are created—a basement level within the chassis I-beams and the upper level within the floor joist system. Floor joists are placed transversely instead of longitudinally. The heating and plumbing occur at two levels. The main distribution occurs longitudinally at the basement level. The horizontal distribution connects to this main spine at the upper level with heating ducts and plumbing parallel to the transverse floor joists. Consequently, the horizontal distribution no longer involves cutting openings through structural members. Other features such as the decking and joists remain the same as in the standard floor system. The floor is rigidly secured to the chassis at each floor joist along the full length of the two main beams and bolted through $2'' \times 6''$ perimeter plates to chassis cross-members.

18.5 Walls (Figure 18.3)

a The wood framing technology of the floor system is extended to the sidewall framing. It consists of a top and bottom plate with studs spaced $16''$ on centers reinforced by horizontal belt rails. Studs $2'' \times 3''$ are used in contrast to $2'' \times 4''$ studs usual in conventional construction.

b At sidewall openings, the studs are doubled, either with another $2'' \times 3''$ or $2'' \times 4''$. Headers for doors and windows consist of $2'' \times 3''$ studs.

c For reinforcement, $3/4''$ diagonal steel strapping ties the floor, walls, and roof into one complete unit. The primary function of the metal strapping is to aid in resistance against sliding caused by loading or by overturning effects caused by wind forces.

d To insure further continuity between the floor system and exterior wall, some manufacturers design their exterior walls to tie down to the floor system; others design their exterior walls to tie into the floor system by extending their wall studs past the bottom plate and lap the longitudinal floor side members. The stud extensions are then nailed to the floor sidemembers.

e Horizontal bracing is provided by $1'' \times 3''$ dadoed belt rails extending over the full length of the wall. Three belt rails are standard, although two or four are also used. All wood framing members are both glued and mechanically nailed at the joints. At critical points, metal splice plates are nailed to both members. To provide adequate continuity between the floor and the wall system, the bottom plate is either glue-nailed or bolted to the floor. Steel plates reinforce the connection between the wall and the floor

Dwelling House Construction 422

system. For added rigidity some manufacturers use steel gusset plates at the corners.

f. Insulation Standard fiberglass insulation thickness varies from $1\frac{1}{8}''$ to $2''$ blanket types to $2\frac{5}{8}''$ batt types. The batt types are $16''$ in width and fit tightly between the studs of the wall sections. Very few walls have vapor barriers.

g. Exterior Finish The standard exterior finish is crimped $0.024''$ prefinished aluminum. The color is painted, baked acrylic, or both, preapplied to both sides of the aluminum. The number of colors used varies from one to three. The aluminum is made rigid by means of machine-pressed or rolled crimping, applied either horizontally or vertically.

h To weatherproof the unit, all joints are sealed with a non-hardening, rubberized, vulcanizing-type seal for protection against leakage. A drip rail is installed the entire length of the unit on each side. In addition, all sidewall openings (other than windows and doors) are protected from leakage by the installation of an aluminum visor drip cap. The siding is attached to the studs or asphalt insulation board by rustproof screws.

i. Interior Finish The interior walls are finished with pre-finished plywood, $\frac{1}{4}''$ in thickness. Most are grooved to simulate random-plank paneling and are fastened to the wall studs with glue and nails. To insure

Figure 18.3 Positioning of Exterior Wall Sections.

good appearance, all nails used to fasten the plywood to the wall studs are set and the hole is filled with matching putty.

j The ¼" plywood walls serve to give added strength by utilizing the stressed-skin principle. By combining the structural strength of the wall frame and the interior sheathing, a wall using smaller studs and thinner sheathing is achieved than would otherwise be possible (see Chapters 5 and 18).

18.6 Roof (Figure 18.4)

a The roof design is one of the unique features of the mobile home. It is a lightweight component using a minimum of building materials. In traditional homes, there are two general types of roofs. One uses a rafter system to support the roof and a separate joist system for the ceiling. In this construction, 2" x 6"'s and 2" x 8"'s are most-commonly used for rafters and joists. Another method uses triangular trusses for the roof structural system. The top chord of the truss supports the roofing while the ceiling can be hung from the bottom chord. Here 2" x 4"'s or 2" x 3"'s are generally used as members of the truss (Chapter 5).

b The mobile home roof construction consists of a structural system covered by an exterior roof and subroof on top and a ceiling on the bottom. Insulation and vapor barriers are inserted in the roof unit.

c The mobile home also uses a truss for its structural system, but it is a bow-string truss instead of triangular, and is jig-fabricated using small structural members. It consists of a 2" x 2" bottom chord and a 1" x 2" cambered top chord. The truss is 2" high at its ends and increases gradually to 6" to 8" high at the center. Plywood plates ¼ thick are glued and nailed on each side for reinforcement. The result is a lightweight prestressed curved form that is efficient for withstanding snow loads and which also provides proper water drainage.

d To form the structural system of the roof, the trusses are arranged 16" on center and are tied together by two longitudinal 1" x 2" side members. These side members extend the length of the roof and are placed in position to allow ½" bearing for the roof truss on the top plate of the side-wall section. Additional longitudinal members are sometimes used for reinforcement.

e On the tops of the trusses, the exterior subroof is installed. Many kinds of materials are used, but the standard is ⅜" rigid insulation board.

Figure 18.4 Positioning of Roof Section.

Manufacturers usually specify that all joints of the exterior subroof occur at the center line of a roof truss.

f The exterior roofing of 26- or 30-gauge galvanized steel or aluminum decking is attached to the subroof. For weather protection, this roof decking is either painted or coated with a rubberized and fibered aluminum preservative. A coating of rubberized and fibered aluminum will last longer than one which is simply painted. This is due to the thickness of the coat and the water resistance of rubber.

g The ceilings are usually made up of $\frac{1}{4}''$ plywood and serve to enclose the roof system. In addition, they are used as a backup board to which acoustical tile, planks, or custom-textured ceiling panels are attached. The ceiling board is fastened to the roof truss.

h Insulation Fiberglass blanket insulation is the most common. Thicknesses range between $\frac{1}{2}''$ and $3\frac{7}{8}''$ and extend the full width of the roof. Single or double vapor barriers are placed below or above the trusses, or in both positions. Often the blanket has an aluminum foil vapor barrier which faces toward the inside of the roof structure.

i Besides the insulation board used for the exterior subroof, some manufacturers use an additional insulation board above the finished ceiling.

Mobile Homes

This double insulation lessens sound transmission from outside and deadens sound that originates inside. It also lessens heat loss in winter. Heat generated by the furnace is retained longer, resulting in lower heating costs. In summer, the interior is cooler because less heat penetrates the roof.

j It is important to ventilate the roof space with open cavities and ventilation holes. This avoids trapping warm moist air under a cold roof and causing condensation, ruining the insulation, staining the ceiling, and possibly decaying the wood trusses. When two vapor barriers are used, the top vapor barrier above the truss is designed to stop penetration by condensed moisture into the insulation. The bottom vapor barrier prevents warm, moist air from the room below condensing in the cold outside air above the ceiling.

k Interior Finish The types of ceiling finishes show a large variation among manufacturers. The more expensive models use a textured acoustical ceiling or acoustical planks with $\frac{1}{4}''$ plywood backup. Less expensive models substitute $\frac{1}{4}''$ thick wood pulp board or other insulation board applied to the roof truss with chrome head drive screws or false beams which match the interior decor.

18.7 Windows and Doors

a. Windows The most frequently-used window is an all-aluminum awning type—usually 12" high. When possible, manufacturers try to place windows to provide cross-ventilation. A bay window is frequently used at the front. All windows are installed in their rough openings with a layer of non-hardening, rubberized sealant between the exterior flange of the window and the rough opening. Double caulking is usually used on the top edge. Each window is secured to the framing with rust-resistant, positive grip screws. The screen unit is usually removable. The interior garnish is anodized. To exclude water, strip caulking is applied to window and door frames, and to wall openings generally.

b. Doors Exterior doors, commonly 32" wide by 72" high, are prehung in extruded aluminum frames. Doors are usually all aluminum insulated in the core with fiberglass or foamed polystyrene (Chapter 16). A jalousie-type ventilating sash equipped with a crank-type operator is generally found in the lower portion.

c Interior doors, commonly $1\frac{3}{8}''$ thick, are usually finished with a $\frac{1}{4}''$

plywood to match the interior paneling. Most closet and cabinet doors are built and finished to match the decor. Closet doors are louvered to provide ventilation.

18.8 Expandable and Double-Wide Units

a In addition to the standard coach design already described, mobile homes come in two other variations: the expandables and double-wides. These versions allow for the accommodation of greater floor area than the standard coach units while still exploiting the technical advantages of mobile home production.

b The expandable is a direct extension of the coach design. It consists of attaching to a standard unit additional walls, floors, and roof sections. These added structural elements are collapsible so that the expandable unit may be transported as a standard coach. When the unit is on site, these structural elements can be pulled or swung out from the unit and fastened together to form additional rooms projecting from the main body of the home.

c Expandables come in three different types: the pull-out, the flip-out, and the tip-out. In the pull-out, the added elements are already fastened together into one component within the home. This entire component is simply pulled out to form the house extension. In the flip-out, the additional elements are attached directly to the unit and are flipped out and fastened together to form the added room. The tip-out is similar to the flip-out, but the additional roof and the outermost exterior wall of the added room are constructed in one piece. Like the flip-out model, the elements swing out and are fastened together when the unit is erected on-site.

d The double-wides accommodate even greater floor area than the expandables. Two separately transported coaches, once on the site, are bolted together through the floor and roof assemblies, creating a home of twice the width of one coach. One coach unit called the "A" unit is the heavier of the two, containing the bathroom and kitchen, while the other unit, the "B" unit, generally contains either bedrooms or living room. For double-wides, tapered roofs are most common (as opposed to the bow-string truss used in single-wides). The roofs are tapered upwards toward a common ridge.

e To insure close tolerances, double-wides are assembled on the assembly line with the two units attached together. When a double-wide is to be transported, the two units are separated and transported independently to the site. While in transit, dummy walls are erected to cover openings in the wall the two units share. These dummy walls are of stud-and-plywood construction or may consist simply of plastic sheets. When the two units are fastened together, the dummy walls are taken down. The joints between the units are water-sealed at the site. Roof shingling and wall siding are completed as required. A metal strip covers the junction of the two floors and is carpeted or tiled over. Appropriate baseboards and molding are installed at the joints.

Index

Abutting properties, 3
Acrylics, 392, 393, 396, 398, 400
Adhesives, 94, 201, 202, 267, 270-272, 274-277, 310, 393, 396
Aggregate, 47, 295-297
Aluminum, 164-185, 226-229, 239, 240, 356
  boarding, 239, 240
  gutters, 226-229
  nails, 356
  pigment, 381, 384, 386
  windows. See Windows
American National Standards, 310
Anchors, 47, 55, 56, 71, 72, 82, 87, 98, 99, 5, 137-141
Areaway, 12, 36, 101, 114
Ashlar, 58, 242
Asphalt flooring, 352-355
ASTM, 288, 289, 295, 306

Bacteria, aerobic, 26, 27
  anaerobic, 26, 27
Backer board, 264, 276, 302
Backfill, 22, 23
Balloon frame, 82-86
  anchor bolts, 82
  bracing, 86
  firestopping, 83-86
  joists, 84
  plate, 82, 83, 85
  posts, 82
  ribband, ribbon, 82, 84, 85
  sheathing, 82, 86
  sill, 82, 85
  studs, 82-86
Baluster, 314, 319, 327, 329, 330
Banks, excavation, 17, 20, 21
Barge board, 219
Baseboards, 337-339
Base shoe, 337-339
Batter boards, 10-14, 20
Bead, joint, 316
Beams. See Girders; Post, plank and beam
Bids, 1
Bolts, 47, 55, 56, 71, 72, 82, 87, 98, 99, 101, 114, 362, 364
  anchor. See Anchors
Bookshelves, 342, 345, 346
Boundaries, bounds, 10
Braced frame, 83, 86-91, 101, 121
  anchors, 87
  brace, diagonal, 87-90
  firestopping, 89, 90
  girt, 83, 87, 88, 90, 91, 101, 121
  plate, 87, 88

  post, 87, 88, 91
  sheathing, 87, 89, 90
  sill, 87, 89, 90
  studs, 87-91
Braces, framing. See Balloon, braced; Platform frames.
Bracing, trench, 20, 21
Brick, 40, 58, 151, 159, 167, 175, 182, 187, 218, 231, 242-245, 294
  fire, 159
  veneer, 167, 175, 182, 187, 218, 231, 242-245
Bridging, 92, 100, 103
Building code, 1, 8, 9, 103, 153, 403, 404
Building lines, 10, 11, 12
Building paper, 133, 171, 173, 178-180, 231
Building permits, 9
Built-up roofing, 211
Bulldozer, 19, 20
Bureau of Reclamation, 17
Butt, 166, 364-367
  clearances, 366, 367
  leaves, 364, 365
  pins, 364, 365
  wide, 366, 367
Butt joint, 315, 316

Cabinets, See Millwork
Calcimine, 381
Calking, 23, 406, 409
Carriage, 113, 115, 313, 322, 324, 325, 327-329
Casement windows, 164-167, 170-174, 176, 177, 189. See also Windows
Cellulosics, 392, 398
Cement, 47, 50, 55, 60, 243, 289, 304, 306, 310, 312
Cement-asbestos, 239, 287, 288
Cesspools, 26
Chair rails, 339
Chimney, 36, 37, 38, 92, 99, 100, 147-163
  ash chute, 159-162
  cap, 155
  cleanouts, 147, 152, 153
  cricket, 155, 215, 216
  corbel, 161
  draft, 149
  design, 147
  fire hazard, 147
  fireplace. See Fireplace
  flashing, 155, 212, 214-216
  flue, 147-156, 158, 159, 160-162: circular, 148-150; effective area, 148-150; lined, 148; rectangular, 148-150; size, 148; square, 148-150; turns, 149-151

Chimney (continued)
  flue linings, 147-156, 159, 160-162
  footings, 36-38, 155, 162, 163
  foundations, 162, 163
  framing around, 92, 99, 100, 129, 147, 153-155, 162
  height, 150, 151
  incinerator, 147, 148, 160, 162
  inside, 152
  leakage, 151, 155
  outside, 147
  prefabricated, 161
  rough opening, 148
  smoke pipe, 153, 162
  testing, 155
  thimbles, 147
  tops, 151
  walls, 152
  withes, 147, 152
Clapboards. See Lap siding
Clay, 4, 5, 18, 22, 25, 40
Clearances, lot line, 8
Closers, door, 377, 379
Closets, 345, 347
Coatings, 381-389
  alkyd, 382
  alligatoring, 388-389
  aluminum, 381, 384, 386
  application, 387
  blistering, 389
  calcimine, 381
  chalking, 388
  checking, 388
  concrete, 387
  cracking, 389
  defects, deterioration, 387-389
  disintegration, 388
  drying oils, 381, 382
  dry wall, 386
  enamel, 381
  epoxy, 382
  epoxy-polyester, 382
  flatting, 388
  floor, 387
  fillers, 385
  finish coats, 387
  fissuring, 388
  formulations, 385-386
  intumescent, 387
  lacquer, 381
  latices, 382
  lead sulfate, 385
  masonry, 386
  metals, 386
  mildew, 389
  natural, 386
  oleoresinous, 382
  paint: oil, 381; water, 381
  peeling, 389: water, 381
  pigments: colored; mixed; transparent; white, 383-385
  plaster, 386
  polyester, 382
  polyurethane, 382
  prime coats, 387
  red lead, 384
  sagging, 389
  shellac, 381
  soiling, 388, 389
  spotting, 389
  stain, 386
  thickness, 387
  titanium dioxide, 384
  varnish, 381
  vinyl, 382
  washing, 389
  white lead, 385
  wood, 386, 387
  wrinkling, 389
  zinc: chromate, 385; oxide, 383-384; sulfide, 384
Columns. See Posts
Composite materials, 394, 396
Concrete, 35, 47-51, 57, 60-61, 95, 257, 294
  aggregate, 47
  block. See Concrete block
  bonding, 49
  cement, 47, 48
  curing, 50
  depositing, 44, 48
  finishing, 51, 57
  footings. See Footings
  foundations. See Foundations
  freezing, 50
  grout, 50
  honeycomb, 48
  joints: horizontal, 49; vertical, 42, 49
  laitence, 48, 49, 50
  latex, 50
  machine, 48
  materials, 47
  mixes, 47-48
  mixing, hand, 48
  ready-mixed, 46
  proportions, 47, 48
  rodding, 48, 50
  sand, 47, 48
  set, final, 49
  set, initial, 49
  spading, 48

Concrete *(continued)*
  strength, 47
  temperature,
  vibrating, 48
  water, 47, 48
  water-cement ratio, 47, 48
  watertight, 49
  weight, 41
  workability, 48
  working, 48-49
Concrete block, 35, 51-59
  advantages, 57, 58
  anchors, 55, 56
  coatings, 53
  corners, 53, 54, 55, 56
  coursing, 53, 54
  curing, 53
  disadvantages, 57, 58
  laying, 53
  lintels, 54, 55
  mortar, 55, 57
  nailing blocks, 55
  openings in walls, 54, 55
  piers, 55, 56
  pilasters, 55, 56
  reinforced, 58, 59
  returns, 55, 56
  sizes, 52-53
  types, 52-53
  use, 51, 53
  wall heights, 53
  waterproofing, 57
Concrete floor, 35, 60-61, 95, 257
  drains, 61
  finish, 61
  joint, wall, 61
  material, 60, 61
  preparation, 60-61
Condensation, 247, 248, 259-262
  damage, 261
  remedy, 261
Conductance, 246, 252, 253
Conduction, 248-253
Conductors, 23, 24, 229
Convection, 248-253
Copper, 212, 213. See Flashing; Roofing
Corbel, 161
Core board, 264
Cores, structural sandwich. See Structural sandwich
Corner beads, 279, 293
Corner boards, 222, 232-234, 236, 237
Cornerite, 289-292, 294
Cornice, 195, 196, 218-223, 225, 337, 339-340

  barge board, 219
  boxed, 220, 222, 223
  brackets, 220
  closed, 218, 220
  drip, 221
  eaves, 218, 220-223
  edging strip, 219, 220, 222
  fascia, 218-223, 225
  false rafters, 218, 219
  flying rafter, 219
  frieze, 219, 220-223
  lookouts, 222, 223
  molding, 219-223
  open, 218
  plancia, 220, 222, 223
  rake, 219, 222, 223
  return, 222, 223
  roof boards, 219-222
  shingles, 195, 196, 220
  simple, 220, 221
  verge board, 219
Cork flooring, 352
Coverage, lot, 8
Cricket, 155, 215, 216
Curbs, 3

Damper, 156-159, 161-163
Deed restrictions, 8
Dew point, 260, 261
Distribution box, 27, 30
Door closers, 377-379
Doors, 330-338
  batten, 330-331
  casing, 336-338
  flush, 331-333, 337; hollow, 331-333; solid, 331, 332
  frames, 335-337: exterior, 335-337; interior, 335-336; thresholds, 335, 337
  panelled, 332-334: flat, 333, 334; lights, 333, 337; moldings, 333, 334; plywood, 333, 334; rails, 333, 334; raised, 333, 334; solid, 333, 334; stiles, 333, 334; veneered, 333, 334, 337
  plinth blocks, 338
  sizes, 335
  special, 335
  stock, 335
Dormers, 129-131
  flashing, 215, 216
Double glazing. See Windows, multiple; Glass, insulating.
Double-hung windows. See Windows, double hung
Dovetail, 315-317, 329
Dowel, 315, 359, 363

Downspouts. *See* Leaders
Drainage, 2, 4, 8, 35
Drain pipes, 22, 23
Drains, 5, 21, 23, 40, 157, 161, 229, 230
  encircling, 23, 24, 40
  floor, 61
  pitch, 25
  roof, 23, 229, 230
  tile, 22, 23, 32, 34, 57
  wall, 21, 22, 23, 40
Dry wall construction. *See* Wallboards, Plywood
Dry wells, 6, 20-25, 57, 223

Engineers, Corps of, 17
Easing, 320, 329
Eastern frame. *See* Braced frame
Eaves, 117, 196, 204, 205, 218, 220-223, 235, 242
Eavetroughs. *See* Gutters
Effluent, 26, 27, 31
Elbow, 227, 229, 230
Electricity, lines, 5, 6, 8
Enamel, 381
Epoxies, 382, 393, 400
Excavation, 5, 16-21
  banks, 17, 20, 21
  bracing, 20, 21
  equipment, 19
  hand, 17, 19
  lagging, 20, 21
  main, 19
  miscellaneous, 20
  power, 17
  procedure, 19-21
  sheeting, 20, 21
  staging, 20
  trenches, 16, 19, 20, 21
Expansion joints, gutters, 228, 230
Exterior finish, 74, 171, 218, 231-245

Facia, fascia, 218-223, 225
Fibers, 391, 394-395, 400
Fiberboard, 286
Fillers, and close- and open-grained woods, 385
Fireplace, 147, 154, 156-163, 190, 342
  ash dump, 159-162
  combustion air, 190
  damper, 156-159, 161, 163
  draft, 158
  fire brick, 147, 159
  flue. *See* Chimneys, flues
  framing around. *See* Chimneys
  hearth, 147, 156, 159, 161-163
  lip, 149, 157, 158, 161-163
  opening, 156
  proportions, 148, 154, 156
  smoke chamber, 148, 154, 156, 158-163
  smoke shelf, 154, 156-163
  throat, 154, 157
  trimmer arch, 159, 160, 162
  underfire, 147, 154, 156, 159, 161-163
Firestopping, 70, 73, 83-86, 89, 90, 112, 247, 249
  blocking, 84, 86, 89, 90
  concrete, 84, 86
  masonry, 86, 89, 90
  tee sill, 73, 85, 86
Flashing, 135, 155, 172-174, 180, 197-199, 202, 205, 206, 212-217, 235, 242-244, 397, 398
  apron, 214-216
  base, 214-216
  brick veneer, 243, 244
  cap, 214-216
  changes in pitch, 216, 217
  chimney, 155, 212, 214-216
  counter, 212
  cricket, 215, 216
  dormer, 215, 216
  expansion, 213
  materials, 212-214
  principles, 212, 213
  rakes, 212, 216, 217
  roof. *See* Roofing; Shingles; Slate; Tile
  roofing temper, 213
  sheet materials, 212-214
  vents, 216, 217
  window. *See* Window
Flight, 318, 320, 323, 324
Floor, bathroom. *See* Tile, interior
Flooring, 347-355
  block, 350, 351
  end-matched, 348, 351
  grades, 347
  hardwoods, 347, 348
  laying, 349
  nails, 348, 349
  parquetry, 351
  patterns, 349, 350
  plastics, 396-397
  precautions, 351
  preparation for, 348
  random width, 352
  ready-finished, 352
  resilient, 352-355: asphalt, 352-355; below grade, 353, 354; colors, 353; cork, 352; coves, 355; installation, 353-355; linoleum, 352, 353; sheet, 352; substrates,

Index

Flooring *(continued)*
  353, 354; textures, 353; tile, 352-353; vinyl, 352-355, 396-397
  sanding, 352
  scraping, 352
  softwoods, 347
  wood block, 105
Floors, framing, 92-110, 133-136. *See also* Balloon, Braced, Platform, Post, plank and beam framing
  bathroom, 108
  bridging, 92, 100, 103
  cantilever, 97
  combination, 110-111
  floor beam, 92
  girders. *See* Girders
  in-line, 97, 133-136
  joists, 90, 92-99, 101-107
  openings, 98-99
  partitions under, 100
  piers, 92, 93, 95
  posts, 92, 93-95
  spans, 93
  subfloor, 92, 101, 103-110
  2-4-1, 106, 107
Flues. *See* Chimneys
Fluorocarbons, 393
Foams, 258, 394, 397-399
Footings, 1, 19, 35-39, 92, 93, 95, 162-163
  chimney, 36-38
  design, 35-39
  forms, 39
  loads, 36-38
  posts, 35, 37, 38
  reinforcing, 38
  soft soil, 35
Forest Products Laboratory, U.S., 403
Forms, 40-47, 49, 51
  anchor bolt, 47
  braces, 40-44, 46
  bulkheads, 42, 49
  footing, 39
  materials, 40, 42-44, 46
  openings, 40, 46-47
  pressure, 40, 44
  rangers, 40, 43, 44, 46
  runways, 46
  sectional, 42, 43, 44
  shoe, 41-45
  simple stud, 41, 45
  sleeves, 40, 46-47
  spreaders, 44-45
  stripping, 51
  studs, 40-46
  stud and ranger, 42-44
  ties, 40
  tops, 46
  wall, 40-47, 51
  wire, 44-45
Foundation, 1, 35-61, 162-163, 242-245
  concrete block. *See* Concrete block
  concrete, cast. *See* Concrete
  forms. *See* Forms
  masonry, 242-245
  materials, 35
  rubble. *See* Rubble
  settlement, 35-38
Framing
  balloon. *See* Balloon framing
  braced, eastern. *See* Braced framing
  floor. *See* Floor framing
  in-line. *See* In-line framing
  partition. *See* Partition framing
  platform, western. *See* Platform framing
  post-and-lintel. *See* Post, plank and beam
  roof. *See* Roof
  stair. *See* Stair framing
  wall types, 69
Frieze, 219, 220-223, 235, 242
Frost, 1, 4, 35
Furring, 265, 268, 269, 284, 489, 305, 308

Gas, 5, 6, 8, 19
Gaskets, 170, 185, 187, 188, 406
Girders, 63, 92-99, 135, 138, 142, 145. *See also* Floors; Framing
  bearing plates, 93, 95, 96
  built up, 94, 95
  fire protection
  laminated, 94, 96, 135, 138, 142
  ledgers, 95, 97
  nailers, 94, 95, 98
  plywood, 96
  reinforced concrete, 94
  spacing, 97
  steel, 94-96
  wood, 94, 95
Girt, 83, 87, 88, 90, 91, 101, 121
  dropped, 87, 88, 90, 91
  function, 87, 88
  joints: mortise and tenon; pins; step, 88-91
  raised, 87, 88, 90, 91
Glass, 164, 167-170, 185, 187, 190-192, 246-248
  absorptive, 192
  figured, 167
  flat drawn, 167
  float, 167
  insulating, 168, 185, 187, 190-192

Glass *(continued)*
  multiple, 190-192
  plate, 167
  sheet, 168
  strength, 167, 168
  reflective, 192
  weight, 168
  wind, 168
Glaziers' points, 169, 170
Glazing. *See* Glass; Windows
Glue. *See* Adhesives
Glued plywood joist, 106-109
Goose neck, 230
Grades, 2, 3, 15, 19
  finish, 2, 3, 15, 19
  natural, 2, 3
Gravel, 18, 22, 25, 32, 34, 60
Grease trap, 27
Grounds, 74, 172-175
Grub hoe, 19
Gutter, 117, 220, 223-230, 397, 399
  expansion joints, 228, 230
  hung, 226, 227-229
  materials, 223, 226
  metal, 223, 226, 227: attached, 228; clearance, 228; hangers, 227-229; molded, 228, 229; pitch, 228
  plastic, 223, 226, 227, 230, 397, 399
  profiles, 225, 227-229
  requirements, 223, 224
  wood, 223, 224: cement, 224; clearance, 225, 226; connection, 225; details, 224-226; fastening, 225; length, 226; molded, 224, 225; pitch, 226; species, 224; splicing, 224, 225; strainers, 226, 228; vee, 224, 225
Gypsum, 263, 264, 288, 290-292, 297, 299, 302

Hands, doors and locks, 377, 378
Hangers, 98, 99, 114, 125, 137-141
Hardboard, 236, 238-240, 286, 287, 361
Hardware, finish, 364-380
  butts. *See* Butts
  closers, door, 377-379
  finishes, 379-380
  hands, 377, 378
  hinges. *See* Hinges
  latches. *See* Locks and latches
  locks. *See* Locks and latches
  materials, 379-380
  miscellaneous, 378
  pivots, 364, 366, 368
  springs, 377
Headroom, 322, 327

Heat loss, 246-252
  comfort, 248
  condensation, 247, 248, 259-262
  conductance, 246, 252, 253
  conduction, 248-253
  convection, 248-253
  gradient, 259-261
  heaters, 246, 248
  infiltration, 246, 252
  radiation, 248, 249, 251-254
  resistance, 246
  structure, 246, 247
  temperature, 248-251
Heaving, 4, 35
Hinges, 166, 364-369
  checking, 366
  double-acting, 366
  floor, 366, 367
  flush, 366, 368, 369
  gravity, 366, 368
  invisible, 367
  knuckle, 364
  leaves, 364
  lipped, 366, 368, 369
  overlay, 366
  pins, 364
  spring, 364, 367, 377
  strap, 364, 365
  wide-throw, 366
Hip. *See* Roof; Shingles
Horizontal Boarding, 231, 240, 241
  application, 240, 241
  water table, 240, 241
Horse, 113, 322
Housed joint, 316
Humidity, 196, 247, 248, 259-262
  absolute, 260
  condensation, 190, 247, 248, 259-262
  dew point, 260, 261
  difference, 260
  gradient, 259-261
  relative, 248
  vapor pressure, 260-262

Incinerator, 147, 148, 160, 162
Infiltration, 246, 252
In-line framing, 133-134, 136
  frame, 133
  joists, cantilevered, 133, 134
  sheathing, 133, 136
  subfloors, 133, 136
  studs, 133, 136
  trussed rafters, 133, 136
Insulation, 135, 246-262, 291, 302, 394, 397-399, 403, 418, 419, 422-424. *See also*

Insulation *(continued)*
  Heat losses
    bats, 256, 257, 262
    blanket, 251, 254-257, 262
    bulk, 250, 254, 257
    conduction, 252-253
    convection, 249-253
    fill, 254, 256, 257
    foam, 256-258, 394, 397-399
    lath, 291
    materials, 252-259
    metal, 254, 257-259, 262
    sprayed, 254, 258
    wallboard, 251-255, 262, 264, 302
    windows, 190-192
Interior finish, 62, 313-355. *See also* Doors; Flooring; Joints; Millwork; Moldings; Stairs
  back priming, 315
  care, 314-315
  grades, 62
  priming, 314
  species, 314

Joints, 62, 80, 90, 91, 315-317, 330, 341, 407, 408
  beaded, 316
  butt, 315, 316
  butterfly, 316
  coped, 315
  corner, 316
  dovetail, 315-317, 329, 330, 341
  dowel, 315
  housed, 316
  lap, 315
  miter, 315, 316
  mortise and tenon, 90, 91, 316
  pinned, 90, 91
  quirk, 316, 317
  sheet materials. *See* Wallboard
  shiplap, 62, 80, 315-317
  shoulder, 315, 316
  spline, 315
  step, 91
  tongue and groove, 62, 80, 316, 317
  wallboard. *See* Wallboard
Joist, 62, 73-76, 84, 92, 95, 97-99, 101-103, 108, 109, 114, 115, 120, 122-126, 142, 144, 145, 255. *See also* Lumber; Balloon, Braced, Platform, Floor framing
  chamfering, 108
  deflection, 101, 102
  fiber stresses, 101, 102
  framing anchors, 98, 99, 114
  framing into supports, 97
  grades, 62
  hangers, 98, 99, 114
  headers, 92, 95, 98, 99, 114, 115
  holes for piping, 103, 108, 109
  ledgers, 95, 97
  modulus of elasticity, 101, 102
  notches, 95, 97, 98, 108
  shrinkage, 97, 98
  sizes, 101-103
  spacing, 101-103
  spans, 101-103
  species, 62, 65
  stiffness, 101, 102
  stirrups. *See* Hangers; Framing anchors
  strength, 101
  structural grades, 62, 63
  tall joists, 98, 99, 114, 115
  trimmers, 92, 98, 99, 114, 115

Keene's Cement, 297, 301
Keys, 370, 375, 376
Knob, door, 370, 372, 375

Lacquer, 381
Lagging, 20, 21
Laminates, 395, 397
Landing. *See* Stairs
Latches. *See* Locks and latches
Latices, 382
Lath, 289-297, 302, 303, 305, 306, 308, 309, 312, 357, 361
  application, 291, 293-294
  beads, 292, 302
  cornerite, 289-292, 294
  deflection, 289
  diamond, 292, 293
  fasteners, 291
  flat, 292, 293, 295
  framing, 289, 290
  furring, 289
  gypsum, 288, 290-292, 297, 299, 302
  insulating, 291
  masonry, plaster base, 294, 296, 297, 299, 303
  metal, 291-295, 305, 306, 308, 309, 312
  molding, 292
  nails, 291, 357, 361
  partition tile, 294
  perforated, 291
  plain, 291
  popping, nail, 291
  reinforcement, 289
  rib, 292-295
  screws, 291
  self-furring, 292, 293

Lath *(continued)*
  sheet metal, 295
  staples, 291, 292
  stucco, 292, 305, 308
  supports, 289, 293
  tile, clay, 294
  veneer plaster, 302, 303
  weight, 295
  wire, 292, 295
  wire cloth, 292, 293, 295, 305, 306, 308
  wood, 291
Layout, 10-15
Lead sulfate, basic, 385
Leaders, 23, 223, 226, 227, 229, 230
  elbow, 227, 229, 230
  head, 227, 229, 230
  shoe, 229, 230
  strap, 227, 229, 230
  strainers, 226, 228
Ledge, 4
Ledgers, 95, 97, 99
Legal requirements, 2, 3, 8-9
Level. surveyor's, 15
Lights, glass, 74, 168-170, 177, 191
Lime, 55, 297, 301, 304, 306, 308
Lines, 10-12
  building, 10-12
  lot, 10-12, 14
Linoleum, 352, 353
Lintels, 54, 55
Liquid-applied roofing, 210-211
Locks and latches, 370-376
  bolt: dead, 370, 372, 375; guard, 372; latch, 370, 372, 375
  bored, 372, 374, 375
  cylinders, 370, 372, 375, 376
  cylindrical, 370, 372, 374, 375
  dead, 370, 375
  drivers, 375, 376
  functions, 372
  integral, 370, 372, 373
  jimmy-proof, 375
  keys, 370, 375, 376
  knobs, 370, 372, 375
  lever, 370
  lock, dead
  mortise, 370-373
  operating trim, 370, 372, 375
  pins, 375, 376
  rim, 375
  sliding door, 375
  stops, 372
  strike, 370, 373, 374
  tubes: aligning, 372; latch, 372
  tubular, 370

  unit, 370, 372, 373
Lookouts, 75, 222, 223
Lumber, 40, 42-44, 62-65, 73, 80, 82, 92, 97, 98, 134, 135, 137-141, 145, 315, 317
  beams, 63
  boards, 62, 63
  common, 62, 63
  dimension, 62, 65, 67
  dimensions, 62, 63, 65: actual, 63, 65; nominal, 63, 65
  dressed, 62
  dry, 62, 97
  factory and shop, 62
  finish. See Millwork; Trim
  form, 40, 42-44
  framing, 62-64
  girders, 63, 92-98, 145
  grades, 62
  matched, 62
  patterned, 62
  planks, 134, 135, 137-141
  rough, 62
  select, 62
  shiplap, 62, 80, 315-317
  softwood, 62
  species, 62, 65
  square-edge, 62
  strength, 63
  stringers, 63
  strips, 63
  structural, 62, 63
  surfaced, 63
  timbers, 62
  tongue and groove, 62
  treated, 73, 82
  unseasoned, 62
  worked, 62
  yard, 62, 63-64

Manufactured housing, 401-414
  boxes, 401, 402, 409
  caulking, 406, 409
  combined, 402
  components, 401
  composite, 404
  downlap, 405, 408
  erection, 408-412
  flexibility, 409, 414
  frame and infill, 401, 402, 404
  foundation, 408
  gasket, 406
  horizontal-rail, 404-406
  joints, 407, 408
  modules, 401, 402, 409
  panel, 401-405, 407, 408-411

Manufactured housing *(continued)*
  partitions, 407, 409
  platform, 403, 405
  post and beam, 401, 402, 404, 406, 408
  precutting, 401, 403, 410
  road permits, 403
  sandwich, 404, 406, 409, 413
  stressed-skin, 145-146, 403-405, 409
  trusses, 407, 410, 411
  volume elements, 401, 402, 409
  wall sections, 407, 408
Mantels, 340-342
Masonry, 35, 36, 40, 58, 182, 231, 242-245, 294, 305, 306. *See also* Chimneys; Foundations; Piers; Pilasters; Plaster
  ashlar, 58, 242
  mortar, *See* Mortar
  rubble, 35, 58-60, 242, 243
Masonry veneer, 40, 182, 187, 231, 242-245, 305, 306
  brick, 40, 242-245
  construction, 243-244
  flashing, 243, 244
  foundations, 242
  materials, 242
  mortar, 243
  paper, 243
  ties, 243
  water table, 244, 245
  weep holes, 244
  windows, 182, 187
Mattock, 19
Melamine formaldehyde, 393, 398
Metal foil, 254, 257-262
Millwork, 74, 171, 313-347
  baseboards, 337-339
  base shoe, 337-339
  bookshelves, 342, 345, 346
  cabinets, 341, 343-346: counters, 343-346; doors, 343, 344; drawers, 341, 343, 344; kitchen, 341, 343-345; metal, 345; shelves, 341, 343, 346; toe space, 341, 343, 346
  chair rails, 339
  closets, 345, 347
  cornices, 337, 339-340
  definition, 338
  mantels, 340-342
  wood species, 314
Miter joint, 315, 316
Mobile homes, 415-427
  assembly, 416
  cabinets, 418
  ceilings, 418, 425
  chassis, 416-418, 420

  cost, 415
  delivery, 415
  details, 418
  dimensions, 416
  doors, 416, 418, 426-427
  double wide, 426-427
  expandable, 426-427
  exploded view, 417, 418
  flexibility, 416
  floor assembly, 416-421: cavity, 420; standard, 419
  flooring, 418, 420
  growth, 415
  heating, 417, 418, 421
  insulation, 418, 419, 422-424
  materials, 416, 418-427
  organization, 415
  plumbing, 417, 421
  prefabrication, 415
  production, 415
  rationale, 416
  road restrictions, 416
  roof assembly, 416, 418, 423-425
  roofing, 424
  standardization, 416
  transportability, 416
  trusses, 418, 423
  vapor barriers, 423
  volume, 416
  wall assembly, 416, 421-423: exterior finish, 422; interior finish, 422-423
  weights, 416
  windows, 416, 418, 425
Module, 401, 402, 409
Modulus of elasticity, 101, 102
Moldings, 317-318, 219-223, 283, 285-287, 317, 318, 337, 339, 340. *See also* Doors; Millwork; Stairs; Windows
  cornice, 337, 339-340
  profiles, 317
Mortar, 55, 56, 60, 243, 310, 312, 400
Mortise and tenon, 90, 91, 316
Mullions, 171, 173, 182, 183

Nails, nailing, 356-359
  aluminum, 356
  annular, 358-359
  asbestosboard, 360
  box, 357, 360
  brad, 360
  casing, 357, 360
  clinch, 360
  common, 68, 357, 360
  concrete, 360
  copper, 356

Nails, nailing *(continued)*
  cut, 356, 357
  dating, 361
  double-headed, 357, 361
  fiberboard, 361
  finish, 357, 360
  flooring, 348, 349, 357, 360
  gypsum lath, 361
  gypsum wallboard, 357, 361
  hardboard, 361
  heads, 357-359
  hinge, 361
  insulation board, 361
  lath, 391, 357, 361
  masonry, 361
  metals, 359
  penny, 358
  points, 356-358
  ring, 358-359
  roofing, 357, 361
  schedule, 68
  shake, 361
  shanks, 358-359
  shingle, 357, 361
  siding, 233, 357, 361
  sinkers, 358
  sizes, 358-361: spikes, 358; steel, 356
  slating, 357, 361
  surface treatments, 359
  threaded, 358-359
  underlayment, 361
  wallboard, 361
  wire, 356-359
Neoprene, 400
Newel post, 319, 320, 322, 329, 330
Nosing, 320, 322, 325, 329
Nylon, 392, 399

Offset stakes, 10, 12, 13, 14, 20
Oil, drying, 381, 382
Openings, framing. *See* Balloon, Braced, Platform, Floor, Roof, Special framing
Overhangs, 75, 141, 142, 144, 145

Paint. *See* Coatings
Panels, 401-405, 407, 408-411. *See also* Manufactured housing
  partition, 407, 409
  stressed-skin, 145-146, 403-405, 409
  structural sandwich, 404, 406, 409, 413
  wall, 407, 408
Panels, door, 332-334. *See also* Doors
Parquetry, 351
Particle board, 239
Partition, 73, 97, 100, 101, 110-112, 115, 116, 128. *See also* Balloon, Braced, Platform framing.
  bearing, 100, 101, 110
  blocking, 112
  continuity, 110, 111
  corners, 112
  firestopping, 112
  framing under, 100, 101
  nonbearing, 100, 101, 110
  openings, 110, 111
  plates, 110-112
  shrinkage, 110, 112
  soles, 100, 101, 110-112
  studs, 100, 101, 110-112
  thickness, 110
Peat, 18
Pentachlorphenol, 73
Perlite, 295, 296, 298
Phenolics, 393, 398, 399
Picks, 19
Piers, 19, 35-36, 55, 56, 92, 93, 95
  concrete, 93
  masonry, 35, 36, 55, 56, 93, 95
Pigments, 383-385
  aluminum, 381, 384, 386
  colored, 384
  lead sulfate, basic, 385
  mixed, 385
  red lead, 384
  titanium dioxide, 384
  transparent, 384
  white lead, 385
  zinc: chromate, 385; oxide, 383-384; sulfide, 384
Pilasters, 35, 36, 55, 56
Pin
  butts and hinges, 364, 365
  wood, 90, 91
Planks, 134, 135, 137-141
Plaster, 389-304
  acoustical, 296, 301
  aggregates, 295-297
  application, 298-304
  base coat, 299, 302
  bond, 294, 299
  brown, 296, 298, 299
  cement. *See* Stucco
  corner heads, 293
  cracks, 292
  doubled back, 299
  drying, 300
  finish, 296, 298, 300-303
  float, 296, 298, 301
  gauging, 301
  grounds, 292, 293, 299, 303

Plaster (continued)
  gypsum, 297
  Keene's Cement, 297, 301
  lime, 297, 301
  neat, 299
  mixes, 296, 299-303
  Paris, plaster of, 297
  perlite, 295, 296, 298
  proprietary, 302
  putty, 301
  ready-mix, 299
  sand, 295-298
  scratch, 296, 298, 299
  screeds, 299
  single, 303
  smooth, 301
  speed of set, 297, 304
  three-coat, 296, 298
  trowel, 296, 301
  two coat, 296, 294, 303
  veneer, 301-304
  vermiculite, 295, 296, 298
  wood-fibered, 299
Plasterboard. See Wallboard
Plaster grounds, 292, 293, 299, 303
Plastic, 182-184, 192, 193, 223, 226, 227, 230, 258, 390-400
  ABS, 399
  acrylic, 392, 398, 400
  adhesives, 393, 396
  applications, 396-400
  bathtubs, 397, 400
  cellulosics, 392, 398
  classes, 390-394
  combustibility, 391, 398-399
  composite materials, 394-396
  concrete, 394
  copolymers, 399
  counters, 395-397
  durability, 391
  epoxies, 393, 400
  expansion, 400
  fatigue, 392
  fibers, 391, 394-395, 400
  flame, fire resistant and retardant, 391, 398-399
  flashing, 397, 399
  flexibility, 392
  floors, 396-397
  fluorocarbons, 393
  gasses, 391
  gel coat, 400
  glass fiber, 391, 394, 395, 400
  glazing, 397, 398
  gutters, 223, 226, 227, 230, 397, 399
  hardware, 397, 399
  laminates, 395, 397
  lighting, 397, 398
  melamine formaldehyde, 393, 398
  mortar, 400
  neoprene, 400
  nylon 392, 399
  phenolics, 393, 398, 399
  piping, 397, 399
  plumbing, 397, 400
  polyesters, 393, 394, 400
  polyethylene, 392, 398, 399
  polycarbonate, 393, 398
  polymers, 390
  polypropylene, 393
  polystyrene, 392, 398-399
  polysulfide, 400
  polyurethane, 258, 394, 398-400
  polyvinyl butyral, 393
  polyvinyl chloride (PVC), 392, 396-400
  polyvinylidine chloride, 399, 400
  properties, 390-392
  reinforced, 393-395, 400
  reinforcing, 391
  sandwiches, 395-396: cores, 395-396; facings, 395, 396
  sealants, 397, 400
  siding, 236-238
  silicones, 394, 400
  skylights, 192-193, 396-397
  strength, 391
  smoke, 391
  stiffness, 391
  stucco, 309, 400
  thermoplastic, 390, 392-393
  thermosetting, 390, 393-394
  toughness, 391
  upholstery, 394
  urea formaldehyde, 393, 398
  vapor barrier, 397, 399
  walls, 396-397
  wear, 392
  weathering, 391
  windows, 166, 182-184, 192, 193
Plate. See Balloon, Braced, Platform, Roof framing
Platform frame, 69-82, 403, 405
  anchors, 71, 72
  bands, 69, 73
  bracing, 81, 82
  firestopping, 70
  headers, 69, 70, 73
  joists, 73-76
  panels, 403, 405
  plates, 69-71, 73, 75-76

Platform frame *(continued)*
  platform, 69, 70, 71
  posts, 69, 70, 71, 73, 81
  sheathing, 69, 70, 74, 77-82
  sill, 69-73, 81
  soles or shoes, 69-71
  studs, 69-71, 73-76
  subflooring, 70, 71, 73, 75-77
Plinth block, 338
Plot plan, 3, 6, 7, 8, 11, 12
Plywood, 42-44, 64-67, 78-79, 96, 103-111, 131-133, 236, 238, 239, 283-285, 287
  boarding, 234
  combination floor, 110, 111
  concrete form, 42-44, 67
  dimensions, 64, 66
  exterior, 64, 65
  forms, 42-44
  girders, 96
  glued plywood joist, 106-109
  grades, 64-66
  groups, 65, 66
  identification index, 65, 66, 106
  interior, 65, 67
  joints, 283-285, 287
  nailing schedule, 78-79, 104-106
  roof decking, 131-133
  sheathing, 77-79
  siding, 236, 238
  species, 65, 66, 109
  standard, 66, 67, 108, 109
  structural, 66, 67, 108, 109
  subfloor, 67, 103-110
  2-4-1, 106, 107
  underlayment, 67, 105, 108, 109
  veneers, 64
Polyesters, 382, 393, 394, 400
Polyethylene, 392, 398, 399
Polycarbonate, 393, 398
Polymers, 390
Polypropylene, 393
Polystyrene, 392, 398-399
Polysulfide, 400
Polyurethane, 258, 382, 394, 398-400
polyvinyl butyral, 393
Polyvinyl chloride (PVC), 382, 392, 396-400
Polyvinylidene chloride, 399, 400
Ponding, 120
Popping, nail, 274, 276, 279, 280 291
Post, plank and beam, 134-135, 137-141, 401, 402, 404, 406, 408
  beams, 134, 135, 137, 139
  design, 134, 140
  fastenings, 135, 137-141

  fire resistance, 135
  foundations, 135
  insulation, 135
  loads, 135, 140
  plank, 134, 135, 137-141
  posts, 134, 135, 137, 140
  roof framing, 137, 140-141
  subfloor, 139
  studs, 134, 137
  wiring, 135
Post, 37, 38, 71, 73, 84, 91-95, 114, 134, 135, 137, 140, 142, 145, 319, 320, 329, 330. *See also* Balloon, Braced, Platform, frames: Post, plank and beam
  basement, 92-95
  bases, 93, 94
  built up, 71, 73, 84
  caps, 93-96
  stair, 319, 320, 329, 330
  steel, 93, 95, 142
  wood, 93, 95, 114, 142
Power shovel, 19
Precutting, 401, 403, 410
Prefabrication. *See* Manufactured housing
Presses, hot-plate, 395
Pulp boards, 286
Putty and puttying, 168-170, 177

Quirk, 316, 317

Radiation, 248, 249, 251-254
  bright surfaces, 253-254
  dull surfaces, 253, 254
  emission, 254
  metal foil, 254, 257-261
Rafter, 84, 85, 90, 116-133, 142, 144, 145, 196, 202, 218-222, 255. *See also* Roofs; Cornices
  bird's mouth joint, 120, 121, 123, 124, 130, 218, 222
  common, 118, 119, 122, 123, 126, 128
  connectors, 125-127
  false, 218, 219
  flying, 219
  header, 129-131
  hip, 118, 119, 123
  jack, 118, 119, 123, 126, 128, 142, 144, 145, 218-222, 225
  lookout, 125
  sizes, 125
  spans, 117
  species, 117
  tail, 129, 130
  trimmer, 129-131
  trussed, 118, 125-127

Rafter *(continued)*
  valley, 119, 128, 129-131
Rail, stair, 314, 319-321, 329, 330
Ramp, 320, 329
Red lead, 384
Ribband, Ribbon. See Balloon framing
Ridge. See Roofs; shingles; slate; tile
Right triangle. See Triangle, 3-4-5
Rise, 118, 318-320, 323
Riser, 113, 116, 313, 318, 320-325, 327-329
Roads, 2
Rock, 4, 5
Roll roofing, 202, 203
Roof boards, 77, 131-133, 196, 202, 204, 205, 208, 219-222
Roof drains. See Gutters, Leaders
Roofing. See Shingles, slate; Roll, Built-up sheet, Liquid roofing; Tile
Roof, 85, 117-133, 196, 219
  analysis, 118-129
  collar beams, 122, 124-126
  cornice. See Cornice
  deck, 117, 128
  deflection, 117
  dormers, 129-131
  eaves, 117
  ells, 128-129
  flat, 118
  framing, 116-133
  gable, 85, 117, 119, 120, 122, 125, 126, 219
  gambrel, 117, 119, 124, 126
  hip, 117, 118, 119, 125, 126, 129
  lean-to, 117, 121
  loads, 117, 119, 120, 132
  mansard, 117, 128, 219
  openings, 129-131
  pitch, 117, 118
  plate, 117-125
  ponding, 120
  purlin, 118, 128
  rafters. See Rafter
  reactions, 119, 125
  ridge, 117, 118, 126, 129
  ridge board, 117, 120, 122, 126, 128
  ridge pole, 117
  rise, 118
  roof boards. See Roof boards
  roofing paper, 133
  run, 118
  shed, 117-121, 219
  shingle strips, 133, 196
  valley, 118, 128
  wind loads, 117, 124, 126
Rubble, 35, 58-60, 242, 243

bond stones, 60
coursed, 60
details, 60
foundations, 60
mortar, 60
random, 58, 60
stability, 60
stones, 60
Run, 118, 318, 319, 320, 323

Sand, 18, 25, 32, 417, 418, 289, 295-298, 306, 307
Sash, 165, 166, 168, 169, 172-175, 177-188, 191, 192. See also Windows
Screws, 291, 358, 359, 362, 363
  drive, 358, 359, 363
  dowel, 359, 363
  heads, 359, 363
  knob, 363
  lag, 362, 363
  lath, 291
  metals, 362
  sizes, 359, 362
  types, 359, 362
  wood, 359, 362, 363
Septic tank, 6, 8, 20, 21, 25-34
  absorption area, 32-34
  bacteria, 26, 27
  capacity, 27
  detergents, 26
  distribution box, 27, 30
  effluent, 26, 27, 31
  grease trap, 27
  materials, 27
  scum, 26, 27, 31
  sludge, 26, 27, 31
  soil, relative absorption, 26
  soil test, 26, 31-32
  tile field, 26, 27, 28, 30, 31, 32, 33, 34
  washer, automatic, 27
Service, 20, 21
  trenches, 16, 19, 20
  underground, 20
Set-back, 8
Sewers, 5, 6, 8, 17, 19, 20, 23, 24, 26, 27, 28, 31
Shadowing, 282, 284
Shakes, 194, 199, 200, 361
Sheathing. See Balloon, Braced, Platform framing; Forms; Plywood; Wallboard
  matched, 62, 77
  nailing, 80
Sheeting, 20, 21
Sheet roofing, 202-204
  laying, 203, 204

Sheet roofing *(continued)*
  materials, 203
  substrates, 204
Shellac, 381
Shingle, asbestos, 198, 199, 205, 207, 231, 240
  composition, 207
  hip, 198, 205, 207
  laying, 198, 199, 205, 207
  nailing, 207, 357, 361
  ridge, 198, 205, 207
  shapes, 207
  starting, 198, 205
  weather, 198, 205, 207
Shingle, asphalt, 200-203, 231, 240
  adhesive, 201, 202
  composition, 200
  fire resistance, 200
  flashing, 202
  hips, 203
  laying, 201, 203
  ridges, 201-203
  shapes, 200, 203
  sizes, 200
  starting course, 201, 202
  weather, 201, 240
  valleys, 202, 203
Shingle, walls, 240-242
  application, 240, 241
  eaves, 242
  frieze, 242
  materials, 240
  water table, 241, 242
  wood, 240-242
Shingle, wood, 194-200, 231, 240
  Boston hip, ridge, 197, 198
  cement, 197
  flashing, 197-199, 242
  grades, 195, 242
  hips, 197
  laying, 195, 242
  nailing, 196, 357, 361
  ridges, 197, 198
  saddle, 197, 198
  sawn, 194
  shingle strips, 133, 196
  sizes, 195
  snakes, 194, 199, 200
  spacing, 196
  species, 194
  split, 194
  starting course, 195, 196, 220
  straight edge, 196
  valleys, 197, 199
  weather, 195, 196, 240

Shiplap, 62, 80, 315-317
Shoulder joint, 315, 316
Shovels, 19
Siding, 172-174, 231-238
  bevel. *See* Siding, lap
  clapboards. *See* Siding, lap
  drop, 231, 232, 234: corner boards, 232, 234; doors, 231, 232; mitered, 232, 234; shapes, 232, 234; use, 231, 234; water table, 232; windows, 231
  lap, 172, 173, 231, 233-238: aluminum, 236-238; application, 234, 235; corner boards, 233, 236, 237; eaves, 235; frieze, 235; hardboard, 236, 238; miter, 233, 235; nailing, 233, 357, 361; plastic, 236-238; plywood, 236, 238; spacing, 233-235; water table, 233, 235, 236
Sills, house. *See* Balloon, Braced, Platform, Special framing
Silicones, 394, 400
Silt, 18
Site, 1-9
Skylights, 192-193, 396-397
Slate, 197-199, 204-207, 357, 361
  Boston hip, ridge, 205, 206
  cant strip, 204-206
  cement, 206
  colors, 204
  felt, 204, 205
  flashing, 205, 206
  hips, 193, 205, 206
  laying, 204, 206
  nailing, 204, 206, 357, 261
  ridges, 198, 205, 206
  sizes, 204, 206
  sources, 204
  starting course, 205, 206
  valleys, 199, 207
  weather, 205, 206
Sludge, 26, 27, 31
Smoke chamber. *See* Fireplace
Smoke pipes, 133, 162
Smoke shelf. *See* Fireplace
Soil, 3, 4, 16, 17-18, 22, 23, 26, 31-32, 35-38
  classification, 17-18
  expansive, 4
  relative absorption, 25
  test, 17, 26, 31-32
Soil pipe, 22, 23
Soles, shoes. *See* Balloon, Braced, Platform, Partition framing.
Special framing, 141-146
  corner windows, 141
  girders: built-up, 142, 143; plywood, 142,

Special framing *(continued)*
  143; steel, 142, 143; wood, 142, 143
  openings, 141, 142, 143
  overhangs, 141, 142, 144, 145
  stressed cover, 145-146
  structural analysis, 141
Specification clauses
  adjacent properties, protection, 3
  cabinets, 313
  chimneys, 147
  closets, 313
  conductors, drains, 21
  doors, 313
  drains, 16, 22: land, 16, 22; wall, 16, 22
  dry wells, 16, 22
  excavation: backfill, frost, grading, solid bed, work covered, 16
  fill: area, under basement floors, 16
  fireplaces, 147, 313
  flashing, 212
  floors, basement, 16, 60
  forms, 40
  framing, lumber, 62
  gutters, 224, 226, 228
  lath: gypsum, 288; metal, 289
  leaders, 226, 228
  plywood, 62
  rock, blasting, 4, 16
  roof boards, 62, 194
  sheathing paper, 194, 201
  shingles: asphalt, doubling, nails, spacing, weather, wood, 194-201
  site: examination; relation to grades, 1
  staircase, 313
  topsoil, removal, 4, 16
  trim, 313
  underground water, 16
  underlayment, 62
  wallboards, 263-264
  windows: basement, double-hung, glazing, metal, hanging, priming, sash, wood, 164
Spikes, 358
Spindles, 321
Splash strip, 223
Spline, 315
Springs
  hinges, 364, 367, 377
  water, 5
Stains, 386
Staples, 271, 291, 292, 302
Stair, 113-116, 313, 318-330
  angle post, 319, 329, 330
  baluster, 314, 319, 327, 329, 330
  bracket, 330
  built in place, 325, 327
  carriages, 113, 115, 313, 322, 324, 325, 327-329
  clearance, 322
  drop, 319
  easing, 320, 329
  enclosed, 319
  flight, 318, 320, 323, 324
  framing, 113-116
  headers, 114, 115
  headroom, 322, 327
  horses, 113, 322
  landings, 113, 115, 116, 320, 322-324, 327
  laying out, 323
  newel post, 319, 320, 322, 329, 330
  nosing, 320, 322, 325, 329
  open, 319-321, 328
  ornamental, 324
  plans, 323-326
  posts, 320
  rail, 314, 319-321, 329, 330
  ramps, 320, 329
  rise, 318-320, 323
  risers, 113, 116, 313, 318, 320, 325, 327-329
  rough, 113-116
  run, 318, 319, 320, 323
  shop built, 325, 327
  spindle, 321
  spiral, 324
  stringers: curb, open, rough, wall, 313, 318-321, 324, 325, 328, 329
  thrust block, 116, 327
  treads, 113, 116, 313, 318, 320-322, 324, 325, 327-330
  trimmers, 114, 115
  wells, 114, 115
  winder, 324, 326
Stake-out, 10
Stops, lock, 372
Straight edge, 196
Strainers, 226, 228
Stressed cover or skin, 145-146, 403-405, 409
Strikes, 370, 371, 374
Stringers. *See* Stairs
Structural sandwich, 404, 406, 409, 413. *See also* Manufactured housing
Stucco, 292-294, 304-309
  application, 304, 307
  bond, 306
  brown, 306-308
  cement, 294, 304, 306
  colors, 304, 306

Stucco *(continued)*
  concrete, 306
  curing, 307
  epoxide, 309
  finish, 306-308
  framing, 305
  furring, 305, 308
  lime, 304, 306, 307
  masonry, 305, 306
  metal fabric, 305, 306, 308
  mixes, 306-307
  pigments, 307
  reinforcement, 305-306, 308. *See also*
    Lath, metal
  sand, 306, 307
  scratch, 306-308
  stucco mesh, netting, 292, 305, 308
  textures, 307, 308
  three-coat, 304
  two-coat, 304
  wire fabric, 293, 308
Studs. *See* Balloon, Braced, Platform,
  Roof, Special framing; Forms
Subfloor. *See* Balloon, Braced, Platform,
  framing; Floors, framing; Plywood
Subsoil, 4, 8, 25
  examination, 4
Surveyor's
  stakes, 2
  tape, 11, 12, 15

Tee sill, 73, 85, 86, 89
Temperature gradients, 259-261
Termite shields, 71, 72, 82, 235
Test pit, 26, 31-33
Thimbles, 147
Thresholds, 335, 337
Tile, drain. *See* Drains
Tile field. *See* Septic tanks
Tile, floor. *See* Tile, interior
Tile, interior, 108, 209-312
  accessories, 310, 312
  adhesives, 310
  application, 310-312
  buttering, 310, 311
  ceramic, 309, 312
  cove, 311, 312
  dado, 312
  floating, 310, 311
  floor, 309, 311, 312
  glazed, 309
  gypsum board, 309-311
  lath, metal, 309-312
  masonry, 309
  mortar, 310, 312

  mosaic, 311, 312
  studs, 309
  subfloor, 108, 311
  vitreous, 309
  walls, 309, 310
Tile, roof, 204-210
  framing for, 205
  felt, 205
  French, 205, 206
  Greek, 206, 207
  interlocking, 205, 206
  materials, 205
  mission, 206, 207
  Roman, 206, 207
  roof boards, 205
  shingle, 205, 206
  Spanish, 206, 207
Tile, wall. *See* Tile, interior
Titanium dioxide, 384
Tongue and groove joints. *See* Joints
Topography, 3
Topsoil, 3, 10, 19, 20
Tract, 15
Transit, 10, 15
Tread. *See* Stairs
Trees, 5, 6
Trenches, 16, 19, 20, 21
Trenching machine, 19
Triangle, 3-4-5, 11-12
Trim, 313-341. *See also* Millwork
Trusses, 76, 83, 110, 111, 118, 125-127,
  407, 410, 411, 418, 423. *See also* Rafters

Underground water, 5
Underlayment, 62, 67, 105, 108-111
Urea formaldehyde, 393, 398
Utilities, 1, 5

Valley, 118-120, 128, 129-131, 197, 199,
  202, 203, 207
Vapor barrier, 256-258, 260-262, 397, 399
Vapor pressure, 260-262
Varnish, 381
Vehicle, 381
Veneer, masonry. *See* Masonry veneer
  plaster, 301-304
  wood, 64
Vents, 216, 217
Verge board, 219
Vermiculite, 295, 296, 298
Vertical boarding, 238-240
  aluminum, 239, 240
  battens, 238, 239
  cement-asbestos, 239
  flush, 238, 239

Vertical boarding *(continued)*
  hardboard, 239, 240
  particle board, 239
  patterns, 238, 239
  plywood, 239
  square edge, 238, 239
Vinyl flooring, 352-355, 396-397
Volume elements, 401, 402, 409

Wallboard, 62, 73, 77, 245, 251-255, 263-288, 302, 309, 311
  adhesives, 267, 270-272, 274-277
  application, 266-280
  backer, 264, 276, 302
  battens, 285, 288
  beads, 279, 285
  blocking, 285
  cement-asbestos, 287-288
  core, 264
  corners, 278, 279-282, 302
  cracking, 280
  cutting, 272, 273
  double ply, 266, 274, 276, 277, 302
  edges, 278, 279
  fasteners, 267, 270-280
  fiberboard, 286
  framing, 265-269
  furring, 265, 268, 269, 284
  gypsum, 263, 264
  gypsum board, 263-284, 309-311
  hardboard, 286-287
  horizontal, 265, 266, 275, 277, 281
  insulating, 251-255, 262, 264, 302
  joints, 265, 273, 276-279, 283, 285, 286, 288, 302: compounds, 278
  laminating, 276
  moldings, 283, 285-287
  nailing, 270, 271, 273-275, 281, 285, 288, 357, 361
  plywood, 283-285. *See also* Plywood
  popping, nails, 274, 276, 279, 280
  pre-decorated, 274
  pulp boards, 286
  ridges, 282
  runners, 268, 269
  screws, 270, 271, 274
  shadowing, 282, 284
  single layer, 265, 266
  staples, 271, 302
  studs, 265-269, 284
  taping, 265, 277, 278, 302
  temperature, 272
  types, 264, 265
  type X, 264, 302
  ventilation, 272

  vertical, 265, 266, 277, 281
  water: mains, 5, 6, 8, 19; subsurface, 4, 5
Waterproofing, 5, 57
Water table, 1, 231-233, 235, 236, 240-242, 244, 245
Weather, roofing, 195, 196, 198, 200-207, 231, 240
Weatherstripping, 164, 173, 179, 185, 189, 190
Weep holes, 224
Western frame. *See* Platform frame
Wet wall, 74
White lead, 385
Wind loads, 117, 124, 126
Winder, 324, 326
Window, 47, 164-193, 231
  aluminum, 164, 166, 167, 174, 176, 177, 184, 185
  apron, 172-175, 180
  awning, 165, 167
  balances, 165, 177, 178, 180, 185, 187
  bars, 168
  basement, 47, 164, 175, 177
  blind stop, 175, 178, 179, 182
  bow, 166
  bronze, 167
  casement, 164-167, 170-174, 176, 177, 189
  casing, 171-175, 178-180, 187
  clips, 170, 176, 177
  cottage, 180, 182
  double hung, 164-170, 177-185, 189
  drip cap, 171, 172, 174, 180
  fixed, 165, 166, 175, 177, 187, 188
  flashing, 172-174, 180
  frames, 166, 168, 170-171, 176, 178-188
  gaskets, 170, 185, 187, 188
  glass. *See* Glass
  glazing, 164, 168-170, 176, 177, 190-192
  grilles, 168
  ground, 172-175
  head, 165, 170, 171, 178, 179, 183, 187
  hopper, 165-167
  infiltration, 189
  jambs, 165, 170, 171, 178-180, 183, 187
  jalousies, 167
  lights, 168-170, 177, 191
  lintel, 182, 187
  materials, 167
  meeting rails, 169, 178, 179, 183, 189, 190
  metal, 167, 170, 174-177, 183
  mullions, 171, 173, 182, 183
  multiple, 190-192
  muntins, 166
  parting strip, 175, 178, 179

picture, 166
Window *(continued)*
  pivoted, 165, 166
  plastic, 166, 182-184, 192, 193
  points, glazier's, 169, 170
  projected, 165-167, 176, 188
  putty, 168-170, 177
  rails, 166, 168, 169, 171-173, 178, 179, 183
  reversible, 166
  sash, 165, 166, 168, 169, 172-175, 177-188, 191, 192
  sash weights, 165, 177-185
  screens, 171-173, 177-179, 182, 187
  sealants, 170, 187
  secondary sash, 191
  sills, 165, 170-173, 177-182
  single-hung, 165, 185
  stationary, 165, 166, 175, 177, 187, 188
  steel, 164, 166, 167, 174-177
  stiles, 165, 166, 168, 178, 179, 182, 183
  stool, 172, 174, 175, 180, 182, 187
  stop, 175, 180
  storm sash, 171, 177-180, 182, 190, 191
  traverse, 165, 166, 185-187
  types, 165
  veneer masonry, 182, 187
  ventilating, 165
  weatherstripping, 164, 173, 179, 185, 189, 190
  wood, 164, 167, 168, 177-183
  yoke, 178
  zipper, 188
Withes, 147, 152

Zinc
  chloride, 43
  chromate, 385
  oxide, 383-384
  sulfide, 384
Zoning ordinances, 1, 8